**4th Edition**

# Essential Mathematics with Applications

**Vernon C. Barker**
***Palomar College***

**Richard N. Aufmann**
***Palomar College***

**HOUGHTON MIFFLIN COMPANY** Boston Toronto
Geneva, Illinois Palo Alto Princeton, New Jersey

Senior Sponsoring Editor: *Maureen O'Connor*
Senior Associate Editor: *Robert Hupp*
Editorial Production Manager: *Nancy Doherty Schmitt*
Production/Design Coordinator: *Carol Merrigan*
Senior Manufacturing Coordinator: *Marie Barnes*
Marketing Manager: *Charles Baker*

**Credits**

Cover design by Len Massiglia, LMA Communications.
Cover image by Herman Cherry, BRIEF ENCOUNTER 1990. Oil on linen, 66 × 60 inches. Courtesy Luise Ross Gallery, NY. Photo credit: Regina Cherry.

Chapter opening art by Patrice Rossi. Interior art by Network Graphics and Nugraphic Design Studio.

Page 20, table sources, *USA Today,* August 24, 1993 and *The Merger Yearbook,* U.S./International Ed. (New York: Securities Data Publishing Co., 1994). Page 39, table sources, *USA Today,* September 15, 1993 and Craig Carter, *Complete Baseball Record Book—1994* (The Sporting News). Page 55, art, copyright 1993, USA TODAY. Reprinted with permission. Page 124, table sources, *USA Today,* August 23, 1993 and Lee Wolverton, "Rooney, Steelers Expect Prosperity Despite Cost Spike," *Pittsburgh Business Times,* June 7, 1993. Page 126, art, middle of the page, copyright 1993, USA TODAY. Reprinted with permission. Page 130, table at the bottom of the page source, Bureau of Labor Statistics. Page 196, art, copyright 1993, USA TODAY. Reprinted with permission. Page 208, art, copyright 1993, USA TODAY. Reprinted with permission. Page 250, table source, *Money* Magazine, March 1994.



Printed in the U.S.A.

ISBN 0-395-71229-7

3456789-B-98 97 96

# Contents

# Preface

The fourth edition of *Essential Mathematics with Applications* provides mathematically sound and comprehensive coverage of the topics considered essential in a basic college mathematics course. The text has been designed not only to meet the needs of the traditional college student but also to serve the needs of returning students whose mathematical proficiency may have declined during years away from formal education.

In this new edition of *Essential Mathematics with Applications,* careful attention has been given to implementing the standards suggested by NCTM. Each chapter begins with a mathematical vignette in which there may be a historical note, application, or curiosity related to mathematics. At the end of each section there are "Applying the Concepts" exercises that include writing, synthesis, critical thinking, and challenge problems. The chapter ends with "Projects in Mathematics" that can be used for cooperative learning activities.

## INSTRUCTIONAL FEATURES

### Interactive Approach

*Essential Mathematics with Applications* uses an interactive style that provides a student with an opportunity to try a skill as it is presented. Each section is divided into objectives, and every objective contains one or more sets of matched-pair examples. The first example in each set is worked out; the second example, called "You Try It," is for the student to work. By solving this problem, the student practices concepts as they are presented in the text. There are *complete* worked-out solutions to these examples in an appendix at the end of the book. By comparing their solution to the solution in the appendix, students are able to obtain immediate feedback on and reinforcement of the concept.

### Emphasis on Problem-Solving Strategies

*Essential Mathematics with Applications* features a carefully developed approach to problem-solving that emphasizes developing strategies to solve problems. Students are encouraged to develop their own strategies, to draw diagrams, and to write strategies as part of their solution to a problem. In each case, model strategies are presented as guides for students to follow as they attempt the "You Try It" problem. Having students provide strategies is a natural way to incorporate writing into the math curriculum.

### Emphasis on Applications

The traditional approach to teaching mathematics covers only the straightforward manipulation of numbers and symbols and thereby fails to teach students the practical value of mathematics. By contrast, *Essential Mathematics with Applications* contains an extensive collection of contemporary application problems. Wherever appropriate, the last objective of a section presents applications that require the student to use the skills covered in that section to solve practical problems. This carefully integrated applied approach generates student awareness of the value of mathematics as a real-life tool.

### Completely Integrated Learning System Organized by Objectives

Each chapter begins with a list of the learning objectives included within that chapter. Each of the objectives is then restated in the chapter to remind the student of the current topic of discussion. The same objectives that organize the text are also used as the structure for exercises and the Computer Tutor. For each objective in the text, there is a corresponding computer tutorial.

## AN INTERACTIVE APPROACH

Instructors have long realized the need for a text that requires students to use a skill as it is being taught. *Essential Mathematics with Applications* uses an interactive technique that meets this need. Every objective, including the one shown on the facsimile pages below, contains at least one pair of examples. One of the examples is worked. The second example in the pair (You Try It) is not worked so that students may "interact" with the text by solving it. To provide immediate feedback, a complete solution to this example is provided in the answer section at the end of the book. The benefit of this interactive style is that students can check that a new skill has been learned before attempting a homework assignment.

Section 2.7 / Division of Fractions and Mixed Numbers **97**

## 2.7 Division of Fractions and Mixed Numbers

**An explanatory passage begins each skill objective.**

*Objective A* ***To divide fractions***

The **reciprocal** of a fraction is the fraction with the numerator and denominator interchanged. For instance, the reciprocal of $\frac{2}{3}$ is $\frac{3}{2}$. The process of interchanging the numerator and denominator of a fraction is called **inverting** the fraction.

To find the reciprocal of a whole number, first write the whole number as a fraction with a denominator of 1; then find the reciprocal of that fraction.

The reciprocal of 5 is $\frac{1}{5}$. $\left(\text{Think } 5 = \frac{5}{1}\right)$

Reciprocals are used to rewrite division problems as related multiplication problems. Look at the following two problems:

**Paired examples follow the explanatory passage.**

$8 \div 2 = 4$ — 8 divided by 2 is 4.

$8 \times \frac{1}{2} = 4$ — 8 times the reciprocal of 2 is 4.

"Divided by" means the same as "times the reciprocal of." Thus "÷ 2" can be replaced with "$\times \frac{1}{2}$," and the answer will be the same. Fractions are divided by making this replacement.

➡ Divide: $\frac{2}{3} \div \frac{3}{4}$ $\qquad \frac{2}{3} \div \frac{3}{4} = \frac{2}{3} \times \frac{4}{3} = \frac{2 \cdot 4}{3 \cdot 3} = \frac{2 \cdot 2 \cdot 2}{3 \cdot 3} = \frac{8}{9}$

**The interactive key is the You Try It. It has not been worked so that the student may practice the skill, referring to the worked example at the left, if necessary.**

**Example 1** Divide: $\frac{5}{8} \div \frac{4}{9}$

**Solution** $\frac{5}{8} \div \frac{4}{9} = \frac{5}{8} \times \frac{9}{4} = \frac{5 \cdot 9}{8 \cdot 4}$

$= \frac{5 \cdot 3 \cdot 3}{2 \cdot 2 \cdot 2 \cdot 2 \cdot 2} = \frac{45}{32} = 1\frac{13}{32}$

**You Try It 1** Divide: $\frac{3}{7} \div \frac{2}{3}$

**Your solution**

**Example 2** Find the quotient of $\frac{3}{5}$ and $\frac{12}{25}$.

**Solution** $\frac{3}{5} \div \frac{12}{25} = \frac{3}{5} \times \frac{25}{12} = \frac{3 \cdot 25}{5 \cdot 12}$

$= \frac{\overset{1}{\cancel{3}} \cdot \overset{1}{\cancel{5}} \cdot 5}{\underset{1}{\cancel{5}} \cdot 2 \cdot 2 \cdot \underset{1}{\cancel{3}}} = \frac{5}{4} = 1\frac{1}{4}$

**You Try It 2** Divide: $\frac{3}{4} \div \frac{9}{10}$

**Your solution**

*Solutions on p. A16*

**Reference to the Answer Section allows the student to check solutions immediately.**

*Objective B* ***To divide whole numbers, mixed numbers, and fractions***

To divide a fraction and a whole number, first write the whole number as a fraction with a denominator of 1.

➡ Divide: $\frac{3}{7} \div 5$

$\frac{3}{7} \div \boxed{5} = \frac{3}{7} \div \boxed{\frac{5}{1}} = \frac{3}{7} \times \frac{1}{5} = \frac{3 \cdot 1}{7 \cdot 5} = \frac{3}{35}$

• Write 5 with a denominator of 1; then divide the fractions.


Content and Format © 1995 HMCo.

The solution of an application problem in *Essential Mathematics with Applications* is always accompanied by two parts: **Strategy** and **Solution.** The strategy is a written description of the steps that are necessary to solve the problem; the solution is the implementation of the strategy. This format provides students with a structure for problem solving. It also encourages students to write strategies for solving problems which, in turn, fosters organizing problem-solving strategies in a logical way.

**198** Chapter 5 / Percents

***Objective B To solve application problems***

To solve percent problems, remember that it is necessary to identify the percent, base, and amount. Usually the base follows the phrase "percent of."

**Example 4**
The Kaminski family has an income of \$1500 each month and makes a car payment of \$180 a month. What percent of the monthly income is the car payment?

**A strategy which the student may use in solving an application problem is stated.**

**Strategy**
To find what percent of the income the car payment is, write and solve the basic percent equation using $n$ to represent the unknown percent. The base is \$1500 and the amount is \$180.

**This strategy is used in the solution of the worked example.**

**Solution** $n \times \$1500 = \$180$
$n = \$180 \div \$1500$
$n = 0.12 = 12\%$

The car payment is 12% of the monthly income.

**You Try It 4**
Tomo Nagata had an income of \$20,000 and paid \$3000 in income tax. What percent of the income is the income tax?

**Your strategy**

**Your solution**

**Example 5**
Claudia Sloan missed 25 questions out of 200 on a state nursing license exam. What percent of the questions did she answer correctly?

**Strategy**
To find what percent of the questions she answered correctly:

- Find the number of questions the nurse answered correctly (200 − 25).
- Write and solve a basic percent equation using $n$ to represent the unknown percent. The base is 200 and the amount is the number of questions answered correctly.

**Solution** 200 − 25 = 175 questions answered correctly.

$n \times 200 = 175$
$n = 175 \div 200$
$n = 0.875 = 87.5\%$

Claudia answered 87.5% of the questions correctly.

**You Try It 5**
A survey of 1000 people showed that 667 people favored Minna Oliva for governor of the state. What percent of the people surveyed did not favor her?

**Students are encouraged to write a strategy for the application problem they try to solve.**

**Your strategy**

**Your solution**

*Solutions on p. A27*



Content and Format © 1995 HMCo.

## OBJECTIVE-SPECIFIC APPROACH

Many mathematics texts are not organized in a manner that facilitates management of learning. Typically, students are left to wander through a maze of apparently unrelated lessons, exercise sets, and tests. *Essential Mathematics with Applications* solves this problem by organizing all lessons, exercise sets, and computer tutorials around a carefully constructed hierarchy of objectives. The advantage of this objective-by-objective organization is that it enables the student who is uncertain at any step in the learning process to refer easily to the original presentation and review that material. The Objective-Specific Approach also gives the instructor greater control over the management of student progress. The Computer Tutor is also organized around the objectives of the text. As a result, supplemental instruction is available for any objectives that are troublesome for a student.

A numbered objective statement names the topic of each lesson.

> **2.1 The Least Common Multiple and Greatest Common Factor**
>
> *Objective A* *To find the least common multiple (LCM)*

The exercise sets correspond to the objectives in the text.

> **2.1 Exercises**
>
> *Objective A*
>
> Find the LCM.
>
> **1.** 5, 8 **2.** 3, 6 **3.** 3, 8 **4.** 2, 5 **5.** 5, 6

The answers to the Chapter Test show the objective to study if the student incorrectly answers the exercise.

> CHAPTER TEST *pages 113–114*
>
> **1.** 120 (2.1A) **2.** 8 (2.1B) **3.** $\frac{11}{4}$ (2.2A) **4.** $3\frac{3}{5}$ (2.2B) **5.** $\frac{49}{5}$ (2.2B)
>
> **6.** $\frac{5}{8} = \frac{45}{72}$ (2.3A) **7.** $\frac{5}{8}$ (2.3B) **8.** $1\frac{11}{12}$ (2.4A) **9.** $1\frac{61}{90}$ (2.4B) **10.** $22\frac{4}{15}$ (2.4C)

The answers to the Cumulative Review exercises also show the objective that relates to the exercise.

> CUMULATIVE REVIEW *pages 115–116*
>
> **1.** 290,000 (1.1D) **2.** 291,278 (1.3B) **3.** 73,154 (1.4B) **4.** 540 R2 (1.5C)
>
> **5.** 1 (1.6B) **6.** $2 \cdot 2 \cdot 11$ (1.7B) **7.** 210 (2.1A) **8.** 20 (2.1B) **9.** $\frac{23}{3}$ (2.2B)

## ADDITIONAL LEARNING AIDS

### Projects in Mathematics

The Projects in Mathematics feature occurs at the end of each chapter. These projects can be used as extra credit or cooperative learning activities. The projects cover various aspects of mathematics including the use of calculators, extended applications, additional problem-solving strategies, and other topics related to mathematics.

### Chapter Summaries

At the end of each chapter there is a Chapter Summary that includes Key Words and Essential Rules that were covered in the chapter. These chapter summaries provide a single point of reference as the student prepares for a test.

### Study Skills

The To the Student passage provides suggestions for using this text and approaches to creating good study habits.

### Computer Tutor

The Computer Tutor is a networkable, interactive, algorithmically-driven software package. This powerful ancillary features full-color graphics, a glossary, extensive hints, animated solution steps, and a comprehensive class management system.

### Glossary

New to this edition, a Glossary at the end of the book includes definitions of terms used in the text.

## EXERCISES

### End-of-Section Exercises

*Essential Mathematics with Applications* contains more than **3000** exercises. At the end of each section there are exercise sets that are keyed to the corresponding learning objective. The exercises are carefully developed to ensure that students can apply the concepts in the section to a variety of problem situations.

### Applying the Concepts Exercises

The End-of-Section Exercises are followed by Applying the Concepts Exercises. These sections contain a variety of exercise types, including:

- challenge problems
- problems that require that the student determine if a statement is always true, sometimes true, or never true
- problems that ask students to determine incorrect procedures
- problems that require a calculator to determine a solution

### Writing Exercises

Within the "Applying the Concepts Exercises," there are Writing Exercises denoted by [W]. These exercises ask students to write about a topic in the section or to research and report on a related topic.

### Chapter Review Exercises

Review Exercises are found at the end of each chapter. These exercises are selected to help the student integrate all of the topics presented in the chapter. The answers to all review exercises are given in the answer section at the end of the book.

### Cumulative Review Exercises

Cumulative Review Exercises, which appear at the end of each chapter (beginning with Chapter 2), help students maintain skills learned in previous chapters. The answers to all Cumulative Review Exercises are given in the answer section. Along with the answer, there is a reference to the objective that pertains to each exercise.

## NEW TO THIS EDITION

### Topical Coverage

The material in Chapter 6, *Applications for Business and Consumers,* has been updated to reflect current interest rates and prices.

One of the main challenges for students is the ability to translate verbal phrases into mathematical expressions. One reason for this difficulty is that students are not exposed to verbal phrases until later in most texts. In *Essential Mathematics with Applications,* we introduce verbal phrases for operations as we introduce the operation. For instance, after addition concepts have been presented, we provide exercises which say "Find the sum of...." or "What is 6 more than 7?" In this way, students are constantly confronted with verbal phrases and must make a mathematical connection between the phrase and a mathematical operation.

*Point of Interest* notes are interspersed throughout the text. These notes are interesting sidelights of the topic being discussed.

### Computer Tutor

The Computer Tutor has been completely revised. It is now an algorithmically-based tutor that includes color and animation. The algorithmic feature essentially provides an infinite number of practice problems for students to attempt. The algorithms have been carefully crafted to present a variety of problem types from easy to difficult. A complete solution to each problem is available.

There is an interactive feature of the Computer Tutor that requires students to respond to questions about the topic in the current lesson. In this way, students can assess their understanding of concepts as they are presented. There is a Glossary which can be accessed at any time so that students can look up words whose definitions that they may have forgotten.

When the student completes a lesson, a printed report is available. This optional report gives the student's name, the objective studied, the number of problems attempted, the number of problems correct, and the percent correct. The instructor also has the option of creating a cumulative report via the Tutor's new class management system.

### Applying the Concepts

The end-of-section exercises are followed by Applying the Concepts Exercises. These exercises include challenge problems and applications of the concepts presented in the section in a way that requires a student to combine several skills to solve a problem. These exercises also include Writing Exercises, denoted by [W], that ask students to write about a topic in the section or to research and report on a related topic.

### Projects in Mathematics

Through the Projects in Mathematics feature, some of the standards suggested by the NCTM can be implemented. These Projects offer an opportunity for students to explore several topics relating to calculators, applications of math, logic, problem solving, and related subjects.

### Glossary

The Glossary contains the definitions or the descriptions of all the key words highlighted in the text.

## SUPPLEMENT FOR THE STUDENT

### Computer Tutor

The Computer Tutor is an interactive instructional computer program for student use. Each objective of the text is supported by a tutorial in the Computer Tutor. These tutorials contain an interactive lesson that covers the material in the objective. Following the lesson are randomly generated exercises for the student to attempt. A record of the student's progress is available.

The Computer Tutor can be used in several ways: (1) to cover material the student missed because of an absence; (2) to reinforce instruction on a concept that the student has not yet mastered; (3) to review material in preparation for exams. This tutorial is available for the IBM PC and compatible computers running Windows and the Macintosh.

## SUPPLEMENTS FOR THE INSTRUCTOR

### Instructor's Answer Booklet

The Instructor's Answer Booklet contains the answers to all end-of-section exercise sets, Chapter Reviews, Chapter Tests, and Cumulative Reviews.

### Additional Supplements

The following are available to accompany *Essential Mathematics with Applications*. Please ask your Houghton Mifflin sales representative for details.

*Instructor's Resource Manual with Chapter Tests*
*Computer Test Generator*
*Printed Test Bank*
*Solutions Manual*
*Student Solutions Manual*
*Videotapes*

## ACKNOWLEDGMENTS

The authors would like to thank the people who reviewed this manuscript and provided many valuable suggestions:

Ronald J. Bautch, *Cook County College/Denton Center, TX*

Rebecca C. Benson-Beaver, *Valencia Community College, FL*

Martha J. Eagle, *Longview Community College, MO*

Diane Fariss, *McLennan Community College, TX*

Nancy C. Hall, *Muskingham Area Technical College, OH*

William C. Hoston

Laura L. Hoye, *Trident Technical College, SC*

Robert L. Jones, *IVY Technical College/Terre Haute, IN*

JoAnne Kennedy, *LaGuardia Community College, NY*

Gerald L. LaPage, *Bristol Community College, MA*

Lynda D. MacLeod, *Robeson Community College, NC*

James J. Meissner, *Butler County Community College, PA*

Linda J. Murphy, *Northern Essex Community College, MA*

Nancy K. Nickerson, *Northern Essex Community College, MA*

Elizabeth Ogilvie, *Horry-Georgetown Technical College, SC*

Rose L. Pugh, *Bellevue Community College, SC*

James P. Ryan, *Madera Community College, CA*

William D. Summons, *Shelby State Community College, TN*

# To the Student

Many students feel that they will never understand math, while others appear to do very well with little effort. Oftentimes what makes the difference is that successful students take an active role in the learning process.

Learning mathematics requires your *active* participation. Although doing homework is one way you can actively participate, it is not the only way. First, you must attend class regularly and become an active participant in class. Secondly, you must become actively involved with the textbook.

*Essential Mathematics with Applications* was written and designed with you in mind as a participant. Here are some suggestions on how to use the features of this textbook.

There are 6 chapters in this text. Each chapter is divided into sections, and each section is subdivided into learning objectives. Each learning objective is labeled with a letter from A–D.

First, read each objective statement carefully so you will understand the learning goal that is being presented. Next, read the objective material carefully, being sure to note each bold word. These words indicate important concepts that you should familiarize yourself with. Study each in-text example carefully, noting the techniques and strategies used to solve the example.

You will then come to the key learning feature of this text, the *boxed examples*. These examples have been designed to aid you in a very specific way. Notice that in each example box, the example on the left is completely worked out and the "You Try It" example on the right is not. *You* are expected to work the right-hand example (in the space provided) in order to immediately test your understanding of the material you have just studied.

You should study the worked-out example carefully by working through each step presented. This allows you to focus on each step and reinforces the technique for solving that type of problem. You can then use the worked-out example as a model for solving similar problems.

Next, try to solve the "You Try It" example using the problem-solving techniques that you have just studied. When you have completed your solution, check your work by turning to the page at the end of the book where the complete solution can be found. The page number on which the solution appears is printed at the bottom of the example box in the right-hand corner. By checking your solution, you will know immediately whether or not you fully understand the skill you just studied.

When you have completed studying an objective, do all of the exercises in the exercise set that corresponds with that objective. The exercises will be labeled with the same letter as the objective. Math is a subject that needs to be learned in small sections and practiced continually in order to be mastered. Doing all of the exercises in each exercise set will help you master the problem-solving techniques necessary for success.

Once you have completed the exercises to an objective, you should check your answers to the odd-numbered exercises with those found in the back of the book.

After completing a chapter, read the Chapter Summary. This summary highlights the important topics covered in the chapter. Following the Chapter Summary are Chapter Review Exercises, a Chapter Test, and a Cumulative Review (beginning with Chapter 2). Doing the review exercises is an important way of testing your understanding of the chapter. The answer to each review exercise is given at the back of the book. Each answer to the Chapter Test and Cumulative Review is followed by a reference that tells which objective that exercise was taken from. For example, (4.2B) means Section 4.2, Objective B. After checking your answers, restudy any objective that you missed. It may be very helpful to retry some of the exercises for that objective to reinforce your problem-solving techniques.

The Chapter Test should be used to prepare for an exam. We suggest that you try the Chapter Test a few days before your actual exam. Take the test in a quiet place and try to complete the test in the same amount of time you will be allowed for your exam. When taking the Chapter Test, practice the strategies of successful test takers: 1) scan the entire test to get a feel for the questions; 2) read the directions carefully; 3) work the problems that are easiest for you first; and, perhaps most importantly, 4) try to stay calm.

When you have completed the Chapter Test, check your answers. If you missed a question, review the material in that objective and rework some of the exercises from that objective. This will strengthen your ability to perform the skills outlined in that objective.

The Cumulative Review allows you to refresh the skills you learned in previous chapters. This is very important in mathematics. By consistently reviewing previous material, you will retain the skills already learned as you build new ones.

Remember, to be successful: attend class regularly; read the textbook carefully; actively participate in class; work with your textbook using the "You Try It" examples for immediate feedback and reinforcement of each skill; do all the homework assignments; review constantly; and work carefully.

# Index of Applications

Chapter

# 1

# Whole Numbers

## Objectives

### Section 1.1

To identify the order relation between two numbers

To write whole numbers in words and in standard form

To write whole numbers in expanded form

To round a whole number to a given place value

### Section 1.2

To add whole numbers

To solve application problems

### Section 1.3

To subtract whole numbers without borrowing

To subtract whole numbers with borrowing

To solve application problems

### Section 1.4

To multiply a number by a single digit

To multiply larger whole numbers

To solve application problems

### Section 1.5

To divide by a single digit with no remainder in the quotient

To divide by a single digit with a remainder in the quotient

To divide by larger whole numbers

To solve application problems

### Section 1.6

To simplify expressions that contain exponents

To use the Order of Operations Agreement to simplify expressions

### Section 1.7

To factor numbers

To find the prime factorization of a number

# Family Tree for Numbers

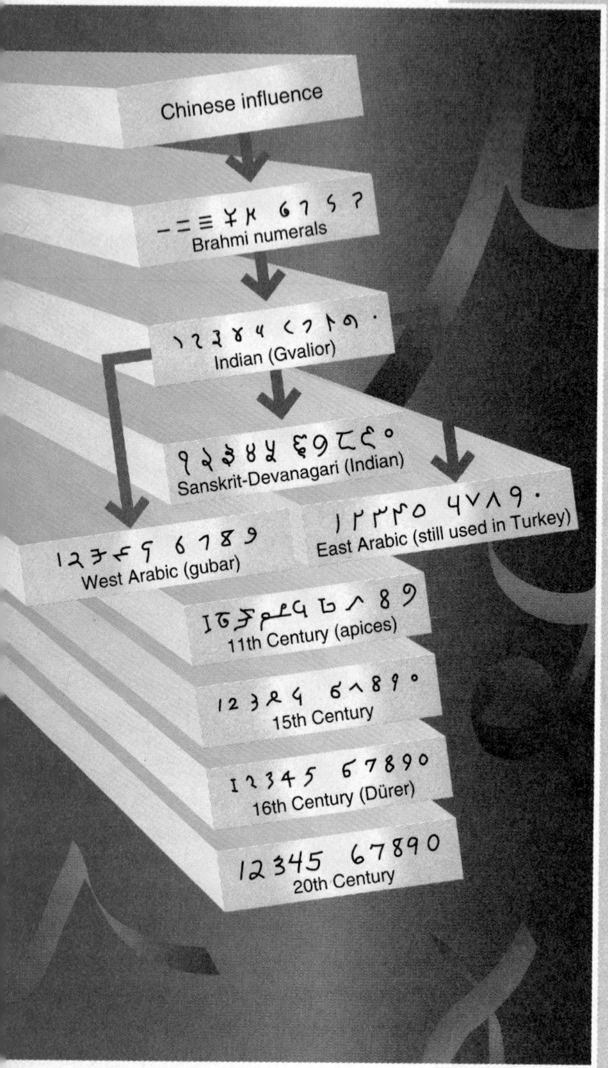


Our number system is called the Hindu-Arabic system because it has its ancestry in India and was refined by the Arabs. But despite the influence of these cultures on our system, there is some evidence that our system may have originated in China around 1400 B.C. That is 34 centuries ago.

The family tree below illustrates the most widely believed account of the history of our number system. In the 16th century, with Gutenberg's invention of the printing press, symbols for our numbers started to become standardized.

# 1.1 Introduction to Whole Numbers

## Objective A To identify the order relation between two numbers

**POINT OF INTEREST**

The numerals that we use have not always appeared as they do today. In the second century, the symbols for 1 through 9 were

By the eighth century, the symbols 2 and 3 were used instead of horizontal lines, but five looked like ㄣ and 9 was shown as ɛ.

The **whole numbers** are 0, 1, 2, 3, 4, 5, 6, 7, 8, 9, 10, 11, 12, 13, 14, . . . .

The three dots mean that the list continues on and on and that there is no largest whole number.

Just as distances are associated with the markings on the edge of a ruler, the whole numbers can be associated with points on a line. This line is called the **number line.** The arrow on the number line indicates that there is no largest whole number.

The Number Line

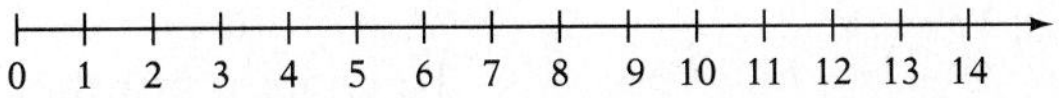


The **graph of a whole number** is shown by placing a heavy dot on the number line directly above the number. Here is the graph of 7 on the number line:

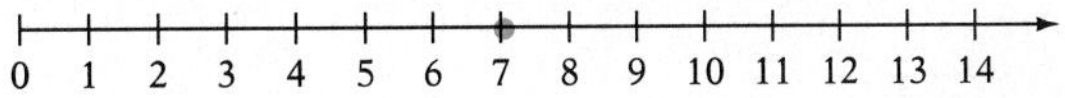


The number line can be used to show the order of whole numbers. A number that appears to the left of a given number is **less than** the given number. The symbol for "is less than" is <. A number that appears to the right of a given number is **greater than** the given number. The symbol for "is greater than" is >.

Four is less than seven.
$4 < 7$

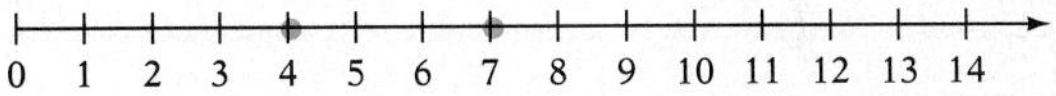


Twelve is greater than seven.
$12 > 7$

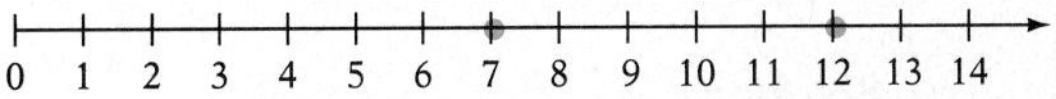


**Example 1** Graph 11 on the number line.

**Solution**

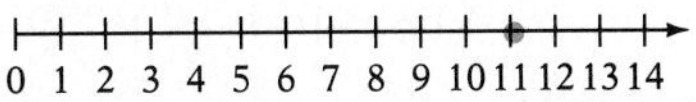


**You Try It 1** Graph 9 on the number line.

**Your solution**

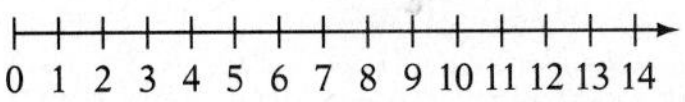


**Example 2** Place the correct symbol, < or >, between the two numbers.

**a.** 39 24

**b.** 0 51

**Solution** **a.** $39 > 24$

**b.** $0 < 51$

**You Try It 2** Place the correct symbol, < or >, between the two numbers.

**a.** 45 29

**b.** 27 0

**Your solution** **a.**

**b.**

*Solutions on p. A7*

## *Objective B To write whole numbers in words and in standard form*

**POINT OF INTEREST**
The Babylonians had a place-value system based on 60. Its influence is still with us in angle measurement and time: 60 seconds in 1 minute, 60 minutes in 1 hour. It appears that the earliest record of a base 10 place-value system for natural numbers was developed in the eighth century.

When a whole number is written using the digits 0, 1, 2, 3, 4, 5, 6, 7, 8, and 9, it is said to be in **standard form.** The position of each digit in the number determines the digit's **place value.** The diagram below shows a **place-value chart** naming the first twelve place values. The number 37,462 is in standard form and has been entered in the chart.

In the number 37,462, the position of the digit 3 determines that its place value is ten-thousands.

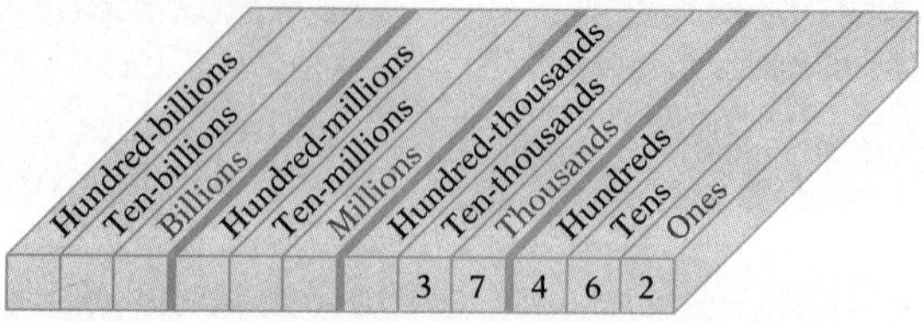


When a number is written in standard form, each group of digits separated by a comma is called a **period.** The number 3,786,451,294 has four periods. The period names are shown in red in the place-value chart above.

To write a number in words, start from the left. Name the number in each period. Then write the period name in place of the comma.

3,786,451,294 is read "three billion seven hundred eighty-six million four hundred fifty-one thousand two hundred ninety-four."

To write a whole number in standard form, write the number named in each period, and replace each period name with a comma.

Four million sixty-two thousand five hundred eighty-four is written 4,062,584. The zero is used as a place holder for the hundred-thousands' place.

**Example 3** Write 25,478,083 in words.

**Solution** Twenty-five million four hundred seventy-eight thousand eighty-three

**You Try It 3** Write 36,462,075 in words.

**Your solution**

**Example 4** Write three hundred three thousand three in standard form.

**Solution** 303,003

**You Try It 4** Write four hundred fifty-two thousand seven in standard form.

**Your solution**

*Solutions on p. A7*

## *Objective C To write whole numbers in expanded form*

The whole number 26,429 can be written in **expanded form** as

20,000 + 6000 + 400 + 20 + 9

The place-value chart can be used to find the expanded form of a number.

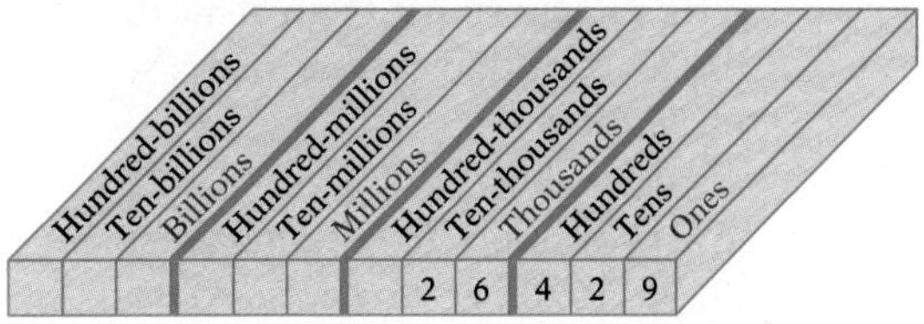


| 2 Ten-thousands | + | 6 Thousands | + | 4 Hundreds | + | 2 Tens | + | 9 Ones |
|---|---|---|---|---|---|---|---|---|
| 20,000 | + | 6000 | + | 400 | + | 20 | + | 9 |

Write the number 420,806 in expanded form.

Note the effect of having zeros in the number.

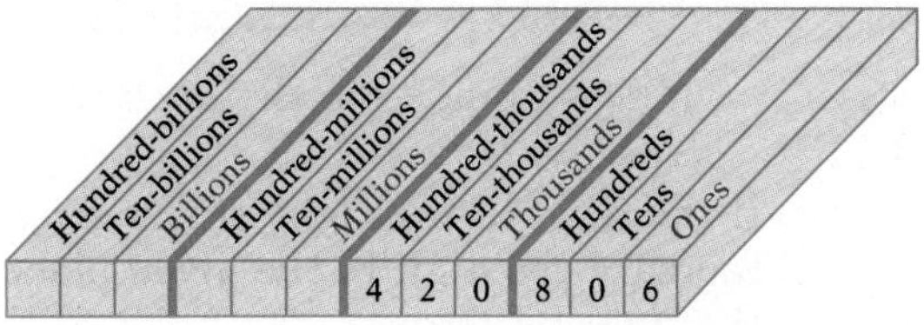


| 4 Hundred-thousands | + | 2 Ten-thousands | + | 0 Thousands | + | 8 Hundreds | + | 0 Tens | + | 6 Ones |
|---|---|---|---|---|---|---|---|---|---|---|
| 400,000 | + | 20,000 | + | 0 | + | 800 | + | 0 | + | 6 |

or simply 400,000 + 20,000 + 800 + 6

**Example 5** Write 23,859 in expanded form.

**Solution** 20,000 + 3000 + 800 + 50 + 9

**You Try It 5** Write 68,281 in expanded form.

**Your solution**

**Example 6** Write 709,542 in expanded form

**Solution** 700,000 + 9000 + 500 + 40 + 2

**You Try It 6** Write 109,207 in expanded form.

**Your solution**

*Solutions on p. A7*

## *Objective D To round a whole number to a given place value*

When the distance to the moon is given as 240,000 miles, the number represents an approximation to the true distance. Giving an approximate value for an exact number is called **rounding.** A number is always rounded to a given place value.

37 is closer to 40 than it is to 30. 37 rounded to the nearest ten is 40.

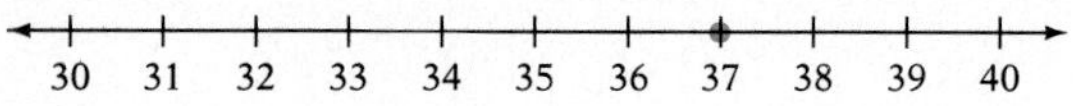


3673 rounded to the nearest ten is 3670. 3673 rounded to the nearest hundred is 3700.

A whole number is rounded to a given place value without using the number line by looking at the first digit to the right of the given place value.

If the digit to the right of the given place value is less than 5, that digit and all digits to the right are replaced by zeros.

➡ Round 13,834 to the nearest hundred.

Given place value

13,834

$3 < 5$

13,834 rounded to the nearest hundred is 13,800.

If the digit to the right of the given place value is greater than or equal to 5, increase the digit in the given place value by 1, and replace all other digits to the right by zeros.

➡ Round 386,217 to the nearest ten-thousand.

Given place value

386,217

$6 > 5$

386,217 rounded to the nearest ten-thousand is 390,000.

**Example 7** Round 525,453 to the nearest ten-thousand.

**Solution**

Given place value

525,453

$5 = 5$

525,453 rounded to the nearest ten-thousand is 530,000.

**You Try It 7** Round 368,492 to the nearest ten-thousand.

**Your solution**

**Example 8** Round 1972 to the nearest hundred.

**Solution**

Given place value

1972

$7 > 5$

1972 rounded to the nearest hundred is 2000.

**You Try It 8** Round 3962 to the nearest hundred.

**Your solution**

*Solutions on p. A7*

## 1.1 Exercises

### *Objective A*

Graph the number on the number line.

**1.** 3

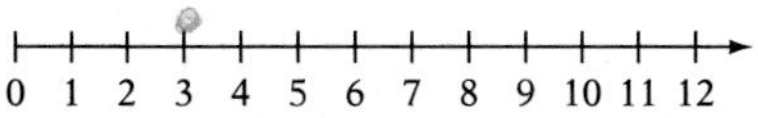


**2.** 5

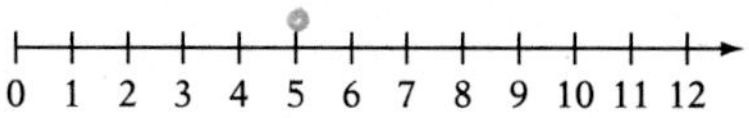


**3.** 9

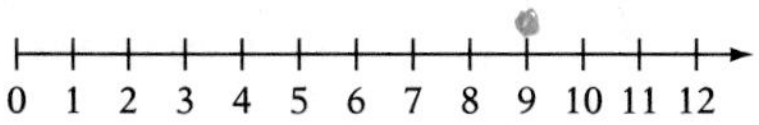


**4.** 0

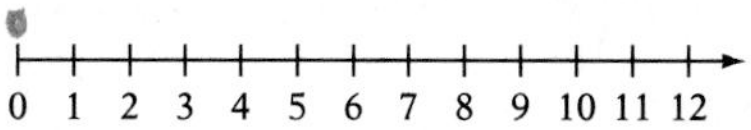


Place the correct symbol, < or >, between the two numbers.

**5.** 37 < 49 **6.** 58 > 21 **7.** 101 > 87

**8.** 16 > 5 **9.** 245 > 158 **10.** 2701 > 2071

**11.** 0 < 45 **12.** 107 > 0 **13.** 815 < 928

### *Objective B*

Write the number in words.

**14.** 2675 **15.** 3790 **16.** 42,928 **17.** 58,473

**18.** 356,943 **19.** 498,512 **20.** 3,697,483 **21.** 6,842,715

Write the number in standard form.

**22.** Eighty-five

**23.** Three hundred fifty-seven

**24.** Three thousand four hundred fifty-six

**25.** Sixty-three thousand seven hundred eighty

**26.** Six hundred nine thousand nine hundred forty-eight

**27.** Seven million twenty-four thousand seven hundred nine

### *Objective C*

Write the number in expanded form.

**28.** 5287

**29.** 6295

**30.** 58,943

**31.** 453,921

**32.** 200,583

**33.** 301,809

**34.** 403,705

**35.** 3,000,642

### *Objective D*

Round the number to the given place value.

**36.** 926 Tens

**37.** 845 Tens

**38.** 1439 Hundreds

**39.** 3973 Hundreds

**40.** 43,607 Thousands

**41.** 52,715 Thousands

**42.** 647,989 Ten-thousands

**43.** 253,678 Ten-thousands

### *APPLYING THE CONCEPTS*

Answer true or false for Exercise 44 and 45. If the answer is false, give an example to show that it is false.

**44.** If you are given two distinct whole numbers, then one of the numbers is always greater than the other number.

**45.** A rounded-off number is always less than its exact value.

**46.** What is the largest three-digit whole number. What is the smallest five-digit whole number?

**47.** In the roman numeral system, IV = 4 and VI = 6. Does the position of the I in this system change the value of the number it represents? Determine the value of IX and XI.

**48.** If 3846 is rounded off to the nearest ten and then that number is rounded to the nearest hundred, is the result the same as what you get when you round 3846 to the nearest hundred? If not, which of the two methods is correct for rounding to the nearest hundred?

# 1.2 Addition of Whole Numbers

## *Objective A* *To add whole numbers*

**Addition** is the process of finding the total of two or more numbers.

**POINT OF INTEREST**

The first use of the plus sign appeared in 1489 in *Mercantile Arithmetic.* It was used to indicate a surplus and not as the symbol for addition. That use did not appear until around 1515.

By counting, we see that the total of \$3 and \$4 is \$7.

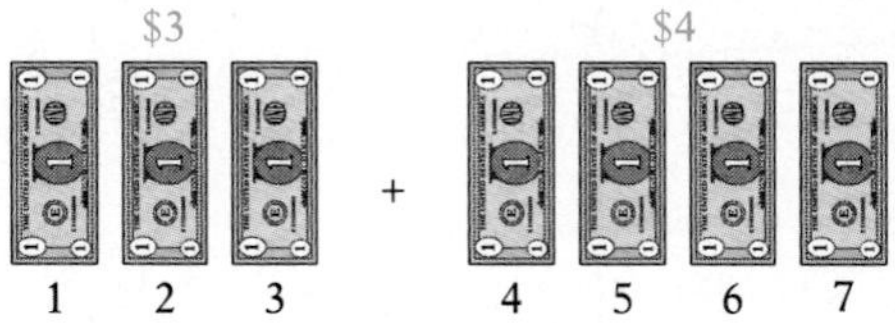


| \$3 | + | \$4 | = | \$7 |
|---|---|---|---|---|
| **Addend** | | **Addend** | | **Sum** |

Addition can be illustrated on the number line by using arrows to represent the addends. The size, or magnitude, of a number can be represented on the number line by an arrow.

The number 3 can be represented anywhere on the number line by an arrow that is 3 units in length.

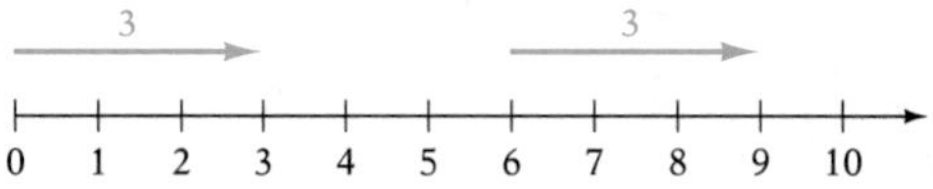


To add on the number line, place the arrows representing the addends head to tail, with the first arrow starting at zero. The sum is represented by an arrow starting at zero and stopping at the tip of the last arrow.

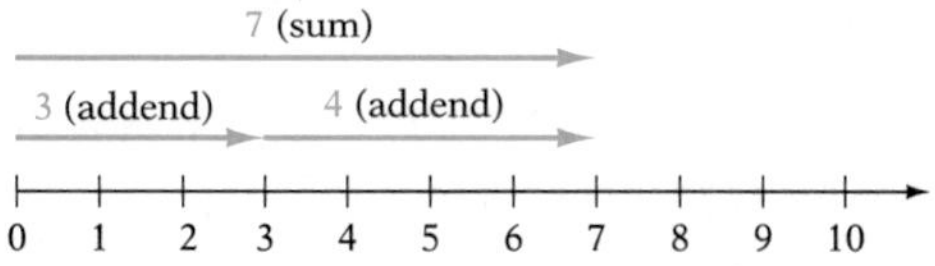


$$3 + 4 = 7$$

More than two numbers can be added on the number line.

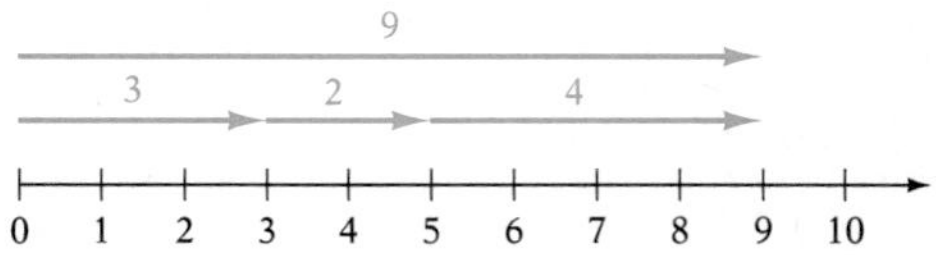


$$3 + 2 + 4 = 9$$

Some special properties of addition that are used frequently are given below.

**Addition Property of Zero**

Zero added to a number does not change the number.

$$4 + 0 = 4$$
$$0 + 7 = 7$$

**Commutative Property of Addition**

Two numbers can be added in either order; the sum will be the same.

$$4 + 8 = 8 + 4$$
$$12 = 12$$

**Associative Property of Addition**

Grouping the addition in any order gives the same result. The parentheses are grouping symbols and have the meaning "do the operations inside the parentheses first."

$$\underbrace{(4 + 2)}_{6} + 3 = 4 + \underbrace{(2 + 3)}_{5}$$
$$6 + 3 = 4 + 5$$
$$9 = 9$$

The number line is not useful for adding large numbers. The basic addition facts for adding one digit to one digit are listed in the Appendix on page A2 for your review. Addition of larger numbers requires the repeated use of the basic addition facts.

To add large numbers, begin by arranging the numbers vertically, keeping the digits of the same place value in the same column.

➡ Add: 321 + 6472

| | THOUSANDS | HUNDREDS | TENS | ONES |
|---|---|---|---|---|
| | | 3 | 2 | 1 |
| + | 6 | 4 | 7 | 2 |
| | 6 | 7 | 9 | 3 |

• Add the digits in each column.

There are several words or phrases in English that indicate the operation of addition. Here are some examples.

| | | |
|---|---|---|
| added to | 3 added to 5 | 5 + 3 |
| more than | 7 more than 5 | 5 + 7 |
| the sum of | the sum of 3 and 9 | 3 + 9 |
| increased by | 4 increased by 6 | 4 + 6 |
| the total of | the total of 8 and 3 | 8 + 3 |
| plus | 5 plus 10 | 5 + 10 |

➡ What is the sum of 24 and 71?

The phrase *the sum of* means to add.

$$\begin{array}{r} 24 \\ +\ 71 \\ \hline 95 \end{array}$$

The sum of 24 and 71 is 95.

When the sum of the digits in a column exceeds 9, the addition will involve "carrying."

➡ Add: 487 + 369

| | HUNDREDS | TENS | ONES |
|---|---|---|---|
| | | 1 | |
| | 4 | 8 | 7 |
| + | 3 | 6 | 9 |
| | | | 6 |

• Add the ones' column.
7 + 9 = 16 (1 ten + 6 ones).
Write the 6 in the ones' column and carry the 1 ten to the tens' column.

$$\begin{array}{r} {\scriptstyle 1\ 1\ \ } \\ 4\ 8\ 7 \\ +\ 3\ 6\ 9 \\ \hline 5\ 6 \end{array}$$

• Add the tens' column.
1 + 8 + 6 = 15 (1 hundred + 5 tens).
Write the 5 in the tens' column and carry the 1 hundred to the hundreds' column.

$$\begin{array}{r} {\scriptstyle 1\ 1\ \ } \\ 4\ 8\ 7 \\ +\ 3\ 6\ 9 \\ \hline 8\ 5\ 6 \end{array}$$

• Add the hundreds' column.
1 + 4 + 3 = 8 (8 hundreds).
Write the 8 in the hundreds' column

**Example 1** Find the total of 17, 103, and 8.

**Solution**

$$\begin{array}{r} 17 \\ 103 \\ +\quad 8 \\ \hline 128 \end{array}$$

**You Try It 1** What is 347 increased by 12,453?

**Your solution**

**Example 2** Add: 89 + 36 + 98

**Solution**

$$\begin{array}{r} {}^{2}\phantom{9} \\ 89 \\ 36 \\ +\ 98 \\ \hline 223 \end{array}$$

**You Try It 2** Add: 95 + 88 + 67

**Your solution**

**Example 3** Add:

$$\begin{array}{r} 41{,}395 \\ 4{,}327 \\ 497{,}625 \\ +\quad 32{,}991 \\ \hline \end{array}$$

**Solution**

$$\begin{array}{r} {}^{1\,1\,2\;\,2\,1} \\ 41{,}395 \\ 4{,}327 \\ 497{,}625 \\ +\quad 32{,}991 \\ \hline 576{,}338 \end{array}$$

**You Try It 3** Add:

$$\begin{array}{r} 392 \\ 4{,}079 \\ 89{,}035 \\ +\quad 4{,}992 \\ \hline \end{array}$$

**Your solution**

*Solutions on p. A7*

## ESTIMATION

### Estimation and Calculators

At some places in the text, you will be asked to use your calculator. Effective use of a calculator requires that you estimate the answer to the problem. This helps ensure that you have entered the numbers correctly and pressed the correct keys.

For example, if you use your calculator to find 22,347 + 5896 and the answer in the calculator's display is 131,757,912, you should realize that you have entered some part of the calculation incorrectly. In this case, you pressed [×] instead of [+]. By estimating the answer to a problem, you can help ensure the accuracy of your calculations. The symbol ≈ is used to denote **approximately equal.**

For example, to estimate the answer to 22,347 + 5896, round off each number to the same place value. In this case we will round to the nearest thousand. Then add.

$$\begin{array}{rcr} 22{,}347 & \approx & 22{,}000 \\ +\quad 5{,}896 & \approx & +\quad 6{,}000 \\ \hline & & 28{,}000 \end{array}$$

The sum 22,347 + 5896 is approximately 28,000. Knowing this, you would know that 131,757,912 is much too large and is therefore incorrect.

To estimate the sum of two numbers, first round each whole number to the same place value and then add. Compare this answer with the calculator's answer.

## *Objective B To solve application problems*

To solve an application problem, first read the problem carefully. The **Strategy** involves identifying the quantity to be found and planning the steps that are necessary to find that quantity. The **Solution** involves performing each operation stated in the Strategy and writing the answer.

A table displaying the number of pounds of the best selling ready-to-eat cereals for the year ending in 1992 is shown at the right.

| *Company* | *Cereal* | *Amount (lb)* |
|---|---|---|
| General Mills | Cheerios | 201,600,000 |
| Kellogg's | Corn Flakes | 128,200,000 |
| Kellogg's | Frosted Flakes | 113,300,000 |
| Kellogg's | Raisin Bran | 81,100,000 |
| Kellogg's | Rice Krispies | 74,900,000 |
| Kellogg's | Mini-Wheats | 67,100,000 |
| General Mills | Total | 52,900,000 |

➡ Find the total amount of ready-to-eat cereal sold for Kellogg's five best-selling cereals for the year ending in 1992.

**Strategy** To find the total amount of cereal sold by Kellogg's, read the table to find the amount of each kind of cereal sold by Kellogg's. Then add the amounts.

**Solution**

$$\begin{array}{r} 128{,}200{,}000 \\ 113{,}300{,}000 \\ 81{,}100{,}000 \\ 74{,}900{,}000 \\ +\ \ 67{,}100{,}000 \\ \hline 464{,}600{,}000 \end{array}$$

For the year ended in 1992, Kellogg's sold 464,600,000 pounds of ready-to-eat cereal.

**Example 4**
Your paycheck shows deductions of \$225 for savings, \$98 for taxes, and \$27 for insurance. Find the total of the three deductions.

**Strategy**
To find the total of the deductions, add the three amounts (\$225, \$98, and \$27).

**Solution**

$$\begin{array}{r} \$225 \\ 98 \\ +\ \ 27 \\ \hline \$350 \end{array}$$

The total of the three deductions is \$350.

**You Try It 4**
Anna Barrera has a monthly budget of \$475 for food, \$275 for car expenses, and \$120 for entertainment. Find the total amount budgeted for the three items each month.

**Your strategy**

**Your solution**

*Solution on p. A7*

## 1.2 Exercises

### *Objective A*

Add.

1. $\begin{array}{r} 17 \\ +\ 11 \\ \hline \end{array}$
2. $\begin{array}{r} 25 \\ +\ 63 \\ \hline \end{array}$
3. $\begin{array}{r} 83 \\ +\ 42 \\ \hline \end{array}$
4. $\begin{array}{r} 63 \\ +\ 94 \\ \hline \end{array}$
5. $\begin{array}{r} 77 \\ +\ 25 \\ \hline \end{array}$
6. $\begin{array}{r} 63 \\ +\ 49 \\ \hline \end{array}$
7. $\begin{array}{r} 56 \\ +\ 98 \\ \hline \end{array}$
8. $\begin{array}{r} 86 \\ +\ 68 \\ \hline \end{array}$
9. $\begin{array}{r} 658 \\ +\ 831 \\ \hline \end{array}$
10. $\begin{array}{r} 842 \\ +\ 936 \\ \hline \end{array}$
11. $\begin{array}{r} 735 \\ +\ 93 \\ \hline \end{array}$
12. $\begin{array}{r} 189 \\ +\ 50 \\ \hline \end{array}$
13. $\begin{array}{r} 859 \\ +\ 725 \\ \hline \end{array}$
14. $\begin{array}{r} 637 \\ +\ 829 \\ \hline \end{array}$
15. $\begin{array}{r} 470 \\ +\ 749 \\ \hline \end{array}$
16. $\begin{array}{r} 427 \\ +\ 690 \\ \hline \end{array}$
17. $\begin{array}{r} 36{,}925 \\ +\ 65{,}392 \\ \hline \end{array}$
18. $\begin{array}{r} 56{,}772 \\ +\ 51{,}239 \\ \hline \end{array}$
19. $\begin{array}{r} 50{,}873 \\ +\ 28{,}453 \\ \hline \end{array}$
20. $\begin{array}{r} 34{,}872 \\ +\ 46{,}079 \\ \hline \end{array}$
21. $\begin{array}{r} 878 \\ 737 \\ +\ 189 \\ \hline \end{array}$
22. $\begin{array}{r} 768 \\ 461 \\ +\ 669 \\ \hline \end{array}$
23. $\begin{array}{r} 319 \\ 348 \\ +\ 912 \\ \hline \end{array}$
24. $\begin{array}{r} 292 \\ 579 \\ +\ 315 \\ \hline \end{array}$
25. $\begin{array}{r} 9409 \\ 3253 \\ +\ 7078 \\ \hline \end{array}$
26. $\begin{array}{r} 8188 \\ 8020 \\ +\ 7104 \\ \hline \end{array}$
27. $\begin{array}{r} 2038 \\ 2243 \\ +\ 3139 \\ \hline \end{array}$
28. $\begin{array}{r} 4252 \\ 6882 \\ +\ 5235 \\ \hline \end{array}$
29. $\begin{array}{r} 67{,}428 \\ 32{,}171 \\ +\ 20{,}971 \\ \hline \end{array}$
30. $\begin{array}{r} 52{,}801 \\ 11{,}664 \\ +\ 89{,}638 \\ \hline \end{array}$
31. $\begin{array}{r} 76{,}290 \\ 43{,}761 \\ +\ 87{,}402 \\ \hline \end{array}$
32. $\begin{array}{r} 43{,}901 \\ 98{,}301 \\ +\ 67{,}943 \\ \hline \end{array}$

Add.

**33.** 20,958 + 3218 + 42

**34.** 80,973 + 5168 + 29

**35.** 392 + 37 + 10,924 + 621

**36.** 694 + 62 + 70,129 + 217

**37.** 294 + 1029 + 7935 + 65

**38.** 692 + 2107 + 3196 + 92

**39.** 97 + 7234 + 69,532 + 276

**40.** 87 + 1698 + 27,317 + 727

**41.** What is 9874 plus 4509?

**42.** What is 7988 plus 5678?

**43.** What is 3487 increased by 5986?

**44.** What is 99,567 added to 126,863?

**45.** What is 23,569 more than 9678?

**46.** What is 7894 more than 45,872?

**47.** What is 479 added to 4579?

**48.** What is 23,902 added to 23,885?

**49.** Find the total of 659, 55, and 1278.

**50.** Find the total of 4561, 56, and 2309.

**51.** Find the sum of 34, 67,892, 329, and 8.

**52.** Find the sum of 32,876, 45, 1289, and 7.

Estimate by rounding to the nearest hundreds. Then use your calculator to add.

53. 1234 + 9780 + 6740

54. 919 + 3642 + 8796

55. 241 + 569 + 390 + 1672

56. 107 + 984 + 1035 + 2904

Estimate by rounding off to the nearest thousands. Then use your calculator to add.

57.
$$\begin{array}{r} 32{,}461 \\ 9{,}844 \\ +\ 59{,}407 \\ \hline \end{array}$$

58.
$$\begin{array}{r} 29{,}036 \\ 22{,}904 \\ +\ 7{,}903 \\ \hline \end{array}$$

59.
$$\begin{array}{r} 25{,}432 \\ 62{,}941 \\ +\ 70{,}390 \\ \hline \end{array}$$

60.
$$\begin{array}{r} 66{,}541 \\ 29{,}365 \\ +\ 98{,}742 \\ \hline \end{array}$$

Estimate by rounding to the nearest ten-thousand. Then use your calculator to add.

| 61. | 62. | 63. | 64. |
|---|---|---|---|
| 67,421 | 21,896 | 281,421 | 542,698 |
| 82,984 | 4,235 | 9,874 | 97,327 |
| 66,361 | 62,544 | 34,394 | 7,235 |
| 10,792 | 21,892 | 526,398 | 73,667 |
| + 34,037 | + 1,334 | + 94,631 | + 173,201 |

Estimate by rounding to the nearest million. Then use your calculator to add.

| 65. | 66. | 67. | 68. |
|---|---|---|---|
| 28,627,052 | 1,792,085 | 12,377,491 | 46,751,070 |
| 983,073 | 29,919,301 | 3,409,723 | 6,095,832 |
| + 3,081,496 | + 3,406,882 | 7,928,026 | 280,011 |
| | | + 10,705,682 | + 1,563,897 |

## *Objective B* *Application Problems*

**69.** The attendance at a Friday night rock concert was 2114, and the attendance at the Saturday night concert was 3678. Find the total attendance at the two concerts.

**70.** The attendance at a Friday night San Diego Padres game was 16,542. Attendance at the Saturday night game was 20,763. Find the total attendance for the two games.

**71.** Dan Marino threw 5 passes for 42 yards in the first quarter, 7 passes for 117 yards in the second quarter, 9 passes for 66 yards in the third quarter, and 8 passes for 82 yards in the fourth quarter. Find the total number of yards gained by passing.

**72.** The high school basketball team scored 17 points in the first quarter, 26 points in the second quarter, 24 points in the third quarter, and 22 points in the fourth quarter. Find the total number of points scored.

**73.** Ken Owens, an account executive, received commissions of $3,168 during October, $2,986 during November, and $819 during December. Find the total commission Ken received for the three-month period.

**74.** The attendance figures for the first four games of the 1992 National League playoffs between the Atlanta Braves and the Pittsburg Pirates were 51,971, 51,975, 56,610 and 57,164. Find the total attendance for the four games.

**75.** There were 207,429 paid admissions for the first four games of the 1992 World Series. There were 52,268 and 51,763 paid admissions for the fifth and sixth games of the series.
**a.** Find the total paid attendance for the fifth and sixth games.
**b.** Find the total paid attendance for the entire series.

**76.** A computer software store in the Westside shopping center sold 3578 compact disks during the first 9 months of the year. The store sold 359 compact disks in October, 259 in November, and 1027 in December.
a. How many compact disks were sold during the last three months of the year?
b. What was the number of compact disks sold during the year?

**77.** A student has $2135 in a checking account to be used for the fall semester. During the summer the student makes deposits of $518, $678, and $468.
a. Find the total amount deposited.
b. Find the new checking account balance, assuming there are no withdrawals.

**78.** The odometer on a moving van reads 68,692. The driver plans to drive 515 miles the first day, 492 miles the second day, and 278 miles the third day.
a. How many miles will be driven during the three days?
b. What will the odometer reading be at the end of the trip?

### *APPLYING THE CONCEPTS*

**79.** How many two-digit numbers are there? How many three-digit numbers are there?

**80.** If you roll two ordinary six-sided dice and add the two numbers that appear on top, how many different sums are possible?

**81.** If you add two *different* whole numbers, is the sum always greater than either one of the numbers? If not, give an example.

**82.** If you add two whole numbers, is the sum always greater than either one of the numbers? If not, give an example. (Compare this with the previous exercise.)

**83.** [W] Make up a word problem for which the answer is the sum of 34 and 28.

**84.** Call a number "lucky" if it ends in a 7. How many lucky numbers are less than 100?

The accompanying table shows the average amount of money all Americans have invested in selected assets and the amounts invested for Americans between the ages of 16 and 34.

**85.** What is the total average amount for all Americans in checking accounts, savings accounts, and U.S. Savings Bonds?

**86.** What is the total average amount for Americans ages 16 to 34 in checking accounts, savings accounts, and U.S. Savings Bonds?

**87.** What is the total average amount for Americans ages 16 to 34 in all categories except home equity and retirement?

**88.** Is the sum of the average amounts invested in home equity and retirement for all Americans greater than or less than that same sum for Americans between the ages of 16 and 34?

| | *All Americans* | *Ages 16 to 34* |
|---|---|---|
| *Checking Accounts* | $487 | $375 |
| *Savings Accounts* | 3,494 | 1,155 |
| *U.S. Savings Bonds* | 546 | 266 |
| *Money Market* | 10,911 | 4,427 |
| *Stocks/mutual Funds* | 4,510 | 1,615 |
| *Home Equity* | 43,070 | 17,184 |
| *Retirement* | 9,016 | 4,298 |

# 1.3 Subtraction of Whole Numbers

## *Objective A To subtract whole numbers without borrowing*

**POINT OF INTEREST**

The use of the minus sign dates from the same period as the plus sign, around 1515.

**Subtraction** is the process of finding the difference between two numbers.

By counting, we see that the difference between \$8 and \$5 is \$3.

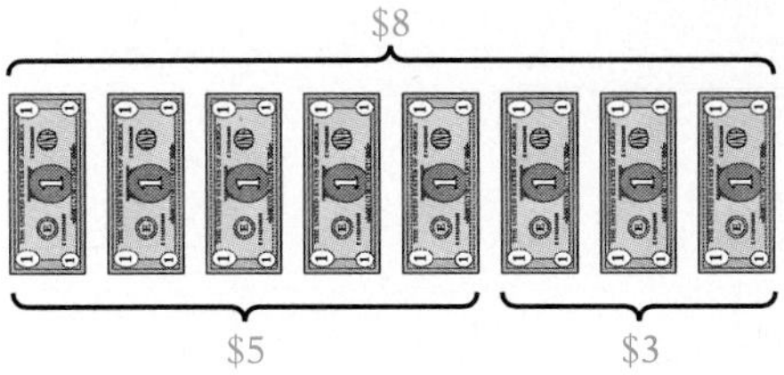


| \$8 | − | \$5 | = | \$3 |
|---|---|---|---|---|
| **Minuend** | | **Subtrahend** | | **Difference** |

The difference 8 − 5 can be shown on the number line.

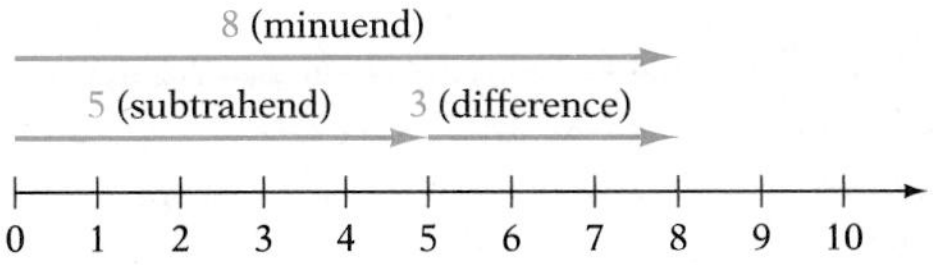


Note from the number line that addition and subtraction are related.

| | |
|---|---|
| Subtrahend | 5 |
| + Difference | + 3 |
| = Minuend | 8 |

The fact that the sum of the subtrahend and the difference equals the minuend can be used to check subtraction.

To subtract large numbers, begin by arranging the numbers vertically, keeping the digits that have the same place value in the same column. Then subtract the digits in each column.

➡ Subtract 8955 − 2432 and check.

| THOUSANDS | HUNDREDS | TENS | ONES |
|---|---|---|---|
| 8 | 9 | 5 | 5 |
| − 2 | 4 | 3 | 2 |
| 6 | 5 | 2 | 3 |

Check:

| | |
|---|---|
| Subtrahend | 2432 |
| + Difference | + 6523 |
| = Minuend | 8955 |

**Example 1** Subtract 6594 − 3271 and check.

**Solution**

```
  6594    Check:   3271
− 3271           + 3323
  3323             6594
```

**You Try It 1** Subtract 8925 − 6413 and check.

**Your solution**

**Example 2** Subtract 15,762 − 7541 and check.

**Solution**

```
  15,762    Check:   7,541
−  7,541           + 8,221
   8,221            15,762
```

**You Try It 2** Subtract 17,504 − 9302 and check.

**Your solution**

*Solutions on p. A8*

## *Objective B To subtract whole numbers with borrowing*

In all the subtraction problems in the previous objective, for each place value the lower digit was not larger than the upper digit. When the lower digit is larger than the upper digit, subtraction will involve "borrowing."

➡ Subtract: 692 − 378

| | HUNDREDS | TENS | ONES |
|---|---|---|---|
| | | 8+1 | |
| | 6 | 9 | 2 |
| − | 3 | 7 | 8 |

Because 8 > 2, borrowing is necessary. 9 tens = 8 tens + 1 ten.

| | HUNDREDS | TENS | ONES |
|---|---|---|---|
| | | 8+①→ | 10 |
| | 6 | ~~9~~ | 2 |
| − | 3 | 7 | 8 |

Borrow 1 ten from the tens' column and write 10 in the ones' column.

| | HUNDREDS | TENS | ONES |
|---|---|---|---|
| | | 8 | 12 |
| | 6 | ~~9~~ | ~~2~~ |
| − | 3 | 7 | 8 |

Add the borrowed 10 to 2.

| | HUNDREDS | TENS | ONES |
|---|---|---|---|
| | | 8 | 12 |
| | 6 | ~~9~~ | ~~2~~ |
| − | 3 | 7 | 8 |
| | 3 | 1 | 4 |

Subtract the digits in each column.

The phrases below are used to indicate the operation of subtraction. An example is shown at the right of each phrase.

| | | |
|---|---|---|
| minus | 8 minus 5 | 8 − 5 |
| less | 9 less 3 | 9 − 3 |
| less than | 2 less than 7 | 7 − 2 |
| the difference between | the difference between 8 and 2 | 8 − 2 |
| decreased by | 5 decreased by 1 | 5 − 1 |

➡ Find the difference between 1234 and 485 and check.

From the phrases that indicate subtraction, the difference between 1234 and 485 is 1234 − 485.

| | | | | |
|---|---|---|---|---|
| | | | 2 | 14 |
| | 1 | 2 | ~~3~~ | ~~4~~ |
| − | | 4 | 8 | 5 |
| | | | | 9 |

| | | | | |
|---|---|---|---|---|
| | | 1 | 12 | 14 |
| | 1 | ~~2~~ | ~~3~~ | ~~4~~ |
| − | | 4 | 8 | 5 |
| | | | 4 | 9 |

| | | | | |
|---|---|---|---|---|
| | 0 | 11 | 12 | 14 |
| | ~~1~~ | ~~2~~ | ~~3~~ | ~~4~~ |
| − | | 4 | 8 | 5 |
| | | 7 | 4 | 9 |

Check:

```
   11
  485
+ 749
 1234
```

Subtraction with a zero in the minuend involves repeated borrowing.

➡ Subtract: 3904 − 1775

| | | | | |
|---|---|---|---|---|
| | | 8 | 10 | |
| | 3 | ~~9~~ | ~~0~~ | 4 |
| − | 1 | 7 | 7 | 5 |

5 > 4
There is a 0 in the tens' column. Borrow 1 hundred (= 10 tens) from the hundreds' column and write 10 in the tens' column.

| | | | | |
|---|---|---|---|---|
| | | | 9 | |
| | | 8 | ~~10~~ | 14 |
| | 3 | ~~9~~ | ~~0~~ | ~~4~~ |
| − | 1 | 7 | 7 | 5 |

Borrow 1 ten from the tens' column and add 10 to the 4 in the ones' column.

| | | | | |
|---|---|---|---|---|
| | | | 9 | |
| | | 8 | ~~10~~ | 14 |
| | 3 | ~~9~~ | ~~0~~ | ~~4~~ |
| − | 1 | 7 | 7 | 5 |
| | 2 | 1 | 2 | 9 |

Subtract the digits in each column.

**Example 3** Subtract 4392 − 678 and check.

**Solution**

```
   3 13  8 12
   4  3  9  2    Check:    678
 −    6  7  8            + 3714
 -----------             ------
   3  7  1  4              4392
```

**You Try It 3** Subtract 3481 − 865 and check.

**Your solution**

**Example 4** Find 23,954 less than 63,221 and check.

**Solution**

```
   5 12  11 11 11
   6  3 , 2  2  1   Check:   23,954
 − 2  3 , 9  5  4          + 39,267
 ----------------          --------
   3  9 , 2  6  7            63,221
```

**You Try It 4** Find 54,562 decreased by 14,485 and check.

**Your solution**

**Example 5** Subtract 46,005 − 32,167 and check.

**Solution**

```
    5 10
  4 6,0 0 5
− 3 2,1 6 7
```

- **There are two zeros in the minuend. Borrow 1 thousand from the thousands' column and write 10 in the hundreds' column.**

```
      9
    5 10 10
  4 6,0 0 5
− 3 2,1 6 7
```

- **Borrow 1 hundred from the hundreds' column and write 10 in the tens' column.**

```
      9  9
    5 10 10 15
  4 6,0  0  5
− 3 2,1  6  7
-------------
  1 3,8  3  8
```

- **Borrow 1 ten from the tens' column and add 10 to the 5 in the ones' column.**

```
Check:   32,167
       + 13,838
       --------
         46,005
```

**You Try It 5** Subtract 64,003 − 54,936 and check.

**Your solution**

*Solutions on p. A8*

## ESTIMATION

### Estimating the Difference Between Two Whole Numbers

Estimate and then use your calculator to find 323,502 − 28,912.

To estimate the difference between two numbers, round each number to the same place value. In this case we will round to the nearest ten-thousand. Then subtract. The estimated answer is 290,000.

```
  323,502 ≈    320,000
−  28,912 ≈  −  30,000
             ---------
               290,000
```

Now use your calculator to find the exact result. The exact answer is 294,590.

323502 [−] 28912 [=] 294590

## Objective C To solve application problems

The table at the right shows the number and value of corporate takeovers during the years 1987 through 1993. Use the table for Example 6 and You Try It 6.

Takeover deals are bouncing back this year. Number and value of deals each year through August 20:

| | Number of Deals | Value (billions) |
|---|---|---|
| 1987 | 1891 | $135 |
| 1988 | 2447 | $199 |
| 1989 | 3509 | $199 |
| 1990 | 3609 | $122 |
| 1991 | 3301 | $ 88 |
| 1992 | 3465 | $ 82 |
| 1993 | 3809 | $135 |

**Example 6**
How many more corporate takeovers occurred in 1993 than in 1987?

**Strategy**
To find the number, subtract the number of takeovers that occurred in 1987 (1891) from the number that occurred in 1993 (3809).

**Solution**

$$\begin{array}{r} 3809 \\ -\ 1891 \\ \hline 1918 \end{array}$$

1918 more takeovers occurred in 1993.

**You Try It 6**
Find the difference between the value of the takeovers that occurred in 1992 and those that occurred in 1993.

**Your strategy**

**Your solution**

**Example 7**
You had a balance of $815 in your checking account. You then wrote checks in the amount of $112 for taxes, $57 for food, and $39 for shoes. What is your new checking account balance?

**Strategy**
To find your new checking account balance:

- Add to find the total of the three checks ($112 + $57 + $39).
- Subtract the total of the three checks from the old balance ($815).

**Solution**

$$\begin{array}{r} 112 \\ 57 \\ +\ 39 \\ \hline 208 \end{array} \text{ total of checks} \qquad \begin{array}{r} 815 \\ -\ 208 \\ \hline 607 \end{array}$$

Your new checking account balance is $607.

**You Try It 7**
Your total salary is $638. Deductions of $127 for taxes, $18 for insurance, and $35 for savings are taken from your pay. Find your take-home pay.

**Your strategy**

**Your solution**

*Solutions on p. A8*

## 1.3 Exercises

### *Objective A*

Subtract.

1. $\begin{array}{r} 9 \\ -5 \\ \hline \end{array}$ 2. $\begin{array}{r} 8 \\ -7 \\ \hline \end{array}$ 3. $\begin{array}{r} 8 \\ -4 \\ \hline \end{array}$ 4. $\begin{array}{r} 7 \\ -3 \\ \hline \end{array}$ 5. $\begin{array}{r} 10 \\ -0 \\ \hline \end{array}$ 6. $\begin{array}{r} 7 \\ -0 \\ \hline \end{array}$

7. $\begin{array}{r} 11 \\ -4 \\ \hline \end{array}$ 8. $\begin{array}{r} 12 \\ -8 \\ \hline \end{array}$ 9. $\begin{array}{r} 19 \\ -8 \\ \hline \end{array}$ 10. $\begin{array}{r} 15 \\ -6 \\ \hline \end{array}$ 11. $\begin{array}{r} 16 \\ -7 \\ \hline \end{array}$ 12. $\begin{array}{r} 18 \\ -9 \\ \hline \end{array}$

13. $\begin{array}{r} 25 \\ -3 \\ \hline \end{array}$ 14. $\begin{array}{r} 55 \\ -4 \\ \hline \end{array}$ 15. $\begin{array}{r} 68 \\ -8 \\ \hline \end{array}$ 16. $\begin{array}{r} 77 \\ -3 \\ \hline \end{array}$ 17. $\begin{array}{r} 89 \\ -23 \\ \hline \end{array}$ 18. $\begin{array}{r} 68 \\ -43 \\ \hline \end{array}$

19. $\begin{array}{r} 54 \\ -21 \\ \hline \end{array}$ 20. $\begin{array}{r} 88 \\ -57 \\ \hline \end{array}$ 21. $\begin{array}{r} 1202 \\ -701 \\ \hline \end{array}$ 22. $\begin{array}{r} 1305 \\ -404 \\ \hline \end{array}$ 23. $\begin{array}{r} 1763 \\ -801 \\ \hline \end{array}$

24. $\begin{array}{r} 1497 \\ -706 \\ \hline \end{array}$ 25. $\begin{array}{r} 8974 \\ -3972 \\ \hline \end{array}$ 26. $\begin{array}{r} 2836 \\ -1711 \\ \hline \end{array}$ 27. $\begin{array}{r} 8976 \\ -7463 \\ \hline \end{array}$ 28. $\begin{array}{r} 9273 \\ -6142 \\ \hline \end{array}$

29. $17 - 8$ 30. $15 - 7$ 31. $12 - 5$ 32. $93 - 71$ 33. $83 - 52$

34. $77 - 36$ 35. $129 - 82$ 36. $132 - 61$ 37. $969 - 44$ 38. $1347 - 103$

39. $4865 - 304$ 40. $1525 - 702$ 41. $9999 - 6794$ 42. $7806 - 3405$ 43. $8843 - 7621$

44. What is 3795 minus 1092?

45. What is 9071 minus 6050?

46. Find the difference between 9763 and 541.

47. Find the difference between 6094 and 3072.

48. What is 3701 less than 6932?

49. What is 2031 less than 5071?

50. Find 6509 decreased by 3102.

51. Find 7994 decreased by 7782.

52. Find 23,907 less 12,705.

53. Find 65,986 less 5741.

### *Objective B*

Subtract.

**54.** 71 − 18

**55.** 93 − 28

**56.** 47 − 18

**57.** 44 − 27

**58.** 71 − 67

**59.** 37 − 29

**60.** 50 − 27

**61.** 70 − 33

**62.** 993 − 537

**63.** 681 − 328

**64.** 250 − 192

**65.** 840 − 783

**66.** 768 − 194

**67.** 679 − 519

**68.** 770 − 395

**69.** 674 − 337

**70.** 3526 − 387

**71.** 1712 − 289

**72.** 4350 − 729

**73.** 1702 − 948

**74.** 1607 − 869

**75.** 5933 − 3754

**76.** 7293 − 3748

**77.** 8143 − 2417

**78.** 7236 − 1978

**79.** 5714 − 2367

**80.** 8462 − 3575

**81.** 9407 − 2918

**82.** 3706 − 2957

**83.** 8605 − 7716

**84.** 8052 − 2709

**85.** 80,305 − 9176

**86.** 70,702 − 4239

**87.** 10,004 − 9306

**88.** 80,009 − 63,419

**89.** 70,618 − 41,213

**90.** 80,053 − 27,649

**91.** 70,700 − 21,076

**92.** 80,800 − 42,023

**93.** 2600 − 1972

**94.** 8400 − 3762

**95.** 9003 − 2471

**96.** 6004 − 2392

**97.** 8202 − 3916

**98.** 7050 − 4137

**99.** 7015 − 2973

**100.** 4207 − 1624

**101.** 7005 − 1796

**102.** 8003 − 2735

**103.** 20,005 − 9,627

**104.** 80,004 − 8,237

**105.** Find 10,051 less 9027.

**106.** Find 17,031 less 5792.

**107.** Find the difference between 1003 and 447.

**108.** Find the difference between 9000 and 763.

**109.** What is 223,491 minus 96,987?

**110.** What is 29,874 minus 21,392?

**111.** What is 29,797 less than 68,005?

**112.** What is 69,379 less than 70,004?

**113.** What is 25,432 decreased by 7994?

**114.** What is 86,701 decreased by 9976?

Estimate by rounding to the nearest ten-thousand. Then use your calculator to subtract.

115. $\begin{array}{r} 80{,}032 \\ -\ 19{,}605 \\ \hline \end{array}$

116. $\begin{array}{r} 90{,}765 \\ -\ 60{,}928 \\ \hline \end{array}$

117. $\begin{array}{r} 32{,}574 \\ -\ 10{,}961 \\ \hline \end{array}$

118. $\begin{array}{r} 96{,}430 \\ -\ 59{,}762 \\ \hline \end{array}$

119. $\begin{array}{r} 567{,}423 \\ -\ 208{,}444 \\ \hline \end{array}$

120. $\begin{array}{r} 300{,}712 \\ -\ 198{,}714 \\ \hline \end{array}$

### ***Objective C*** *Application Problems*

**121.** You have $304 in your checking account. If you write a check for $139, how much is left in your checking account?

**122.** Sam Akyol has $521 in a checking account. If Sam writes a check for $247, how much is left in the checking account?

**123.** The tennis coach at a high school purchased a video camera that costs $1079 and makes a down payment of $180. Find the amount that remains to be paid.

**124.** Rod Guerra, an engineer, purchased a used car that cost $5225 and made a down payment of $450. Find the amount that remains to be paid.

**125.** The odometer of a rental car read 3,459 miles after a trip of 297 miles. What was the odometer reading at the start of the trip?

**126.** At the end of a vacation trip, the odometer of your car read 63,459 miles. If the odometer reading at the start of the trip was 62,963, what was the length of the trip?

**127.** Florida had 20,188,506 acres of wetlands 200 years ago. Today Florida has 10,901,793 acres of wetlands. How many acres of wetlands has Florida lost over the last 200 years?

**128.** The seating capacity of the Kingdome in Seattle is 57,748. The seating capacity of Fenway Park in Boston is 34,182. Find the difference between the seating capacity of the Kingdome and that of Fenway Park.

The graph at the right shows the number of months required to recover all interest and principal from the Social Security fund after a person retires. Use this graph for Exercises 129 and 130.

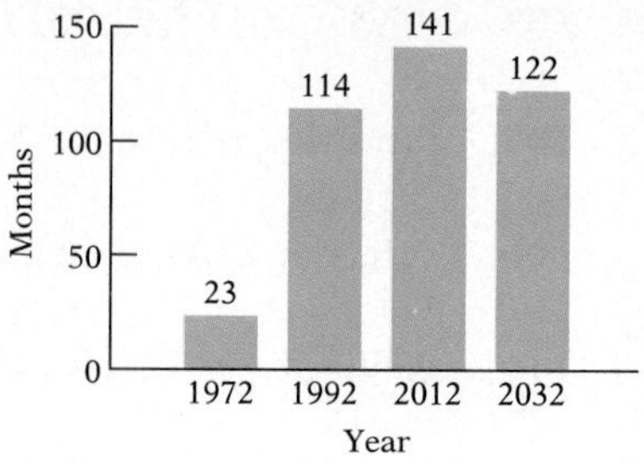


**129.** How much longer did it take to withdraw principal and interest in 1992 than it did in 1972?

**130.** How much longer will it take to withdraw principal and interest in 2012 than it did in 1992?

### *APPLYING THE CONCEPTS*

**131.** Answer true or false.
  **a.** The phrases "the difference between 9 and 5" and "5 less than 9" mean the same thing.
  **b.** $9 - (5 - 3) = (9 - 5) - 3$.
  **c.** Subtraction is an associative operation. *Hint*: See part b of this exercise.

**132.** [W] Explain how you can check the answer to a subtraction problem.

**133.** [W] Make up a word problem for which the difference between 15 and 8 is the answer.

The table at the right shows the cost of two cars and of the various options the car buyer wants to have. Use this table for Exercises 134 and 135.

| | *Ford* | *Chevrolet* |
|---|---|---|
| Basic car | $8589 | $8799 |
| Air | 1025 | 978 |
| Anti-lock brakes | 675 | 533 |
| Stereo AM/FM | 425 | 515 |
| Power windows | 915 | 1025 |

**134.** Which of these two cars, equipped with all the options, costs more?

**135.** What is the difference in cost between the two basic cars?

The data in the table at the right show the approximate total average score on the Scholastic Aptitude Test (SAT) in the United States for a five-year period. Use this table for Exercises 136 and 137.

| *Year* | *Score* |
|---|---|
| 1989 | 903 |
| 1990 | 899 |
| 1991 | 896 |
| 1992 | 898 |
| 1993 | 902 |

**136.** Between which two consecutive years did the SAT scores decrease the most? What was the amount of the decrease?

**137.** Between which two years did the SAT scores increase the most? What was the amount of the increase?

# 1.4 Multiplication of Whole Numbers

## *Objective A* *To multiply a number by a single digit*

Six boxes of toasters are ordered. Each box contains eight toasters. How many toasters are ordered?

This problem can be worked by adding 6 eights.

$8 + 8 + 8 + 8 + 8 + 8 = 48$

This problem involves repeated addition of the same number and can be worked by a shorter process called **multiplication.** Multiplication is the repeated addition of the same number.

$8 + 8 + 8 + 8 + 8 + 8 = 48$

The numbers that are multiplied are called **factors.** The answer is called the **product.**

or

| 6 | × | 8 | = 48 |
|---|---|---|---|
| Factor | | Factor | Product |

The product of 6 × 8 can be represented on the number line. The arrow representing the whole number 8 is repeated 6 times. The result is the arrow representing 48.

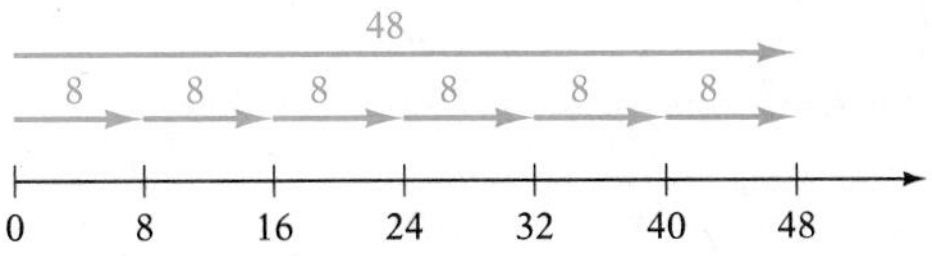


The times sign "×" is one symbol that is used to mean multiplication. Another common symbol used is a dot placed between the numbers.

$7 \times 8 = 56 \qquad 7 \cdot 8 = 56$

As with addition, there are some useful properties of multiplication.

**Multiplication Property of Zero**

The product of a number and zero is zero.

$0 \times 4 = 0$
$7 \times 0 = 0$

**Multiplication Property of One**

The product of a number and one is the number.

$1 \times 6 = 6$
$8 \times 1 = 8$

**Commutative Property of Multiplication**

Two numbers can be multiplied in either order. The product will be the same.

$4 \times 3 = 3 \times 4$
$12 = 12$

**Associative Property of Multiplication**

Grouping the numbers to be multiplied in any order gives the same result. Do the multiplication inside the parentheses first.

$\underbrace{(4 \times 2)}_{8} \times 3 = 4 \times \underbrace{(2 \times 3)}_{6}$
$8 \times 3 = 4 \times 6$
$24 = 24$

The basic facts for multiplying one-digit numbers are listed for your review on page A2 in the Appendix. Multiplication of larger numbers requires the repeated use of the basic multiplication facts.

➡ Multiply: 37 × 4

$$\begin{array}{r} {}^{2}\phantom{7} \\ 37 \\ \times\ \ 4 \\ \hline 8 \end{array}$$

- $4 \times 7 = 28$ (2 tens + 8 ones). Write the 8 in the ones' column and carry the 2 to the tens' column.

$$\begin{array}{r} {}^{2}\phantom{7} \\ 37 \\ \times\ \ 4 \\ \hline 148 \end{array}$$

- The 3 in 37 is 3 tens.

| | | |
|---|---|---|
| | 4 × 3 tens = | 12 tens |
| Add the carry digit. | | + 2 tens |
| | | 14 tens |

- Write the 14.

The phrases below are used to indicate the operation of multiplication. An example is shown at the right of each phrase.

| | | |
|---|---|---|
| times | 7 times 3 | $7 \cdot 3$ |
| the product of | the product of 6 and 9 | $6 \cdot 9$ |
| multiplied by | 8 multiplied by 2 | $2 \cdot 8$ |

**Example 1** Multiply: 735 × 9

**Solution**

$$\begin{array}{r} {}^{34}\phantom{5} \\ 735 \\ \times\ \ \ 9 \\ \hline 6615 \end{array}$$

**You Try It 1** Multiply: 648 × 7

**Your solution**

*Solution on p. A8*

## *Objective B To multiply larger whole numbers*

Note the pattern when the following numbers are multiplied.

Multiply the nonzero part of the factors.

Now attach the same number of zeros to the product as the total number of zeros in the factors.

4 × 2
4 × 20 = 80
1 zero

4 × 2
4 × 200 = 800
2 zeros

4 × 2
40 × 200 = 8000
3 zeros

12 × 5
12 × 5000 = 60,000
3 zeros

➡ Find the product of 47 and 23.

| Multiply by the ones' digit. | Multiply by the tens' digit. | Add. |
| --- | --- | --- |
| 47<br>× 23<br>141 (= 47 × 3) | 47<br>× 23<br>141<br>940 (= 47 × 20)<br><br>Writing the 0 is optional. | 47<br>× 23<br>141<br>94<br>1081 |

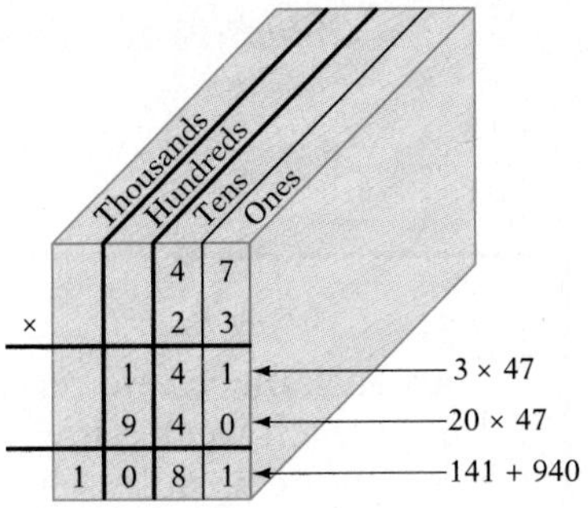


The place-value chart illustrates the placement of the products.

Note the placement of the products when we are multiplying by a factor that contains a zero.

➡ Multiply: 439 × 206

```
  439
× 206
 2634
 000    0 × 439
878
90,434
```

When working the problem, we usually write only one zero. Writing this zero ensures the proper placement of the products.

```
  439
× 206
 2634
8780
90,434
```

**Example 2** Find 829 multiplied by 603.

**Solution**

```
   829
 × 603
  2487
 49740
499,887
```

**You Try It 2** Multiply: 756 × 305

**Your solution**

*Solution on p. A8*

**ESTIMATION**

**Estimating the Product of Two Whole Numbers**

Estimate and then use your calculator to find 3267 × 389.

To estimate a product, round each number so that all the digits are zero except the first digit. Then multiply. The estimated answer is 1,200,000.

$$\begin{array}{rcr} 3267 & \approx & 3000 \\ \times\ 389 & \approx & \times\ \ 400 \\ \hline & & 1{,}200{,}000 \end{array}$$

Now use your calculator to find the exact answer. The exact answer is 1,270,863.

3267 [×] 389 [=] 1270863

### *Objective C To solve application problems*

**Example 3**
An auto mechanic receives a salary of $525 each week. How much does the auto mechanic earn in 4 weeks?

**Strategy**
To find the mechanic's earnings for 4 weeks, multiply the weekly salary ($525) by the number of weeks (4).

**Solution**

$$\begin{array}{r} \$525 \\ \times \quad 4 \\ \hline \$2100 \end{array}$$

The mechanic earns $2100 in 4 weeks.

**You Try It 3**
A new-car dealer receives a shipment of 37 cars each month. Find the number of cars the dealer will receive in 12 months.

**Your strategy**

**Your solution**

**Example 4**
A press operator earns $320 for working a 40-hour week. This week the press operator also worked 7 hours of overtime at $13 an hour. Find the press operator's total pay for the week.

**Strategy**
To find the press operator's income for the week:

- Find the overtime pay by multiplying the hours of overtime (7) by the overtime rate of pay ($13).
- Add the weekly salary ($320) to the overtime pay.

**Solution**

$$\begin{array}{r} \$13 \\ \times \quad 7 \\ \hline \$91 \end{array} \text{ overtime pay} \qquad \begin{array}{r} \$320 \\ + \quad 91 \\ \hline \$411 \end{array}$$

The press operator earned $411 this week.

**You Try It 4**
The buyer for Ross Department Store can buy 80 men's suits for $4800. Each sports jacket will cost the store $23. The manager orders 80 men's suits and 25 sports jackets. What is the total cost of the order?

**Your strategy**

**Your solution**

*Solutions on p. A9*

## 1.4 Exercises

### *Objective A*

Multiply.

| | | | | |
|---|---|---|---|---|
| **1.** $\begin{array}{r} 3 \\ \times 4 \\ \hline \end{array}$ | **2.** $\begin{array}{r} 2 \\ \times 8 \\ \hline \end{array}$ | **3.** $\begin{array}{r} 5 \\ \times 7 \\ \hline \end{array}$ | **4.** $\begin{array}{r} 6 \\ \times 4 \\ \hline \end{array}$ | **5.** $\begin{array}{r} 5 \\ \times 5 \\ \hline \end{array}$ |
| **6.** $\begin{array}{r} 7 \\ \times 7 \\ \hline \end{array}$ | **7.** $\begin{array}{r} 0 \\ \times 7 \\ \hline \end{array}$ | **8.** $\begin{array}{r} 8 \\ \times 0 \\ \hline \end{array}$ | **9.** $\begin{array}{r} 8 \\ \times 9 \\ \hline \end{array}$ | **10.** $\begin{array}{r} 7 \\ \times 6 \\ \hline \end{array}$ |
| **11.** $\begin{array}{r} 66 \\ \times \ 3 \\ \hline \end{array}$ | **12.** $\begin{array}{r} 70 \\ \times \ 4 \\ \hline \end{array}$ | **13.** $\begin{array}{r} 67 \\ \times \ 5 \\ \hline \end{array}$ | **14.** $\begin{array}{r} 127 \\ \times \ 9 \\ \hline \end{array}$ | **15.** $\begin{array}{r} 623 \\ \times \ 4 \\ \hline \end{array}$ |
| **16.** $\begin{array}{r} 802 \\ \times \ 5 \\ \hline \end{array}$ | **17.** $\begin{array}{r} 607 \\ \times \ 9 \\ \hline \end{array}$ | **18.** $\begin{array}{r} 300 \\ \times \ 5 \\ \hline \end{array}$ | **19.** $\begin{array}{r} 600 \\ \times \ 7 \\ \hline \end{array}$ | **20.** $\begin{array}{r} 906 \\ \times \ 8 \\ \hline \end{array}$ |
| **21.** $\begin{array}{r} 703 \\ \times \ 9 \\ \hline \end{array}$ | **22.** $\begin{array}{r} 127 \\ \times \ 5 \\ \hline \end{array}$ | **23.** $\begin{array}{r} 632 \\ \times \ 3 \\ \hline \end{array}$ | **24.** $\begin{array}{r} 559 \\ \times \ 4 \\ \hline \end{array}$ | **25.** $\begin{array}{r} 632 \\ \times \ 8 \\ \hline \end{array}$ |
| **26.** $\begin{array}{r} 780 \\ \times \ 7 \\ \hline \end{array}$ | **27.** $\begin{array}{r} 690 \\ \times \ 5 \\ \hline \end{array}$ | **28.** $\begin{array}{r} 465 \\ \times \ 4 \\ \hline \end{array}$ | **29.** $\begin{array}{r} 382 \\ \times \ 7 \\ \hline \end{array}$ | **30.** $\begin{array}{r} 367 \\ \times \ 3 \\ \hline \end{array}$ |
| **31.** $\begin{array}{r} 524 \\ \times \ 4 \\ \hline \end{array}$ | **32.** $\begin{array}{r} 337 \\ \times \ 5 \\ \hline \end{array}$ | **33.** $\begin{array}{r} 841 \\ \times \ 6 \\ \hline \end{array}$ | **34.** $\begin{array}{r} 6709 \\ \times \ 7 \\ \hline \end{array}$ | **35.** $\begin{array}{r} 3608 \\ \times \ 5 \\ \hline \end{array}$ |
| **36.** $\begin{array}{r} 8568 \\ \times \ 7 \\ \hline \end{array}$ | **37.** $\begin{array}{r} 5495 \\ \times \ 4 \\ \hline \end{array}$ | **38.** $\begin{array}{r} 4780 \\ \times \ 4 \\ \hline \end{array}$ | **39.** $\begin{array}{r} 3690 \\ \times \ 5 \\ \hline \end{array}$ | **40.** $\begin{array}{r} 9895 \\ \times \ 2 \\ \hline \end{array}$ |

**41.** Find the product of 5, 7, and 4.

**42.** Find the product of 6, 2, and 9.

**43.** Find the product of 5304 and 9.

**44.** Find the product of 458 and 8.

**45.** What is 3208 multiplied by 7?

**46.** What is 5009 multiplied by 4?

**47.** What is 3105 times 6?

**48.** What is 8957 times 8?

### Objective B

Multiply.

49. $16 \times 21$

50. $18 \times 24$

51. $35 \times 26$

52. $27 \times 72$

53. $39 \times 46$

54. $37 \times 25$

55. $67 \times 23$

56. $95 \times 33$

57. $693 \times 91$

58. $581 \times 72$

59. $419 \times 80$

60. $727 \times 60$

61. $8279 \times 46$

62. $9577 \times 35$

63. $6938 \times 78$

64. $8875 \times 67$

65. $7035 \times 57$

66. $6702 \times 48$

67. $3009 \times 35$

68. $6003 \times 57$

69. $809 \times 530$

70. $607 \times 460$

71. $800 \times 325$

72. $700 \times 274$

73. $987 \times 349$

74. $688 \times 674$

75. $312 \times 134$

76. $423 \times 427$

77. $379 \times 500$

78. $684 \times 700$

79. $985 \times 408$

80. $758 \times 209$

81. $3407 \times 309$

82. $5207 \times 902$

83. $4258 \times 986$

84. $6327 \times 876$

85. What is 5763 times 45?

86. What is 7349 times 7?

87. Find the product of 2, 19, and 34.

88. Find the product of 6, 73, and 43.

89. What is 376 multiplied by 402?

90. What is 842 multiplied by 309?

91. Find the product of 233,489 times 3005.

92. Find the product of 34,985 times 9007.

Estimate and then use your calculator to multiply.

93. 8745 × 63

94. 4732 × 93

95. 39,246 × 29

96. 64,409 × 67

97. 2937 × 206

98. 8941 × 726

99. 3097 × 1025

100. 6379 × 2936

101. 32,508 × 591

102. 62,504 × 923

103. 81,405 × 902

104. 66,735 × 844

**Objective C** *Application Problems*

**105.** Rob Hupp owns a compact car that averages 43 miles on 1 gallon of gas. How many miles could the car travel on 12 gallons of gas?

**106.** A plane flying from Los Angeles to Boston uses 865 gallons of jet fuel each hour. How many gallons of jet fuel were used on a 6-hour flight?

**107.** Assume that a machine at the Coca Cola Bottling Company can fill and cap 4200 bottles of Coke in 1 hour. How many bottles of Coke can the machine fill and cap in 40 hours?

**108.** A computer can store 368,640 bytes of information on 1 disk. How many bytes of information can be stored on 6 disks?

**109.** A computer graphics screen has 640 rows of pixels, and there are 480 pixels per row. Find the total number of pixels on the screen.

**110.** Anthony Dias owns 350 shares of Public Service of Colorado that has a yearly dividend of $2 a share paid in 4 quarterly payments. Find the yearly income from the 350 shares of stock.

**111.** A food server buys a car by making payments of $235 each month for 48 months. Find the total cost of the car.

A lighting consultant to a bank suggests that the bank lobby contain 43 can lights, 15 high-intensity lights, 20 fire safety lights, and one chandelier. The table at the right gives the costs for each type of light from two companies. Use this table for Exercises 112 and 113.

| | *Company A* | *Company B* |
|---|---|---|
| *Can Lights* | $2 each | $3 each |
| *High-Intensity* | $6 each | $4 each |
| *Fire Safety* | $12 each | $11 each |
| *Chandelier* | $998 each | $1089 each |

**112.** Which company offers the lighting designer the lights for the lower total price?

**113.** How much can the lighting designer save by purchasing the lights from the company that offers the lower total price?

The table at the right shows the hourly wages of four different job classifications at a small construction company. Use this table for Exercises 114 to 117.

| *Type of Work* | *Wage per Hour* |
|---|---|
| Electrician | \$17 |
| Plumber | \$15 |
| Clerk | \$8 |
| Bookkeeper | \$10 |

**114.** The owner of this company wants to provide the electrical installation for a new house. Based on the architectural plans for the house, it is estimated that it will require 3 electricians each working 50 hours to complete the job. What is the estimated cost for the electricians' labor?

**115.** Carlos Vasquez, a plumbing contractor, hires 4 plumbers from this company at the hourly wage given in the table. If each plumber works 23 hours, what are the total wages paid by Carlos?

**116.** The owner of this company estimates that remodeling a kitchen will require 1 electrician working 30 hours and 1 plumber working 33 hours. This project also requires 3 hours of clerical work and 4 hours of bookkeeping. What is the total cost for these four components of this remodeling?

**117.** An estimate to remodel a bathroom included 1 electrician working 5 hours and 1 plumber working 18 hours. There was 1 hour estimated for clerical work and 2 hours for bookkeeping. What is the total cost of these four components of this remodeling?

## APPLYING THE CONCEPTS

**118.** Determine whether each of the following statements is always true, sometimes true, or never true.
a. A whole number times zero is zero.
b. A whole number times one is the whole number.
c. The product of two whole numbers is greater than either one of the whole numbers.

**119.** [W] According to the National Safety Council, in a recent year a death resulting from an accident occurred at the rate of 1 every 5 min. At this rate, how many accidental deaths occurred each hour? Each day? Throughout the year? Explain how you arrived at your answers.

**120.** [W] In Brazil, 713 acres are deforested each hour. At this rate, how many acres of deforestation occur each day? Each month? Throughout the year? Explain how you arrived at your answers.

**121.** [W] Pick your favorite number between 1 and 9. Multiply the number by 3. Multiply that product by 37037. How is the product related to your favorite number? Explain why this works. (Suggestion: Multiply 3 and 37037 first.)

**122.** There are quite a few tricks based on whole numbers. Here's one about birthdays. Write down the month in which you were born. Multiply by 5. Add 7. Multiply by 20. Subtract 100. Add the day of the month on which you were born. Multiply by 4. Subtract 100. Multiply by 25. Add the year you were born. Subtract 3400. The answer is the month/day/year of your birthday.

# 1.5 Division of Whole Numbers

*Objective A* *To divide by a single digit with no remainder in the quotient*

**Division** is used to separate objects into equal groups.

A store manager wants to display 24 new objects equally on 4 shelves. From the diagram, we see that the manager would place 6 objects on each shelf.

The manager's division problem can be written as follows:

Number of shelves **Divisor** → $\begin{array}{r}6\\4\overline{)24}\end{array}$ ← Number on each shelf **Quotient**; ← Number of objects **Dividend**

Note that the quotient multiplied by the divisor equals the dividend.

| | | | | | | |
|---|---|---|---|---|---|---|
| $\begin{array}{r}6\\4\overline{)24}\end{array}$ | because | 6 Quotient | × | 4 Divisor | = | 24 Dividend |
| $\begin{array}{r}6\\9\overline{)54}\end{array}$ | because | 6 | × | 9 | = | 54 |
| $\begin{array}{r}5\\8\overline{)40}\end{array}$ | because | 5 | × | 8 | = | 40 |

Here are some important quotients and the properties of zero in division:

**Important Quotients**

Any whole number, except zero, divided by itself is 1. $\begin{array}{r}1\\8\overline{)8}\end{array}$ $\begin{array}{r}1\\14\overline{)14}\end{array}$ $\begin{array}{r}1\\10\overline{)10}\end{array}$

Any whole number divided by 1 is the whole number. $\begin{array}{r}9\\1\overline{)9}\end{array}$ $\begin{array}{r}27\\1\overline{)27}\end{array}$ $\begin{array}{r}10\\1\overline{)10}\end{array}$

**Properties of Zero in Division**

Zero divided by any other whole number is zero. $\begin{array}{r}0\\7\overline{)0}\end{array}$ $\begin{array}{r}0\\13\overline{)0}\end{array}$ $\begin{array}{r}0\\10\overline{)0}\end{array}$

Division by zero is not allowed. $\begin{array}{r}\boxed{?}\\0\overline{)8}\end{array}$ There is no number whose product with 0 is 8.

When the dividend is a larger whole number, the digits in the quotient are found in steps.

➡ Divide $4\overline{)3192}$ and check.

$$\begin{array}{r} 7\phantom{92} \\ 4\overline{)\,3192} \\ -28\phantom{92} \\ \hline 39\phantom{2} \end{array}$$

- Think $4\overline{)31}$.
- Subtract 7 × 4.
- Bring down the 9.

$$\begin{array}{r} 79\phantom{2} \\ 4\overline{)\,3192} \\ -28\phantom{92} \\ \hline 39\phantom{2} \\ -36\phantom{2} \\ \hline 32 \end{array}$$

- Think $4\overline{)39}$.
- Subtract 9 × 4.
- Bring down the 2.

$$\begin{array}{r} 798 \\ 4\overline{)\,3192} \\ -28\phantom{92} \\ \hline 39\phantom{2} \\ -36\phantom{2} \\ \hline 32 \\ -32 \\ \hline 0 \end{array}$$

- Think $4\overline{)32}$.
- Subtract 8 × 4.

Check:

$$\begin{array}{r} 798 \\ \times\ \ 4 \\ \hline 3192 \end{array}$$

The place-value chart can be used to show why this method works.

| | HUNDREDS | TENS | ONES | |
|---|---|---|---|---|
| | 7 | 9 | 8 | |
| $4\overline{)\,3}$ | 1 | 9 | 2 | |
| −2 | 8 | 0 | 0 | 7 hundreds × 4 |
| | 3 | 9 | 2 | |
| | −3 | 6 | 0 | 9 tens × 4 |
| | | 3 | 2 | |
| | | −3 | 2 | 8 ones × 4 |
| | | | 0 | |

There are other ways of expressing division.

54 divided by 9 equals 6.

54 ÷ 9 equals 6.

$\frac{54}{9}$ equals 6.

**Example 1** Divide $7\overline{)56}$ and check.

**Solution**

$$\begin{array}{r} 8 \\ 7\overline{)56} \end{array}$$

Check: $8 \times 7 = 56$

**You Try It 1** Divide $9\overline{)63}$ and check.

**Your solution**

**Example 2** Divide $2808 \div 8$ and check.

**Solution**

$$\begin{array}{r} 351 \\ 8\overline{)\ 2808} \\ -24\phantom{08} \\ \hline 40\phantom{8} \\ -40\phantom{8} \\ \hline 08 \\ -\ \ 8 \\ \hline 0 \end{array}$$

Check: $351 \times 8 = 2808$

**You Try It 2** Divide $4077 \div 9$ and check.

**Your solution**

**Example 3** Divide $7\overline{)2856}$ and check.

**Solution**

$$\begin{array}{r} 408 \\ 7\overline{)\ 2856} \\ -28\phantom{56} \\ \hline 05\phantom{6} \\ -\ \ 0\phantom{6} \\ \hline 56 \\ -56 \\ \hline 0 \end{array}$$

- Think $7\overline{)5}$.
  Place 0 in quotient.
  Subtract $0 \times 7$.
  Bring down the 6.

Check: $408 \times 7 = 2856$

**You Try It 3** Divide $9\overline{)6345}$ and check.

**Your solution**

*Solutions on p. A9*

## *Objective B* *To divide by a single digit with a remainder in the quotient*

Occasionally it is not possible to separate objects into a whole number of equal groups.

A warehouse clerk must place 14 objects into 3 boxes. From the diagram, we see that the clerk would place 4 objects in each box and have 2 objects left over. The 2 is called the **remainder.**

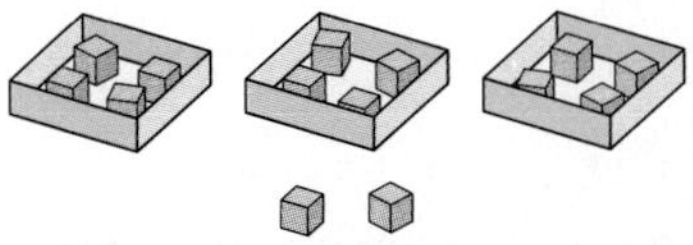

The clerk's division problem could be written

$$\begin{array}{lrl} & 4 \swarrow & \text{Quotient (Number in each box)} \\ \text{Divisor (Number of boxes)} \rightarrow & 3\overline{)\,14} \leftarrow & \text{Dividend (Total number of objects)} \\ & \underline{-12} & \\ & 2 \leftarrow & \text{Remainder (Number left over)} \end{array}$$

The answer to a division problem with a remainder is frequently written

$$\begin{array}{r} 4\ \text{r2} \\ 3\overline{)14} \end{array}$$

Note that $\underset{\text{Quotient}}{4} \times \underset{\text{Divisor}}{3} + \underset{\text{Remainder}}{2} = \underset{\text{Dividend}}{14}$

**Example 4** Divide $4\overline{)2522}$ and check.

**Solution**

$$\begin{array}{r} 630\ \text{r2} \\ 4\overline{)\,2522} \\ \underline{-24\phantom{22}} \\ 12\phantom{2} \\ \underline{-12\phantom{2}} \\ 02 \\ \underline{-\ 0} \\ 2 \end{array}$$

- Think $4\overline{)2}$.
- Place 0 in quotient. Subtract $0 \times 4$.

Check: $(630 \times 4) + 2 =$
$2520 + 2 = 2522$

**You Try It 4** Divide $6\overline{)5225}$ and check.

**Your solution**

**Example 5** Divide $9\overline{)27{,}438}$ and check.

**Solution**

$$\begin{array}{r} 3{,}048\ \text{r6} \\ 9\overline{)\,27{,}438} \\ \underline{-27\phantom{,438}} \\ 0\,4\phantom{38} \\ \underline{-\ \ 0\phantom{38}} \\ 43\phantom{8} \\ \underline{-36\phantom{8}} \\ 78 \\ \underline{-72} \\ 6 \end{array}$$

- Think $9\overline{)4}$.
- Subtract $0 \times 9$.

Check: $(3048 \times 9) + 6 =$
$27{,}432 + 6 = 27{,}438$

**You Try It 5** Divide $7\overline{)21{,}409}$ and check.

**Your solution**

*Solutions on pp. A9 and A10*

### *Objective C To divide by larger whole numbers*

When the divisor has more than one digit, estimate at each step by using the first digit of the divisor. If that product is too large, lower the guess by 1 and try again.

➡ Divide $34\overline{)1598}$ and check.

$$\begin{array}{r} 5 \\ 34\overline{)\,1598} \\ -170 \\ \hline \end{array}$$

- Think $3\overline{)15}$.
- Subtract 5 × 34.

170 is too large. Lower guess by 1 and try again.

$$\begin{array}{r} 4 \\ 34\overline{)\,1598} \\ -136 \\ \hline 238 \end{array}$$

- Subtract 4 × 34.

$$\begin{array}{r} 47 \\ 34\overline{)\,1598} \\ -136 \\ \hline 238 \\ -238 \\ \hline 0 \end{array}$$

- Think $3\overline{)23}$.
- Subtract 7 × 34.

Check:
$$\begin{array}{r} 47 \\ \times 34 \\ \hline 188 \\ 141\phantom{0} \\ \hline 1598 \end{array}$$

The phrases below are used to indicate the operation of division. An example is shown at the right of each phrase.

| | | |
|---|---|---|
| the quotient of | the quotient of 9 and 3 | 9 ÷ 3 |
| divided by | 6 divided by 2 | 6 ÷ 2 |

**Example 6** Find 7077 divided by 34 and check.

**Solution**

$$\begin{array}{r} 208\text{ r5} \\ 34\overline{)\,7077} \\ -68\phantom{0} \\ \hline 27\phantom{0} \\ -\ 0\phantom{0} \\ \hline 277 \\ -272 \\ \hline 5 \end{array}$$

- Think $34\overline{)27}$.
- Place 0 in the quotient.
- Subtract 0 × 34.

Check: (208 × 34) + 5 =
7072 + 5 = 7077

**You Try It 6** Divide 4578 ÷ 42 and check.

**Your solution**

*Solution on p. A10*

**Example 7** Find the quotient of 21,312 and 56 and check.

**Solution**

$$\begin{array}{r} 380 \text{ r}32 \\ 56\overline{)21,312} \\ -16\ 8\phantom{12} \\ \hline 4\ 51\phantom{2} \\ -4\ 48\phantom{2} \\ \hline 32 \\ -\ 0 \\ \hline 32 \end{array}$$

• Think $5\overline{)21}$. $4 \times 56$ is too large. Try 3.

Check: $(380 \times 56) + 32 =$
$21,280 \quad + 32 = 21,312$

**You Try It 7** Divide $18,359 \div 39$ and check.

**Your solution**

**Example 8** Divide $427\overline{)24,782}$ and check.

**Solution**

$$\begin{array}{r} 58 \text{ r}16 \\ 427\overline{)24,782} \\ -21\ 35\phantom{2} \\ \hline 3\ 432 \\ -3\ 416 \\ \hline 16 \end{array}$$

Check: $(58 \times 427) + 16 =$
$24,766 \quad + 16 = 24,782$

**You Try It 8** Divide $534\overline{)33,219}$ and check.

**Your solution**

**Example 9** Divide $386\overline{)206,149}$ and check.

**Solution**

$$\begin{array}{r} 534 \text{ r}25 \\ 386\overline{)206,149} \\ -193\ 0\phantom{49} \\ \hline 13\ 14\phantom{9} \\ -11\ 58\phantom{9} \\ \hline 1\ 569 \\ -1\ 544 \\ \hline 25 \end{array}$$

Check: $(534 \times 386) + 25 =$
$206,124 \quad + 25 = 206,149$

**You Try It 9** Divide $515\overline{)216,848}$ and check.

**Your solution**

*Solutions on p. A10*

## ESTIMATION

### Estimating the Quotient of Two Whole Numbers

Estimate and then use your calculator to find 36,936 ÷ 54.

To estimate a quotient, round each number so that all the digits are zero except the first digit. Then divide.

$36{,}936 \div 54 \approx$
$40{,}000 \div 50 = 800$

The estimated answer is 800.

Now use your calculator to find the exact answer.

36936 [÷] 54 [=] 684

The exact answer is 684.

## *Objective D* *To solve application problems*

The *average* of the total of several events is the quotient of that total and the number of events. For instance, if the total of 6 of your test scores was 498, then

$$\text{average test score} = \frac{498}{6} = 83$$

➡ The table at the right shows the number of home dates and the total home attendance of the baseball teams in the National League. Find the average home attendance for the Chicago team. Round to the nearest whole number.

| Team | Home Dates | Home Att. |
|---|---|---|
| Atlanta | 69 | 3,296,967 |
| Chicago | 74 | 2,446,912 |
| Cincinnati | 69 | 2,189,722 |
| Colorado | 67 | 3,872,511 |
| Florida | 67 | 2,629,117 |
| Houston | 74 | 1,913,944 |
| Los Angeles | 71 | 2,796,973 |
| Montreal | 72 | 1,374,708 |
| New York | 70 | 1,708,606 |
| Philadelphia | 74 | 2,869,808 |
| Pittsburgh | 70 | 1,514,206 |
| St. Louis | 72 | 2,578,292 |
| San Diego | 73 | 1,277,268 |
| San Francisco | 73 | 2,362,524 |

**Strategy**

Determine, from the table, the total attendance and the number of home dates for the Chicago team. To find the average home attendance, divide the total attendance (2,446,912) by the number of home dates (74).

**Solution**

```
       33,066
  74) 2,446,912
     -2 22
        226
       -222
          4 9
        -   0
          4 91
         -4 44
            472
           -444
             28
```

- When rounding to the nearest whole number, compare twice the remainder to the divisor. If twice the remainder is less than the divisor, drop the remainder. If twice the remainder is greater than or equal to the divisor, add 1 to the units digit of the quotient.
- Twice the remainder is $2 \times 28 = 56$. Because $56 < 74$, drop the remainder.

The average attendance to the nearest whole number is 33,066.

**Example 10**
Ngan Hui, a freight supervisor, shipped 192,600 bushels of wheat in 9 railroad cars. Find the amount of wheat shipped in each car?

**Strategy**
To find the amount of wheat shipped in each car, divide the number of bushels (192,600) by the number of cars (9)

**Solution**

$$\begin{array}{r} 21{,}400 \\ 9\overline{)\,192{,}600} \\ \underline{-18\phantom{0{,}600}} \\ 12\phantom{{,}600} \\ \underline{-9\phantom{{,}600}} \\ 36\phantom{00} \\ \underline{-36\phantom{00}} \\ 0\phantom{00} \end{array}$$

Each car carried 21,400 bushels of wheat.

**You Try It 10**
Suppose a Firestone retail outlet can store 270 tires on 15 shelves. How many tires can be stored on each shelf?

**Strategy**

**Your solution**

**Example 11**
The car you are buying costs $11,216. A down payment of $2000 is required. The remaining balance is paid in 48 equal monthly payments. What is the monthly payment?

**Strategy**
To find the monthly payment:

- Find the remaining balance by subtracting the down payment ($2000) from the total cost of the car ($11,216).
- Divide the remaining balance by the number of equal monthly payments (48).

**Solution**

$$\begin{array}{r} 11{,}216 \\ \underline{-\ \ 2{,}000} \\ 9{,}216 \end{array}$$

remaining balance

$$\begin{array}{r} 192 \\ 48\overline{)\,9216} \\ \underline{-48\phantom{16}} \\ 441\phantom{6} \\ \underline{-432\phantom{6}} \\ 96 \\ \underline{-96} \\ 0 \end{array}$$

The monthly payment is $192.

**You Try It 11**
A soft-drink manufacturer produces 12,600 cans of soft drink each hour. Cans are packed 24 to a case. How many cases of soft drink are produced in 8 hours?

**Your strategy**

**Your solution**

*Solutions on p. A11*

## 1.5 Exercises

### *Objective A*

Divide.

**1.** $4\overline{)8}$ **2.** $3\overline{)9}$ **3.** $6\overline{)36}$ **4.** $9\overline{)81}$ **5.** $5\overline{)25}$

**6.** $7\overline{)49}$ **7.** $5\overline{)80}$ **8.** $6\overline{)96}$ **9.** $6\overline{)480}$ **10.** $9\overline{)630}$

**11.** $4\overline{)840}$ **12.** $3\overline{)690}$ **13.** $7\overline{)308}$ **14.** $7\overline{)203}$ **15.** $7\overline{)1442}$

**16.** $9\overline{)6327}$ **17.** $4\overline{)2120}$ **18.** $8\overline{)7280}$ **19.** $9\overline{)8118}$ **20.** $8\overline{)7264}$

**21.** $3\overline{)64{,}680}$ **22.** $4\overline{)50{,}760}$ **23.** $6\overline{)21{,}480}$ **24.** $5\overline{)18{,}050}$

**25.** Find the quotient of 1446 and 3.

**26.** Find the quotient of 4123 and 7.

**27.** What is 7525 divided by 7?

**28.** What is 32,364 divided by 4?

### *Objective B*

Divide.

**29.** $4\overline{)9}$ **30.** $2\overline{)7}$ **31.** $5\overline{)27}$ **32.** $9\overline{)88}$ **33.** $3\overline{)40}$

**34.** $6\overline{)97}$ **35.** $8\overline{)83}$ **36.** $5\overline{)54}$ **37.** $7\overline{)632}$ **38.** $4\overline{)363}$

**39.** $4\overline{)921}$ **40.** $7\overline{)845}$ **41.** $8\overline{)1635}$ **42.** $5\overline{)1548}$ **43.** $7\overline{)9432}$

**44.** $6\overline{)6352}$ **45.** $9\overline{)7004}$ **46.** $7\overline{)6001}$ **47.** $6\overline{)2350}$ **48.** $8\overline{)6370}$

**49.** $7\overline{)8124}$ **50.** $3\overline{)5162}$ **51.** $5\overline{)3542}$ **52.** $8\overline{)3274}$

**53.** $4\overline{)15{,}300}$ **54.** $7\overline{)43{,}500}$ **55.** $8\overline{)72{,}354}$ **56.** $5\overline{)43{,}542}$

**57.** Find the quotient of 3107 and 8.

**58.** Find the quotient of 8642 and 8.

**59.** What is 45,738 divided by 4? Round to the nearest ten.

**60.** What is 37,896 divided by 9? Round to the nearest hundred.

**61.** What is 3572 divided by 7? Round to the nearest ten.

**62.** What is 78,345 divided by 4? Round to the nearest hundred.

### *Objective C*

Divide.

**63.** $27\overline{)96}$ **64.** $44\overline{)82}$ **65.** $42\overline{)87}$ **66.** $67\overline{)93}$

**67.** $41\overline{)897}$ **68.** $32\overline{)693}$ **69.** $23\overline{)784}$ **70.** $25\overline{)772}$

**71.** $74\overline{)600}$ **72.** $92\overline{)500}$ **73.** $70\overline{)329}$ **74.** $50\overline{)467}$

**75.** $36\overline{)7225}$ **76.** $44\overline{)8821}$ **77.** $19\overline{)3859}$ **78.** $32\overline{)9697}$

**79.** $88\overline{)3127}$ **80.** $92\overline{)6177}$ **81.** $33\overline{)8943}$ **82.** $27\overline{)4765}$

**83.** $22\overline{)98{,}654}$ **84.** $77\overline{)83{,}629}$ **85.** $64\overline{)38{,}912}$ **86.** $78\overline{)31{,}434}$

**87.** $206\overline{)3097}$ **88.** $504\overline{)6504}$ **89.** $654\overline{)1217}$ **90.** $546\overline{)2344}$

**91.** $169\overline{)8542}$ **92.** $456\overline{)7723}$ **93.** $223\overline{)8927}$ **94.** $467\overline{)9344}$

**95.** Find the quotient of 5432 and 21.

**96.** Find the quotient of 8507 and 53.

**97.** What is 37,294 divided by 72?

**98.** What is 76,788 divided by 46?

**99.** Find 23,457 divided by 43. Round to the nearest hundred.

**100.** Find 341,781 divided by 43. Round to the nearest ten.

Estimate and then use your calculator to divide.

101. $76\overline{)389,804}$

102. $53\overline{)117,925}$

103. $29\overline{)637,072}$

104. $67\overline{)738,072}$

105. $38\overline{)934,648}$

106. $34\overline{)906,304}$

107. $309\overline{)876,324}$

108. $642\overline{)323,568}$

109. $209\overline{)632,016}$

110. $614\overline{)332,174}$

111. $179\overline{)5,734,444}$

112. $374\overline{)7,712,254}$

### *Objective D* Application Problems

**113.** Assume that the American Red Cross, the Boys Club, the Girls Club, and the North County Crisis Center collected $548,000 to promote and provide community services. What amount did each organization receive if the money was divided evenly?

**114.** An insurance agent drives a car with a 16-gallon gas tank. The agent used 128 gallons of gas in traveling 3456 miles. Find the average number of miles traveled on each gallon of gas.

The graph at the right shows the maximum amount paid by an employee for Social Security benefits for the years shown. Use this graph for Exercises 115 to 118. In each case, assume that the employee makes the maximum contribution for the year given.

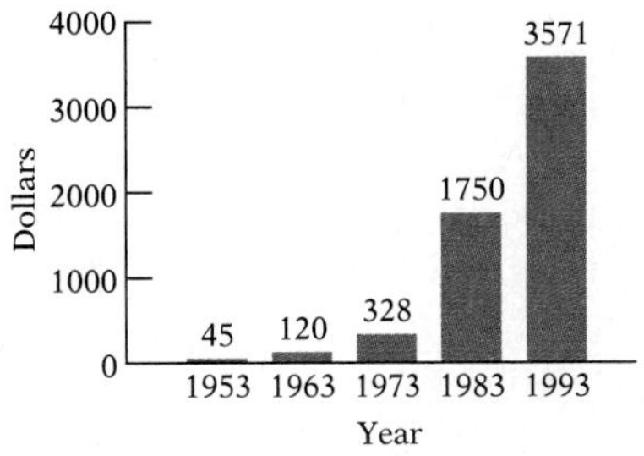


**115.** In 1993, what was the monthly contribution of an employee, to the nearest dollar?

**116.** What is the difference in monthly contribution, to the nearest dollar, between 1993 and 1983?

**117.** To the nearest whole number, how many times greater was the annual contribution in 1993 than the annual contribution in 1953?

**118.** If the maximum Social Security contribution for the years 1993 to 2003 increases by the same amount as it increased from 1983 to 1993, what will the maximum contribution be in 2003?

**119.** A computer can store 2,211,840 bytes of information on 6 disks. How many bytes of information can be stored on 1 disk?

**120.** Ken Martinez, a computer analyst, received $5376 for working 168 hours on a computer consulting project. Find the hourly rate that Ken charged.

**121.** A family bought a living room set for $2856, which includes finance charges. This amount is to be repaid in 24 equal payments. Find the amount of each payment.

**122.** Marie Sarro, a design consultant, made a down payment of $1560 on a car costing $10,536.
a. What is the remaining balance to be paid?
b. The balance is to be paid in 48 equal monthly payments. Find the monthly payment.

Each year *Forbes* magazine publishes what is called the *Forbes* Celebrity list. It is a list of the highest paid entertainers in the United States. Selected entries from the 1993 list are given at the right. Use this list for Exercises 123 to 126.

| *Name* | *Annual Income* |
|---|---|
| Oprah Winfrey | $98,000,000 |
| Steven Spielberg | $72,000,000 |
| Bill Cosby | $66,000,000 |
| Garth Brooks | $48,000,000 |
| Madonna | $36,000,000 |

**123.** The average worker in the United States works approximately 2000 hours per year. Assuming that Oprah Winfrey worked 2000 hours in 1993, find her average hourly wage.

**124.** What is the difference between the average monthly salary of Steven Spielberg and that of Bill Cosby?

**125.** How many times more was Steven Spielberg's annual income than Madonna's annual income?

**126.** If Garth Brooks worked 1200 hours in 1993 and Bill Cosby worked 1500 hours that year, which of the two entertainers had the smaller hourly wage?

### *APPLYING THE CONCEPTS*

**127.** A palindromic number is a whole number that remains unchanged when its digits are written in reverse order. For instance, 292 is a palindromic number. Find the smallest three-digit palindromic number that is divisible by 4.

**128.** Find the smallest four-digit palindromic number (see Exercise 127) that is divisible by 8. (Helpful fact: A whole number whose last three digits are divisible by 8 is itself divisible by 8.)

**129.** The number 10,981 is not divisible by 4. By rearranging the digits, find the largest possible number that is divisible by 4.

**130.** Determine whether each of the following statements is true or false.
a. Any whole number divided by zero is zero.
b. $\frac{0}{0} = 1$
c. Zero divided by any whole number, except zero, is zero.

**131.** Explain how the answer to a division problem can be checked.

# 1.6 Exponential Notation and the Order of Operations Agreement

## Objective A To simplify expressions that contain exponents

Repeated multiplication of the same factor can be written in two ways:

$3 \cdot 3 \cdot 3 \cdot 3 \cdot 3$ or $3^5$ ← **exponent**

The exponent indicates how many times the factor occurs in the multiplication. The expression $3^5$ is in **exponential notation.**

It is important to be able to read numbers written in exponential notation.

$6 = 6^1$ is read "six to the first power" or just "six." Usually the exponent 1 is not written.

$6 \cdot 6 = 6^2$ is read "six squared" or "six to the second power."

$6 \cdot 6 \cdot 6 = 6^3$ is read "six cubed" or "six to the third power."

$6 \cdot 6 \cdot 6 \cdot 6 = 6^4$ is read "six to the fourth power."

$6 \cdot 6 \cdot 6 \cdot 6 \cdot 6 = 6^5$ is read "six to the fifth power."

Each place value in the place-value chart can be expressed as a power of 10.

| | | | |
|---|---|---|---|
| Ten = | 10 = | 10 | $= 10^1$ |
| Hundred = | 100 = | $10 \cdot 10$ | $= 10^2$ |
| Thousand = | 1000 = | $10 \cdot 10 \cdot 10$ | $= 10^3$ |
| Ten-Thousand = | 10,000 = | $10 \cdot 10 \cdot 10 \cdot 10$ | $= 10^4$ |
| Hundred-Thousand = | 100,000 = | $10 \cdot 10 \cdot 10 \cdot 10 \cdot 10$ | $= 10^5$ |
| Million = | 1,000,000 = | $10 \cdot 10 \cdot 10 \cdot 10 \cdot 10 \cdot 10$ | $= 10^6$ |

To simplify a numerical expression containing exponents, write each factor as many times as indicated by the exponent and carry out the indicated multiplication.

$4^3 = 4 \cdot 4 \cdot 4 = 64$

$2^2 \cdot 3^4 = (2 \cdot 2) \cdot (3 \cdot 3 \cdot 3 \cdot 3) = 4 \cdot 81 = 324$

**Example 1** Write $3 \cdot 3 \cdot 3 \cdot 5 \cdot 5$ in exponential notation.

**Solution** $3 \cdot 3 \cdot 3 \cdot 5 \cdot 5 = 3^3 \cdot 5^2$

**You Try It 1** Write $2 \cdot 2 \cdot 2 \cdot 2 \cdot 3 \cdot 3 \cdot 3$ in exponential notation.

**Your solution**

**Example 2** Write $10 \cdot 10 \cdot 10 \cdot 10$ as a power of ten.

**Solution** $10 \cdot 10 \cdot 10 \cdot 10 = 10^4$

**You Try It 2** Write $10 \cdot 10 \cdot 10 \cdot 10 \cdot 10 \cdot 10 \cdot 10$ as a power of 10.

**Your solution**

**Example 3** Simplify $3^2 \cdot 5^3$.

**Solution** $3^2 \cdot 5^3 = (3 \cdot 3) \cdot (5 \cdot 5 \cdot 5) = 9 \cdot 125 = 1125$

**You Try It 3** Simplify $2^3 \cdot 5^2$.

**Your solution**

*Solutions on p. A11*

## *Objective B* *To use the Order of Operations Agreement to simplify expressions*

More than one operation may occur in a numerical expression. The answer may be different, depending on the order in which the operations are performed. For example, consider $3 + 4 \times 5$.

Multiply first, then add.

$$\underbrace{3 + \underbrace{4 \times 5}}$$
$$\underbrace{3 + 20}$$
$$23$$

Add first, then multiply.

$$\underbrace{3 + 4} \times 5$$
$$\underbrace{7 \times 5}$$
$$35$$

An Order of Operations Agreement is used so that only one answer is possible.

**The Order of Operations Agreement**

**Step 1** Do all the operations inside parentheses.
**Step 2** Simplify any number expressions containing exponents.
**Step 3** Do multiplication and division as they occur from left to right.
**Step 4** Do addition and subtraction as they occur from left to right.

Simplify $3 \times (2 + 1) - 2^2 + 4 \div 2$ by using the Order of Operations Agreement.

| | |
|---|---|
| $3 \times \underbrace{(2 + 1)} - 2^2 + 4 \div 2$ | 1. Perform operations in parentheses. |
| $3 \times 3 - \underbrace{2^2} + 4 \div 2$ | 2. Simplify expressions with exponents. |
| $\underbrace{3 \times 3} - 4 + 4 \div 2$ | 3. Do multiplications and divisions as they occur from left to right. |
| $9 - 4 + \underbrace{4 \div 2}$ | |
| $\underbrace{9 - 4} + 2$ | 4. Do additions and subtractions as they occur from left to right. |
| $\underbrace{5 + 2}$ | |
| $7$ | |

One or more of the above steps may not be needed to simplify an expression. In that case, proceed to the next step in the Order of Operations Agreement.

| | |
|---|---|
| Simplify $5 + 8 \div 2$. | No parentheses or exponents. Proceed to step 3 of the Agreement. |
| $5 + \underbrace{8 \div 2}$ | 3. Do multiplication or division. |
| $\underbrace{5 + 4}$ | 4. Do addition or subtraction. |
| $9$ | |

**Example 4** Simplify:
$16 \div (8 - 4) \cdot 9 - 5^2$

**Solution**
$16 \div (8 - 4) \cdot 9 - 5^2$
$16 \div 4 \cdot 9 - 5^2$
$16 \div 4 \cdot 9 - 25$
$4 \cdot 9 - 25$
$36 - 25$
$11$

**You Try It 4** Simplify:
$5 \cdot (8 - 4) \div 4 - 2$

**Your solution**

*Solution on p. A11*

## 1.6 Exercises

### *Objective A*

Write the number in exponential notation.

**1.** $2 \cdot 2 \cdot 2$

**2.** $7 \cdot 7 \cdot 7 \cdot 7 \cdot 7$

**3.** $6 \cdot 6 \cdot 6 \cdot 7 \cdot 7 \cdot 7 \cdot 7$

**4.** $3 \cdot 3 \cdot 3 \cdot 5 \cdot 5 \cdot 5 \cdot 5$

**5.** $6 \cdot 6 \cdot 6 \cdot 11 \cdot 11 \cdot 11 \cdot 11 \cdot 11$

**6.** $2 \cdot 2 \cdot 2 \cdot 2 \cdot 2 \cdot 2 \cdot 3 \cdot 3 \cdot 3$

**7.** $3 \cdot 10 \cdot 10 \cdot 10 \cdot 10$

**8.** $7 \cdot 10 \cdot 10 \cdot 10 \cdot 10 \cdot 10 \cdot 10 \cdot 10$

**9.** $2 \cdot 2 \cdot 3 \cdot 3 \cdot 3 \cdot 5 \cdot 5 \cdot 5 \cdot 5$

**10.** $7 \cdot 7 \cdot 11 \cdot 11 \cdot 11 \cdot 13 \cdot 13 \cdot 13 \cdot 13$

**11.** $2 \cdot 3 \cdot 3 \cdot 7 \cdot 7 \cdot 11$

**12.** $5 \cdot 7 \cdot 7 \cdot 9 \cdot 9 \cdot 11$

**13.** $2 \cdot 2 \cdot 7 \cdot 7 \cdot 7 \cdot 7 \cdot 11 \cdot 11 \cdot 11 \cdot 11$

**14.** $3 \cdot 3 \cdot 7 \cdot 7 \cdot 7 \cdot 7 \cdot 17 \cdot 17 \cdot 17 \cdot 17$

Simplify.

**15.** $2^3$

**16.** $2^6$

**17.** $2^4 \cdot 5^2$

**18.** $2^6 \cdot 3^2$

**19.** $3^2 \cdot 10^2$

**20.** $2^3 \cdot 10^4$

**21.** $6^2 \cdot 3^3$

**22.** $4^3 \cdot 5^2$

**23.** $5 \cdot 2^3 \cdot 3$

**24.** $6 \cdot 3^2 \cdot 4$

**25.** $2^2 \cdot 3^2 \cdot 10$

**26.** $3^2 \cdot 5^2 \cdot 10$

**27.** $0^2 \cdot 4^3$

**28.** $6^2 \cdot 0^3$

**29.** $3^2 \cdot 10^4$

**30.** $5^3 \cdot 10^3$

**31.** $2^2 \cdot 3^3 \cdot 5$

**32.** $5^2 \cdot 7^3 \cdot 2$

**33.** $2 \cdot 3^4 \cdot 5^2$

**34.** $6 \cdot 2^6 \cdot 7^2$

**35.** $5^2 \cdot 3^2 \cdot 7^2$

**36.** $4^2 \cdot 9^2 \cdot 6^2$

**37.** $3^4 \cdot 2^6 \cdot 5$

**38.** $4^3 \cdot 6^3 \cdot 7$

**39.** $4^2 \cdot 3^3 \cdot 10^4$

**40.** $5^2 \cdot 2^3 \cdot 10^3$

**41.** $6^2 \cdot 4^4 \cdot 10$

### *Objective B*

Simplify by using the Order of Operations Agreement.

**42.** $4 - 2 + 3$

**43.** $6 - 3 + 2$

**44.** $6 \div 3 + 2$

**45.** $8 \div 4 + 8$

**46.** $6 \cdot 3 + 5$

**47.** $5 \cdot 9 + 2$

**48.** $3^2 - 4$

**49.** $5^2 - 17$

**50.** $4 \cdot (5 - 3) + 2$

**51.** $3 + (4 + 2) \div 3$

**52.** $5 + (8 + 4) \div 6$

**53.** $8 - 2^2 + 4$

**54.** $16 \cdot (3 + 2) \div 10$

**55.** $12 \cdot (1 + 5) \div 12$

**56.** $10 - 2^3 + 4$

**57.** $5 \cdot 3^2 + 8$

**58.** $16 + 4 \cdot 3^2$

**59.** $12 + 4 \cdot 2^3$

**60.** $16 + (8 - 3) \cdot 2$

**61.** $7 + (9 - 5) \cdot 3$

**62.** $2^2 + 3 \cdot (6 - 2)$

**63.** $3^3 + 5 \cdot (8 - 6)$

**64.** $2^2 \cdot 3^2 + 2 \cdot 3$

**65.** $4 \cdot 6 + 3^2 \cdot 4^2$

**66.** $16 - 2 \cdot 4$

**67.** $12 + 3 \cdot 5$

**68.** $3 \cdot (6 - 2) + 4$

**69.** $5 \cdot (8 - 4) - 6$

**70.** $8 - (8 - 2) \div 3$

**71.** $12 - (12 - 4) \div 4$

**72.** $8 + 2 - 3 \cdot 2 \div 3$

**73.** $10 + 1 - 5 \cdot 2 \div 5$

**74.** $3 \cdot (4 + 2) \div 6$

### *APPLYING THE CONCEPTS*

**75.** Memory in computers is measured in bytes. One kilobyte (1 K) is $2^{10}$ bytes. Write this number in standard form.

**76.** [W] Explain the difference in the order of operations for **a.** $\frac{14 - 2}{2} \div 2 \cdot 3$ and **b.** $\frac{14 - 2}{2} \div (2 \cdot 3)$. Work the two problems. What is the difference between the larger and the smaller answer?

**77.** [W] If a number is divisible by 6 and 10, is the number divisible by 30? If so, explain why. If not, give an example.

# 1.7 Prime Numbers and Factoring

## Objective A To factor numbers

Whole-number factors of a number divide that number evenly (there is no remainder).

1, 2, 3, and 6 are whole-number factors of 6 because they divide 6 evenly.

$1\overline{)6}$ = 6 $\quad 2\overline{)6}$ = 3 $\quad 3\overline{)6}$ = 2 $\quad 6\overline{)6}$ = 1

Note that both the divisor and the quotient are factors of the dividend.

To find the factors of a number, try dividing the number by 1, 2, 3, 4, 5, . . . . Those numbers that divide the number evenly are its factors. Continue this process until the factors start to repeat.

➡ Find all the factors of 42.

$42 \div 1 = 42$ 1 and 42 are factors
$42 \div 2 = 21$ 2 and 21 are factors
$42 \div 3 = 14$ 3 and 14 are factors
$42 \div 4$ Will not divide evenly
$42 \div 5$ Will not divide evenly
$42 \div 6 = 7$ 6 and 7 are factors } Factors are repeating; all the
$42 \div 7 = 6$ 7 and 6 are factors } factors of 42 have been found.

1, 2, 3, 6, 7, 14, 21, and 42 are factors of 42.

The following rules are helpful in finding the factors of a number.

2 is a factor of a number if the last digit of the number is 0, 2, 4, 6, or 8.

436 ends in 6; therefore, 2 is a factor of 436. ($436 \div 2 = 218$)

3 is a factor of a number if the sum of the digits of the number is divisible by 3.

The sum of the digits of 489 is $4 + 8 + 9 = 21$. 21 is divisible by 3. Therefore, 3 is a factor of 489. ($489 \div 3 = 163$)

5 is a factor of a number if the last digit of the number is 0 or 5.

520 ends in 0; therefore, 5 is a factor of 520. ($520 \div 5 = 104$)

**Example 1** Find all the factors of 30.

**Solution**

$30 \div 1 = 30$
$30 \div 2 = 15$
$30 \div 3 = 10$
$30 \div 4$ Will not divide evenly
$30 \div 5 = 6$
$30 \div 6 = 5$

1, 2, 3, 5, 6, 10, 15 and 30 are factors of 30.

**You Try It 1** Find all the factors of 40.

**Your solution**

*Solution on p. A11*

## *Objective B To find the prime factorization of a number*

**POINT OF INTEREST**

Prime numbers are an important part of cryptology, the study of secret codes. To ensure that codes cannot be broken, a cryptologist uses prime numbers that have hundreds of digits.

A number is a **prime number** if its only whole-number factors are 1 and itself. 7 is prime because its only factors are 1 and 7. If a number is not prime, it is called a **composite number.** Because 6 has factors of 2 and 3, 6 is a composite number. The number 1 is not considered a prime number; therefore it is not included in the following list of prime numbers less than 50.

2, 3, 5, 7, 11, 13, 17, 19, 23, 29, 31, 37, 41, 43, 47

The **prime factorization** of a number is the expression of the number as a product of its prime factors. We use a "T-diagram" to find the prime factors of 60. Begin with the smallest prime number as a trial divisor, and continue with prime numbers as trial divisors until the final quotient is 1.

| 60 | | |
|---|---|---|
| 2 | 30 | $60 \div 2 = 30$ |
| 2 | 15 | $30 \div 2 = 15$ |
| 3 | 5 | $15 \div 3 = 5$ |
| 5 | 1 | $5 \div 5 = 1$ |

The prime factorization of 60 is $2 \cdot 2 \cdot 3 \cdot 5$.

Finding the prime factorization of larger numbers can be more difficult. Try each prime number as a trial divisor. Stop when the square of the trial divisor is greater than the number being factored.

➡ Find the prime factorization of 106.

| 106 | |
|---|---|
| 2 | 53 |
| 53 | 1 |

• 53 cannot be divided evenly by 2, 3, 5, 7, or 11. Prime numbers greater than 11 need not be tested because $11^2$ is greater than 53.

The prime factorization of 106 is $2 \cdot 53$.

**Example 2** Find the prime factorization of 315.

**Solution**

| 315 | |
|---|---|
| 3 | 105 |
| 3 | 35 |
| 5 | 7 |
| 7 | 1 |

$315 = 3 \cdot 3 \cdot 5 \cdot 7$

**You Try It 2** Find the prime factorization of 44.

**Your solution**

**Example 3** Find the prime factorization of 201.

**Solution**

| 201 | |
|---|---|
| 3 | 67 |
| 67 | 1 |

• Try only 2, 3, 5, 7, and 11 because $11^2 > 67$.

$201 = 3 \cdot 67$

**You Try It 3** Find the prime factorization of 177.

**Your solution**

*Solutions on p. A11*

## 1.7 Exercises

### *Objective A*

Find all the factors of the number.

| | | | | |
|---|---|---|---|---|
| **1.** 4 | **2.** 6 | **3.** 10 | **4.** 20 | **5.** 5 |
| **6.** 7 | **7.** 12 | **8.** 9 | **9.** 8 | **10.** 16 |
| **11.** 13 | **12.** 17 | **13.** 18 | **14.** 24 | **15.** 25 |
| **16.** 56 | **17.** 36 | **18.** 45 | **19.** 28 | **20.** 32 |
| **21.** 29 | **22.** 33 | **23.** 22 | **24.** 26 | **25.** 44 |
| **26.** 52 | **27.** 49 | **28.** 82 | **29.** 37 | **30.** 79 |
| **31.** 57 | **32.** 69 | **33.** 48 | **34.** 64 | **35.** 87 |
| **36.** 95 | **37.** 46 | **38.** 54 | **39.** 50 | **40.** 75 |
| **41.** 66 | **42.** 77 | **43.** 80 | **44.** 100 | **45.** 84 |
| **46.** 96 | **47.** 85 | **48.** 90 | **49.** 101 | **50.** 123 |

### Objective B

Find the prime factorization.

| | | | | |
|---|---|---|---|---|
| **51.** 6 | **52.** 14 | **53.** 17 | **54.** 83 | **55.** 16 |
| **56.** 24 | **57.** 12 | **58.** 27 | **59.** 9 | **60.** 15 |
| **61.** 36 | **62.** 40 | **63.** 19 | **64.** 37 | **65.** 50 |
| **66.** 90 | **67.** 65 | **68.** 115 | **69.** 80 | **70.** 87 |
| **71.** 18 | **72.** 26 | **73.** 28 | **74.** 49 | **75.** 29 |
| **76.** 31 | **77.** 42 | **78.** 62 | **79.** 81 | **80.** 51 |
| **81.** 22 | **82.** 39 | **83.** 101 | **84.** 89 | **85.** 70 |
| **86.** 66 | **87.** 86 | **88.** 74 | **89.** 95 | **90.** 105 |
| **91.** 67 | **92.** 78 | **93.** 55 | **94.** 46 | **95.** 122 |
| **96.** 130 | **97.** 118 | **98.** 102 | **99.** 150 | **100.** 125 |
| **101.** 120 | **102.** 144 | **103.** 160 | **104.** 175 | **105.** 343 |
| **106.** 216 | **107.** 400 | **108.** 625 | **109.** 225 | **110.** 180 |

### APPLYING THE CONCEPTS

**111.** Twin primes are two prime numbers that differ by 2. For instance, 17 and 19 are twin primes. Find three sets of twin primes, not including 17 and 19.

**112.** In 1742, Christian Goldbach conjectured that every even number greater than 2 could be expressed as the sum of two prime numbers. Show that this conjecture is true for 8, 24, and 72. (*Note*: Mathematicians have not yet been able to determine whether Goldbach's conjecture is true or false.)

**113.** [W] Explain why 2 is the only even prime number.

**114.** [W] Explain the method of finding prime numbers using the Sieve of Erathosthenes.

# Projects in Mathematics

## Order of Operations

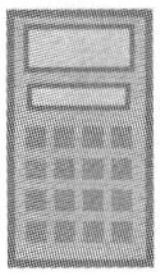

Does your calculator use the Order of Operations Agreement? To find out, try this problem:

$$2 + 4 \cdot 7$$

If your answer is 30, then the calculator uses the Order of Operations Agreement. If your answer is 42, it does not use that agreement.

Even if your calculator does not use the order of Operations Agreement, you can still correctly evaluate numerical expressions. The parentheses keys, [ ( ] and [ ) ], are used for this purpose.

Remember that $2 + 4 \cdot 7$ means $2 + (4 \cdot 7)$ because the multiplication must be completed before the addition. To evaluate this expression, enter the following:

| | | | | | | | | |
|---|---|---|---|---|---|---|---|---|
| Enter: | 2 | [+] | [ ( ] | 4 | [×] | 7 | [ ) ] | [=] |
| Display: | 2 | 2 | [ ( ] | 4 | 4 | 7 | 28 | 30 |

When using your calculator to evaluate numerical expressions, insert parentheses around multiplications or divisions. This has the effect of forcing the calculator to do the operations in the order you want rather than in the order the calculator wants.

***Exercises:***

Evaluate

**1.** $3 \cdot (15 - 2 \cdot 3) - 36 \div 3$

**2.** $4 \cdot 2^2 - (12 + 24 \div 6) - 5$

**3.** $16 \div 4 \cdot 3 + (3 \cdot 4 - 5) + 2$

**4.** $15 \cdot 3 \div 9 + (2 \cdot 6 - 3) + 4$

## Patterns in Mathematics

For the circle at the left, use a straight line to connect each dot on the circle with every other dot on the circle. How many different straight lines are there?

Follow the same procedure for each of the circles shown below. How many different straight lines are there in each?

Find a pattern to describe the number of dots on a circle and the corresponding number of different lines drawn. Use the pattern to determine the number of different lines that would be drawn in a circle with 7 dots and in a circle with 8 dots.

You are arranging a tennis tournament with 9 players. Following the pattern of finding the number of lines drawn on a circle, find the number of singles matches that can be played among the 9 players if each player plays each of the other players only once.

# Chapter Summary

**Key Words**

The *whole numbers* are 0, 1, 2, 3, 4, 5, 6, 7, 8, 9, 10, . . . .

The symbol for "*is less than*" is $<$.

The symbol for "*is greater than*" is $>$.

The position of a digit in a number determines the digit's *place value.*

Giving an approximate value for an exact number is called *rounding.*

An expression of the form $4^2$ is in *exponential notation,* where 4 is the *base* and 2 is the *exponent.*

A number is *prime* if it's only whole-number factors are 1 and itself.

The *prime factorization* of a number is the expression of the number as a product of its prime factors.

**Essential Rules**

| | |
|---|---|
| *The Addition Property of Zero* | Zero added to a number does not change the number.<br>**5 + 0 = 0 + 5 = 5** |
| *The Commutative Property of Addition* | Two numbers can be added in either order.<br>**6 + 5 = 5 + 6 = 11** |
| *The Associative Property of Addition* | Grouping numbers to be added in any order gives the same result.<br>**(2 + 3) + 5 = 2 + (3 + 5) = 10** |
| *The Multiplication Property of Zero* | The product of a number and zero is zero.<br>**0 × 3 = 3 × 0 = 0** |
| *The Multiplication Property of 1* | The product of a number and 1 is the number.<br>**1 × 8 = 8 × 1 = 8** |
| *The Commutative Property of Multiplication* | Two numbers can be multiplied in any order.<br>**5 × 2 = 2 × 5 = 10** |
| *The Associative Property of Multiplication* | Grouping numbers to be multiplied in any order gives the same result.<br>**(3 × 2) × 5 = 3 × (2 × 5) = 30** |
| *The Properties of Zero in Division* | Zero divided by any other number is zero. Division by zero is not allowed. |

*Order of Operations Agreement*

**Step 1** Perform operations inside grouping symbols.
**Step 2** Simplify expressions with exponents.
**Step 3** Do multiplications and divisions as they occur from left to right.
**Step 4** Do additions and subtractions as they occur from left to right.

# Chapter Review

## SECTION 1.1

1. Place the correct symbol, < or >, between the two numbers. 101 87

2. Write 276,057 in words.

3. Write two million eleven thousand forty-four in standard form.

4. Write 10,327 in expanded form.

## SECTION 1.2

5. Add: $\begin{array}{r} 298 \\ 461 \\ +\ 322 \\ \hline \end{array}$

6. Find the sum of 5,894, 6301, and 298.

7. An insurance account executive received commissions of \$723, \$544, \$812, and \$488 during a 4-week period. Find the total income from commissions for the 4 weeks.

8. You had a balance of \$516 in your checking account before making deposits of \$88 and \$213. Find the total amount deposited, and determine your new checking balance.

## SECTION 1.3

9. Subtract: $\begin{array}{r} 4926 \\ -\ 3177 \\ \hline \end{array}$

10. What is 10,134 decreased by 4725?

The graph at the right shows the amount that companies from different countries spend on business trips. Use the graph for Exercises 11–13.

11. How much more does a U.S. company spend for business travel and entertainment than does a company from Italy?

12. How much more does a U.S. company spend for business travel and entertainment than the combined spending of a company from the U.K. and a company from Italy?

13. Does a company from the United States outspend the combined spending of a company from Germany and a company from the U.K.?

## SECTION 1.4

**14.** Multiply: $\begin{array}{r} 843 \\ \times\ 27 \\ \hline \end{array}$

**15.** What is 2019 multiplied by 307?

**16.** You have a car payment of $123 per month. What is the total of the car payments over a 12-month period?

**17.** Vincent Meyers, a sales assistant, earns $240 for working a 40-hour week. Last week Vincent worked an additional 12 hours at $12 an hour. Find Vincent's total pay for last week's work.

## SECTION 1.5

**18.** Divide: $7\overline{)14{,}945}$

**19.** Find the quotient of 109,763 and 84.

**20.** Louis Reyes, a sales executive, drove a car 351 miles on 13 gallons of gas. Find the number of miles driven per gallon of gasoline.

**21.** A car is purchased for $8940, with a down payment of $1500. The balance is paid in 48 equal monthly payments. Find the monthly car payment.

## SECTION 1.6

**22.** Write $2 \cdot 2 \cdot 2 \cdot 2 \cdot 5 \cdot 5 \cdot 5$ in exponential notation.

**23.** Write $5 \cdot 5 \cdot 7 \cdot 7 \cdot 7 \cdot 7 \cdot 7$ in exponential notation.

**24.** Simplify: $3 \cdot 2^3 \cdot 5^2$

**25.** Simplify: $2^3 - 3 \cdot 2$

**26.** Simplify: $3^2 + 2^2 \cdot (5 - 3)$

**27.** Simplify: $8 \cdot (6 - 2) \div 4$

## SECTION 1.7

**28.** Find all the factors of 18.

**29.** Find all the factors of 30.

**30.** Find the prime factorization of 42.

**31.** Find the prime factorization of 72.

# Chapter Test

1. Place the correct symbol, < or >, between the two numbers. 21   19

2. Write 207,068 in words.

3. Write one million two hundred four thousand six in standard form.

4. Write 906,378 in expanded form.

5. Round 74,965 to the nearest hundred.

6. Add: $\begin{array}{r} 25{,}492 \\ +\ 71{,}306 \\ \hline \end{array}$

7. Find the sum of 89,756, 9094, and 37,065.

8. A family drives 425 miles the first day, 187 miles the second day, and 243 miles the third day of their vacation. The odometer read 47,626 miles at the start of the vacation.
   a. How many miles were driven during the 3 days?
   b. What is the odometer reading at the end of the 3 days?

9. Subtract: $\begin{array}{r} 17{,}495 \\ -\ 8{,}162 \\ \hline \end{array}$

10. Find the difference between 29,736 and 9814.

The data in the table at the right are based on information from the California Department of Insurance in 1993. This table shows the annual insurance premiums of some companies for renters insurance. Use it for Exercises 11 and 12.

| | |
|---|---|
| State Farm | $139 |
| Allstate | $129 |
| Farmers | $147 |
| Safeco | $220 |
| Auto Club (AAA) | $136 |
| CNA Insurance Co. | $164 |
| Transamerica | $189 |

11. What is the difference between the annual premium charged by CNA Insurance Co. and that charged by State Farm?

12. What is the difference in premium between the company with the highest premium and the one with the lowest premium?

**13.** Find the product of 90,763 and 8.

**14.** Multiply: $\begin{array}{r} 9736 \\ \times\ 704 \\ \hline \end{array}$

**15.** An investor receives $237 each month from a corporate bond fund. How much will the investor receive over a 12-month period?

**16.** Find the quotient of 5624 and 8.

**17.** Divide: $7\overline{)60{,}972}$

**18.** Divide: $97\overline{)108{,}764}$

**19.** A farmer harvested 48,290 pounds of lemons from one grove and 23,710 pounds of lemons from another grove. The lemons were packed in boxes with 24 pounds of lemons in each box. How many boxes were needed to pack the lemons?

**20.** Write $3 \cdot 3 \cdot 3 \cdot 7 \cdot 7$ in exponential notation.

**21.** Simplify: $3^3 \cdot 4^2$

**22.** Simplify: $4^2 \cdot (4 - 2) \div 8 + 5$

**23.** Simplify: $16 \div 4 \times 2 - (6 - 5)^3$

**24.** Find all the factors of 20.

**25.** Find the prime factorization of 84.

Chapter

# 2 Fractions

## Objectives

### Section 2.1

To find the least common multiple (LCM)

To find the greatest common factor (GCF)

### Section 2.2

To write a fraction that represents part of a whole

To write an improper fraction as a mixed number or a whole number, and a mixed number as an improper fraction

### Section 2.3

To find equivalent fractions by raising to higher terms

To write a fraction in simplest form

### Section 2.4

To add fractions with the same denominator

To add fractions with unlike denominators

To add whole numbers, mixed numbers, and fractions

To solve application problems

### Section 2.5

To subtract fractions with the same denominator

To subtract fractions with unlike denominators

To subtract whole numbers, mixed numbers, and fractions

To solve application problems

### Section 2.6

To multiply fractions

To multiply whole numbers, mixed numbers, and fractions

To solve application problems

### Section 2.7

To divide fractions

To divide whole numbers, mixed numbers, and fractions

To solve application problems

### Section 2.8

To identify the order relation between two fractions

To simplify expressions containing exponents

To use the Order of Operations Agreement to simplify expressions

# Egyptian Fractions

The Rhind papyrus is one of the earliest written accounts of mathematics.[1] In the papyrus, a scribe named Ahmes gives an early account of the concept of fractions. A portion of the Rhind papyrus is rendered below with its hieroglyphic transcription.

The early Egyptians primarily used unit fractions. This is a fraction in which the numerator is a 1. To write a fraction, a small oval was placed above a series of lines. The number of lines indicated the denominator. Some examples of these fractions are

 $= \frac{1}{4}$      $= \frac{1}{2}$

In the first example, each line represents a 1. Because there are 4 lines, the fraction is $\frac{1}{4}$.

The second example is the special symbol that was used for the fraction $\frac{1}{2}$.

[1]Papyrus comes from the stem of a plant. The stem was dried and then pounded thin. The resulting material served as a primitive type of paper.

# 2.1 The Least Common Multiple and Greatest Common Factor

## *Objective A To find the least common multiple (LCM)*

The **multiples** of a number are the products of that number and the numbers 1, 2, 3, 4, 5, ....

$3 \times 1 = 3$
$3 \times 2 = 6$
$3 \times 3 = 9$
$3 \times 4 = 12$ The multiples of 3 are 3, 6, 9, 12, 15, ....
$3 \times 5 = 15$
⋮

A number that is a multiple of two or more numbers is a **common multiple** of those numbers.

The multiples of 4 are 4, 8, 12, 16, 20, 24, 28, 32, 36, ....
The multiples of 6 are 6, 12, 18, 24, 30, 36, 42, ....
Some common multiples of 4 and 6 are 12, 24, and 36.

The **least common multiple** (LCM) is the smallest common multiple of two or more numbers.

The least common multiple of 4 and 6 is 12.

Listing the multiples of each number is one way to find the LCM. Another way to find the LCM uses the prime factorization of each number.

To find the LCM of 450 and 600, find the prime factorization of each number and write the factorization of each number in a table. Circle the largest product in each column. The LCM is the product of the circled numbers.

| | 2 | 3 | 5 |
|---|---|---|---|
| 450 = | 2 | (3 · 3) | 5 · 5 |
| 600 = | (2 · 2 · 2) | 3 | (5 · 5) |

In the column headed by 5, the products are equal. Circle just one product.

The LCM is the product of the circled numbers.
The LCM = $2 \cdot 2 \cdot 2 \cdot 3 \cdot 3 \cdot 5 \cdot 5 = 1800$.

**Example 1** Find the LCM of 24, 36, and 50.

**Solution**

| | 2 | 3 | 5 |
|---|---|---|---|
| 24 = | (2 · 2 · 2) | 3 | |
| 36 = | 2 · 2 | (3 · 3) | |
| 50 = | 2 | | (5 · 5) |

The LCM = $2 \cdot 2 \cdot 2 \cdot 3 \cdot 3 \cdot 5 \cdot 5 = 1800$.

**You Try It 1** Find the LCM of 50, 84, and 135.

**Your solution**

*Solution on p. A12*

## Objective B To find the greatest common factor (GCF)

Recall that a number that divides another number evenly is a factor of that number. 64 can be evenly divided by 1, 2, 4, 8, 16, 32, and 64. 1, 2, 4, 8, 16, 32, and 64 are factors of 64.

A number that is a factor of two or more numbers is a **common factor** of those numbers.

The factors of 30 are 1, 2, 3, 5, 6, 10 15, and 30.
The factors of 105 are 1, 3, 5, 7, 15, 21, 35, and 105.
The common factors of 30 and 105 are 1, 3, 5, and 15.

The **greatest common factor** (GCF) is the largest common factor of two or more numbers.

The greatest common factor of 30 and 105 is 15.

Listing the factors of each number is one way of finding the GCF. Another way to find the GCF uses the prime factorization of each number.

To find the GCF of 126 and 180, find the prime factorization of each number and write the factorization of each number in a table. Circle the smallest product in each column that does not have a blank. The GCF is the product of the circled numbers.

| | 2 | 3 | 5 | 7 |
|---|---|---|---|---|
| 126 = | ② | (3 · 3) circled | | 7 |
| 180 = | 2 · 2 | 3 · 3 | 5 | |

In the column headed by 3, the products are equal. Circle just one product.
Columns 5 and 7 have a blank, so 5 and 7 are not common factors of 126 and 180. Do not circle any number in these columns.

The GCF is the product of the circled numbers.
The GCF = $2 \cdot 3 \cdot 3 = 18$.

**Example 2** Find the GCF of 90, 168, and 420.

**Solution**

| | 2 | 3 | 5 | 7 |
|---|---|---|---|---|
| 90 = | ② | 3 · 3 | 5 | |
| 168 = | 2 · 2 · 2 | ③ | | 7 |
| 420 = | 2 · 2 | 3 | 5 | 7 |

The GCF = $2 \cdot 3 = 6$.

**You Try It 2** Find the GCF of 36, 60, and 72.

**Your solution**

**Example 3** Find the GCF of 7, 12, and 20.

**Solution**

| | 2 | 3 | 5 | 7 |
|---|---|---|---|---|
| 7 = | | | | 7 |
| 12 = | 2 · 2 | 3 | | |
| 20 = | 2 · 2 | | 5 | |

Since no numbers are circled, the GCF = 1.

**You Try It 3** Find the GCF of 11, 24, and 30.

**Your solution**

*Solutions on p. A12*

## 2.1 Exercises

### *Objective A*

Find the LCM.

**1.** 5, 8 **2.** 3, 6 **3.** 3, 8 **4.** 2, 5 **5.** 5, 6

**6.** 5, 7 **7.** 4, 6 **8.** 6, 8 **9.** 8, 12 **10.** 12, 16

**11.** 5, 12 **12.** 3, 16 **13.** 8, 14 **14.** 6, 18 **15.** 3, 9

**16.** 4, 10 **17.** 8, 32 **18.** 7, 21 **19.** 9, 36 **20.** 14, 42

**21.** 44, 60 **22.** 120, 160 **23.** 102, 184 **24.** 123, 234 **25.** 4, 8, 12

**26.** 5, 10, 15 **27.** 3, 5, 10 **28.** 2, 5, 8 **29.** 3, 8, 12 **30.** 5, 12, 18

**31.** 9, 36, 64 **32.** 18, 54, 63 **33.** 16, 30, 84 **34.** 9, 12, 15

### *Objective B*

Find the GCF.

**35.** 3, 5 **36.** 5, 7 **37.** 6, 9 **38.** 18, 24 **39.** 15, 25

**40.** 14, 49 **41.** 25, 100 **42.** 16, 80 **43.** 32, 51 **44.** 21, 44

**45.** 12, 80 **46.** 8, 36 **47.** 16, 140 **48.** 12, 76 **49.** 8, 14

**50.** 24, 30 **51.** 48, 144 **52.** 44, 96 **53.** 18, 32 **54.** 16, 30

**55.** 3, 5, 11 **56.** 6, 8, 10 **57.** 7, 14, 49 **58.** 6, 15, 36 **59.** 8, 12, 16

**60.** 10, 15, 20 **61.** 12, 18, 20 **62.** 24, 40, 72 **63.** 3, 17, 51

**64.** 17, 31, 81 **65.** 14, 42, 84 **66.** 25, 125, 625 **67.** 12, 68, 92

**68.** 28, 35, 70 **69.** 1, 49, 153 **70.** 32, 56, 72 **71.** 24, 36, 48

## APPLYING THE CONCEPTS

**72.** Define the phrase *relatively prime numbers.* List three pairs of relatively prime numbers.

**73.** Joe Salvo, a clerk, works three days and then has a day off. A friend works five days and then has a day off. How many days after Joe and his friend have a day off together will they have another day off together?

**74.** Find the LCM of the following pairs of numbers: 2 and 3, 5 and 7, and 11 and 19. Can you draw a conclusion about the LCM of two prime numbers? Suggest a way of finding the LCM of three prime numbers.

**75.** Find the GCF of the following pairs of numbers: 3 and 5, 7 and 11, and 29 and 43. Can you draw a conclusion about the GCF of two prime numbers. What is the GCF of three prime numbers?

**76.** Is the LCM of two numbers always divisible by the GCF of the two numbers? If so, explain why. If not, give an example.

**77.** Using the pattern for the first two triangles below, determine the center number of the last triangle.

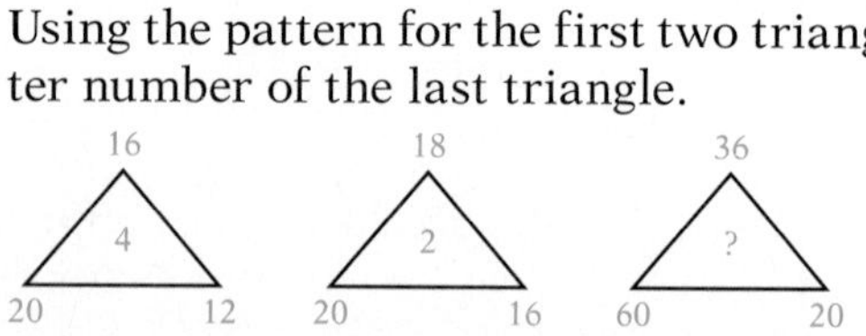


**78.** The ancient Mayans used two calendars, a solar calendar of 365 days and a ritual calendar of 220 days. If a solar year and the ritual year begin on the same day, how many solar years and how many ritual years will pass until this situation occurs again?
(*Mathematics Teacher,* November 1993, page 104)

# 2.2 Introduction to Fractions

## *Objective A* *To write a fraction that represents part of a whole*

A **fraction** can represent the number of equal parts of a whole.

The shaded portion of the circle is represented by the fraction $\frac{4}{7}$. Four-sevenths of the circle are shaded.

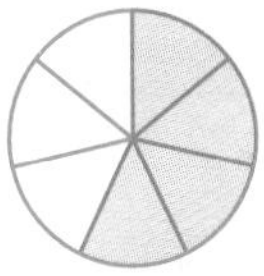

$\frac{4}{7}$

Each part of a fraction has a name.

$$\textbf{Fraction bar} \rightarrow \frac{4}{7} \begin{array}{l} \leftarrow \textbf{Numerator} \\ \leftarrow \textbf{Denominator} \end{array}$$

**POINT OF INTEREST**

The fraction bar was first used in 1050 by al-Hassar. It is also called a vinculum.

A **proper fraction** is a fraction less than 1. The numerator of a proper fraction is smaller than the denominator. The shaded portion of the circle can be represented by the proper fraction $\frac{3}{4}$.

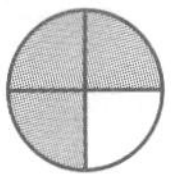

$\frac{3}{4}$

A **mixed number** is a number greater than 1 with a whole-number part and a fractional part. The shaded portion of the circles can be represented by the mixed number $2\frac{1}{4}$.

$2\frac{1}{4}$

An **improper fraction** is a fraction greater than or equal to 1. The numerator of an improper fraction is greater than or equal to the denominator. The shaded portion of the circles can be represented by the improper fraction $\frac{9}{4}$. The shaded portion of the square can be represented by $\frac{4}{4}$.

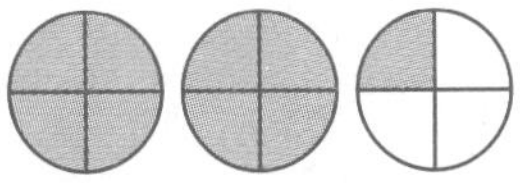

$\frac{9}{4}$

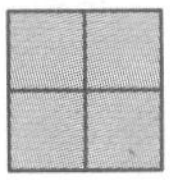

$\frac{4}{4}$

**Example 1**
Express the shaded portion of the circled as a mixed number.

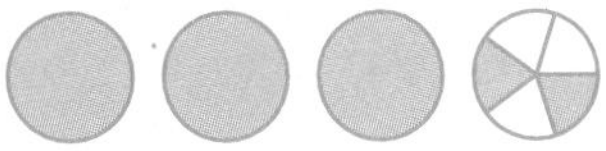

**Solution** $3\frac{2}{5}$

**You Try It 1**
Express the shaded portion of the circles as a mixed number.

**Your solution**

**Example 2**
Express the shaded portion of the circles as an improper fraction.

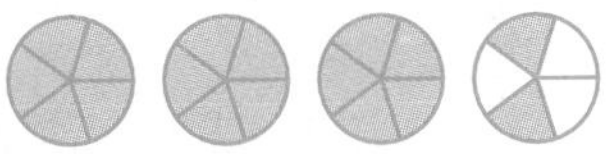

**Solution** $\frac{17}{5}$

**You Try It 2**
Express the shaded portion of the circles as an improper fraction.

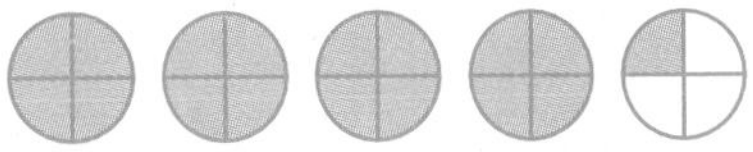

**Your solution**

*Solutions on p. A12*

### *Objective B To write an improper fraction as a mixed number or a whole number, and a mixed number as an improper fraction*

Note from the diagram that the mixed number $2\frac{3}{5}$ and the improper fraction $\frac{13}{5}$ both represent the shaded portion of the circles.

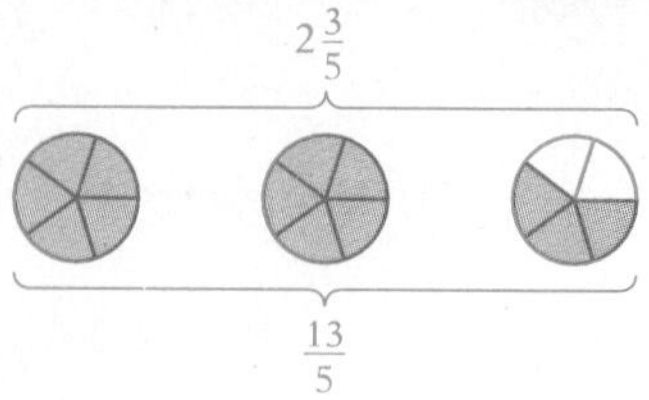


$$2\frac{3}{5} = \frac{13}{5}$$

An improper fraction can be written as a mixed number.

➡ Write $\frac{13}{5}$ as a mixed number.

Divide the numerator by the denominator.

$$\begin{array}{r} 2 \\ 5\overline{)\,13} \\ -10 \\ \hline 3 \end{array}$$

To write the fractional part of the mixed number, write the remainder over the divisor.

$$\begin{array}{r} 2\frac{3}{5} \\ 5\overline{)\,13} \\ -10 \\ \hline 3 \end{array}$$

Write the answer.

$$\frac{13}{5} = 2\frac{3}{5}$$

**POINT OF INTEREST**
Archimedes (c. 287–212 B.C.) is the person who calculated that $\pi \approx 3\frac{1}{7}$. He actually showed that $3\frac{10}{71} < \pi < 3\frac{1}{7}$. The approximation $3\frac{10}{71}$ is more accurate but more difficult to use.

To write a mixed number as an improper fraction, multiply the denominator of the fractional part by the whole-number part. The sum of this product and the numerator of the fractional part is the numerator of the improper fraction. The denominator remains the same.

➡ Write $7\frac{3}{8}$ as an improper fraction.

$$7\frac{3}{8} = \frac{(8 \times 7) + 3}{8} = \frac{56 + 3}{8} = \frac{59}{8} \qquad 7\frac{3}{8} = \frac{59}{8}$$

**Example 3** Write $\frac{21}{4}$ as a mixed number.

**Solution**

$$\begin{array}{r} 5 \\ 4\overline{)\,21} \\ -20 \\ \hline 1 \end{array} \qquad \frac{21}{4} = 5\frac{1}{4}$$

**You Try It 3** Write $\frac{22}{5}$ as a mixed number.

**Your solution**

**Example 4** Write $\frac{18}{6}$ as a whole number.

**Solution**

$$\begin{array}{r} 3 \\ 6\overline{)\,18} \\ -18 \\ \hline 0 \end{array} \qquad \frac{18}{6} = 3$$

Note. The remainder is zero.

**You Try It 4** Write $\frac{28}{7}$ as a whole number.

**Your solution**

**Example 5** Write $21\frac{3}{4}$ as an improper fraction.

**Solution** $21\frac{3}{4} = \frac{84 + 3}{4} = \frac{87}{4}$

**You Try It 5** Write $14\frac{5}{8}$ as an improper fraction.

**Your solution**

*Solutions on p. A12*

## 2.2 Exercises

### *Objective A*

Express the shaded portion of the circle as a fraction.

**1.** 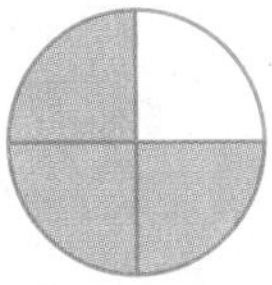

**2.** 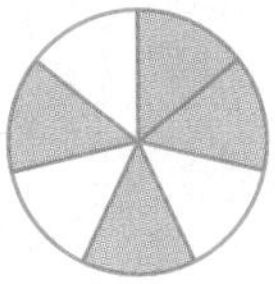

**3.** 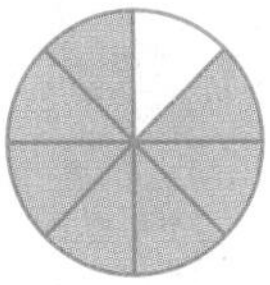

**4.** 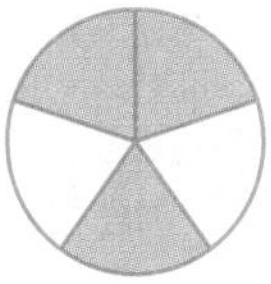

Express the shaded portion of the circles as a mixed number.

**5.** 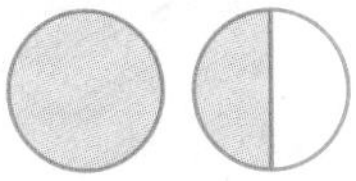

**6.** 

**7.** 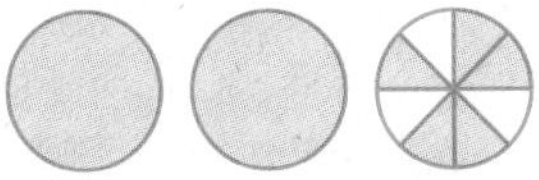

**8.** 

**9.** 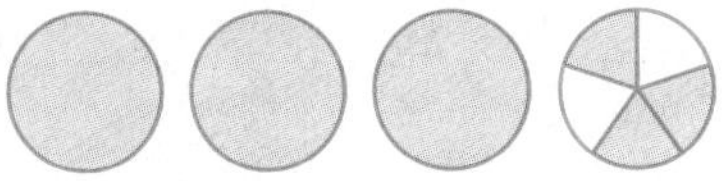

**10.** 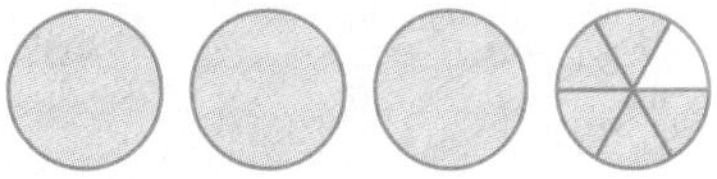

Express the shaded portion of the circles as an improper fraction.

**11.** 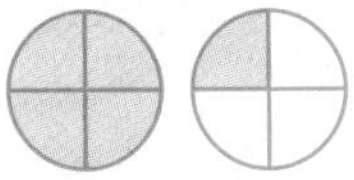

**12.** 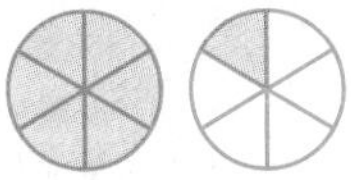

**13.** 

**14.** 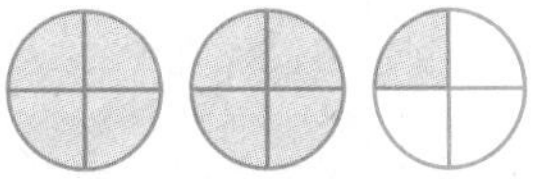

**15.** 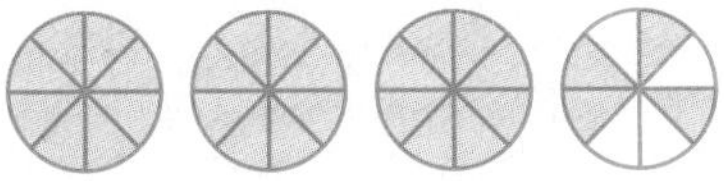

**16.** 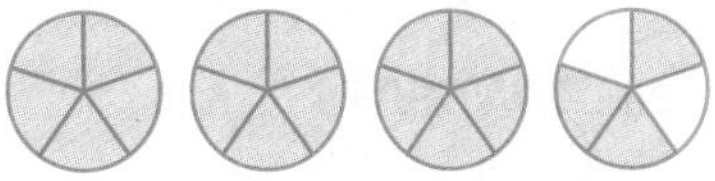

**17.** Shade $\frac{5}{6}$ of 

**18.** Shade $1\frac{2}{5}$ of 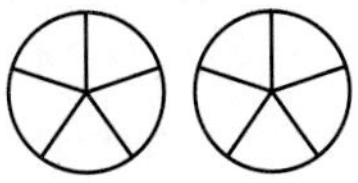

**19.** Shade $\frac{6}{5}$ of 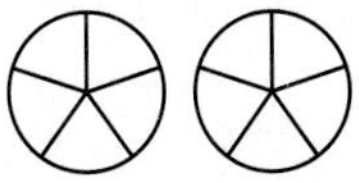

### Objective B

Write the improper fraction as a mixed number or a whole number.

**20.** $\frac{11}{4}$ **21.** $\frac{16}{3}$ **22.** $\frac{20}{4}$ **23.** $\frac{18}{9}$ **24.** $\frac{9}{8}$ **25.** $\frac{13}{4}$

**26.** $\frac{23}{10}$ **27.** $\frac{29}{2}$ **28.** $\frac{48}{16}$ **29.** $\frac{51}{3}$ **30.** $\frac{8}{7}$ **31.** $\frac{16}{9}$

**32.** $\frac{7}{3}$ **33.** $\frac{9}{5}$ **34.** $\frac{16}{1}$ **35.** $\frac{23}{1}$ **36.** $\frac{17}{8}$ **37.** $\frac{31}{16}$

**38.** $\frac{12}{5}$ **39.** $\frac{19}{3}$ **40.** $\frac{9}{9}$ **41.** $\frac{40}{8}$ **42.** $\frac{72}{8}$ **43.** $\frac{3}{3}$

Write the mixed number as an improper fraction.

**44.** $2\frac{1}{3}$ **45.** $4\frac{2}{3}$ **46.** $6\frac{1}{2}$ **47.** $8\frac{2}{3}$ **48.** $6\frac{5}{6}$ **49.** $7\frac{3}{8}$

**50.** $9\frac{1}{4}$ **51.** $6\frac{1}{4}$ **52.** $10\frac{1}{2}$ **53.** $15\frac{1}{8}$ **54.** $8\frac{1}{9}$ **55.** $3\frac{5}{12}$

**56.** $5\frac{3}{11}$ **57.** $3\frac{7}{9}$ **58.** $2\frac{5}{8}$ **59.** $12\frac{2}{3}$ **60.** $1\frac{5}{8}$ **61.** $5\frac{3}{7}$

**62.** $11\frac{1}{9}$ **63.** $12\frac{3}{5}$ **64.** $3\frac{3}{8}$ **65.** $4\frac{5}{9}$ **66.** $6\frac{7}{13}$ **67.** $8\frac{5}{14}$

### APPLYING THE CONCEPTS

**68.** What fraction of the states in the United States of America begins with the letter M? What fraction of the states begins and ends with a vowel?

**69.** [W] Find the business section of your local newspaper. Choose a stock and record the fluctuations in the stock price for one week. Explain the part that fractions play in reporting the price and change in price of the stock.

**70.** [W] Explain in your own words the procedure for rewriting a mixed number as an improper fraction.

# 2.3 Writing Equivalent Fractions

## *Objective A To find equivalent fractions by raising to higher terms*

Equal fractions with different denominators are called **equivalent fractions.**

$\frac{4}{6}$ is equivalent to $\frac{2}{3}$.

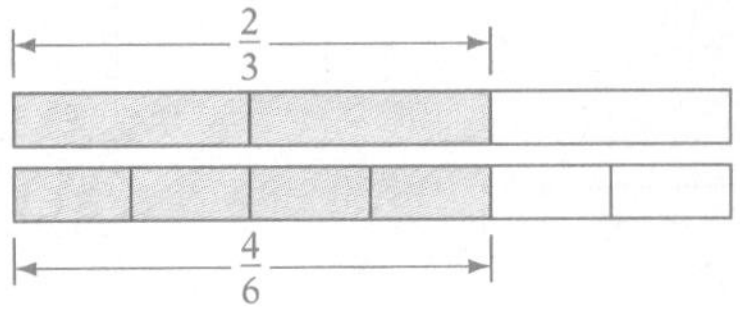


Remember that the Multiplication Property of One stated that the product of a number and one is the number. This is true for fractions as well as whole numbers. This property can be used to write equivalent fractions.

$$\frac{2}{3} \times 1 = \frac{2}{3} \times \frac{1}{1} = \frac{2 \cdot 1}{3 \cdot 1} = \frac{2}{3}$$

$\frac{2}{3} \times 1 = \frac{2}{3} \times \boxed{\frac{2}{2}} = \frac{2 \cdot 2}{3 \cdot 2} = \frac{4}{6}$ $\quad \frac{4}{6}$ is equivalent to $\frac{2}{3}$.

$\frac{2}{3} \times 1 = \frac{2}{3} \times \boxed{\frac{4}{4}} = \frac{2 \cdot 4}{3 \cdot 4} = \frac{8}{12}$ $\quad \frac{8}{12}$ is equivalent to $\frac{2}{3}$.

$\frac{2}{3}$ was rewritten as the equivalent fractions $\frac{4}{6}$ and $\frac{8}{12}$.

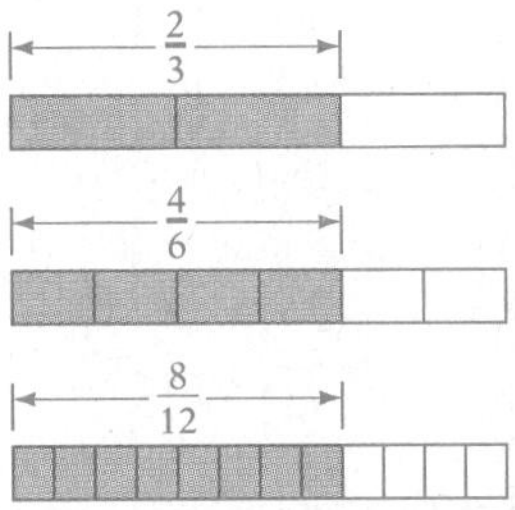


➡ Write a fraction that is equivalent to $\frac{5}{8}$ and has a denominator of 32.

$8\overline{)32}$ = 4 • Divide the larger denominator by the smaller.

$\frac{5}{8} = \frac{5 \cdot 4}{8 \cdot 4} = \frac{20}{32}$ • Multiply the numerator and denominator of the given fraction by the quotient (4).

$\frac{20}{32}$ is equivalent to $\frac{5}{8}$.

---

**Example 1** Write $\frac{2}{3}$ as an equivalent fraction that has a denominator of 42.

**Solution** $3\overline{)42}$ = 14 $\quad \frac{2}{3} = \frac{2 \cdot 14}{3 \cdot 14} = \frac{28}{42}$

$\frac{28}{42}$ is equivalent to $\frac{2}{3}$.

**You Try It 1** Write $\frac{3}{5}$ as an equivalent fraction that has a denominator of 45.

**Your solution**

**Example 2** Write 4 as a fraction that has a denominator of 12.

**Solution** Write 4 as $\frac{4}{1}$.

$1\overline{)12}$ = 12 $\quad 4 = \frac{4 \cdot 12}{1 \cdot 12} = \frac{48}{12}$

$\frac{48}{12}$ is equivalent to 4.

**You Try It 2** Write 6 as a fraction that has a denominator of 18.

**Your solution**

*Solutions on p. A12*

## Objective B To write a fraction in simplest form

A fraction is in **simplest form** when there are no common factors in the numerator and the denominator.

The fractions $\frac{4}{6}$ and $\frac{2}{3}$ are equivalent fractions.

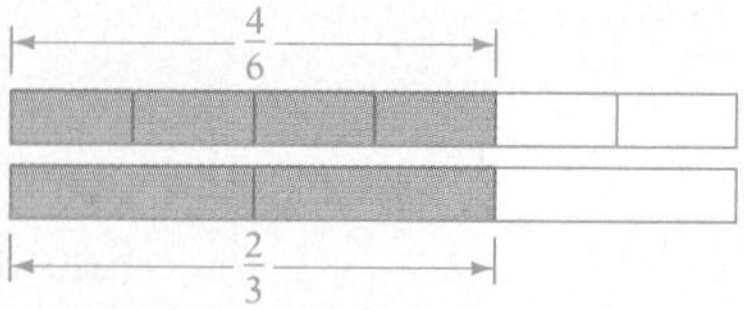


$\frac{4}{6}$ has been written in simplest form as $\frac{2}{3}$.

The Multiplication Property of One can be used to write fractions in simplest form. Write the numerator and denominator of the given fraction as a product of factors. Write factors common to both the numerator and denominator as an improper fraction equivalent to 1.

$$\frac{4}{6} = \frac{2 \cdot 2}{2 \cdot 3} = \frac{\boxed{2} \cdot 2}{\boxed{2} \cdot 3} = \boxed{\frac{2}{2}} \cdot \frac{2}{3} = 1 \cdot \frac{2}{3} = \frac{2}{3}$$

The process of eliminating common factors is displayed with slashes through the common factors as shown at the right.

$$\frac{4}{6} = \frac{\overset{1}{\cancel{2}} \cdot 2}{\underset{1}{\cancel{2}} \cdot 3} = \frac{2}{3}$$

To write a fraction in simplest form, eliminate the common factors.

$$\frac{18}{30} = \frac{\overset{1}{\cancel{2}} \cdot \overset{1}{\cancel{3}} \cdot 3}{\underset{1}{\cancel{2}} \cdot \underset{1}{\cancel{3}} \cdot 5} = \frac{3}{5}$$

An improper fraction can be changed to a mixed number.

$$\frac{22}{6} = \frac{\overset{1}{\cancel{2}} \cdot 11}{\underset{1}{\cancel{2}} \cdot 3} = \frac{11}{3} = 3\frac{2}{3}$$

**Example 3** Write $\frac{15}{40}$ in simplest form.

**Solution** $\frac{15}{40} = \frac{3 \cdot \overset{1}{\cancel{5}}}{2 \cdot 2 \cdot 2 \cdot \underset{1}{\cancel{5}}} = \frac{3}{8}$

**You Try It 3** Write $\frac{16}{24}$ in simplest form.

**Your solution**

**Example 4** Write $\frac{6}{42}$ in simplest form.

**Solution** $\frac{6}{42} = \frac{\overset{1}{\cancel{2}} \cdot \overset{1}{\cancel{3}}}{\underset{1}{\cancel{2}} \cdot \underset{1}{\cancel{3}} \cdot 7} = \frac{1}{7}$

**You Try It 4** Write $\frac{8}{56}$ in simplest form.

**Your solution**

**Example 5** Write $\frac{8}{9}$ in simplest form.

**Solution** $\frac{8}{9} = \frac{2 \cdot 2 \cdot 2}{3 \cdot 3} = \frac{8}{9}$

$\frac{8}{9}$ is already in simplest form because there are no common factors in the numerator and denominator.

**You Try It 5** Write $\frac{15}{32}$ in simplest form.

**Your solution**

**Example 6** Write $\frac{30}{12}$ in simplest form.

**Solution** $\frac{30}{12} = \frac{\overset{1}{\cancel{2}} \cdot \overset{1}{\cancel{3}} \cdot 5}{\underset{1}{\cancel{2}} \cdot 2 \cdot \underset{1}{\cancel{3}}} = \frac{5}{2} = 2\frac{1}{2}$

**You Try It 6** Write $\frac{48}{36}$ in simplest form.

**Your solution**

*Solutions on p. A13*

## 2.3 Exercises

### *Objective A*

Write an equivalent fraction with the given denominator.

1. $\frac{1}{2} = \frac{}{10}$
2. $\frac{1}{4} = \frac{}{16}$
3. $\frac{3}{16} = \frac{}{48}$
4. $\frac{5}{9} = \frac{}{81}$
5. $\frac{3}{8} = \frac{}{32}$
6. $\frac{7}{11} = \frac{}{33}$
7. $\frac{3}{17} = \frac{}{51}$
8. $\frac{7}{10} = \frac{}{90}$
9. $\frac{3}{4} = \frac{}{16}$
10. $\frac{5}{8} = \frac{}{32}$
11. $3 = \frac{}{9}$
12. $5 = \frac{}{25}$
13. $\frac{1}{3} = \frac{}{60}$
14. $\frac{1}{16} = \frac{}{48}$
15. $\frac{11}{15} = \frac{}{60}$
16. $\frac{3}{50} = \frac{}{300}$
17. $\frac{2}{3} = \frac{}{18}$
18. $\frac{5}{9} = \frac{}{36}$
19. $\frac{5}{7} = \frac{}{49}$
20. $\frac{7}{8} = \frac{}{32}$
21. $\frac{5}{9} = \frac{}{18}$
22. $\frac{11}{12} = \frac{}{36}$
23. $7 = \frac{}{3}$
24. $9 = \frac{}{4}$
25. $\frac{7}{9} = \frac{}{45}$
26. $\frac{5}{6} = \frac{}{42}$
27. $\frac{15}{16} = \frac{}{64}$
28. $\frac{11}{18} = \frac{}{54}$
29. $\frac{3}{14} = \frac{}{98}$
30. $\frac{5}{6} = \frac{}{144}$
31. $\frac{5}{8} = \frac{}{48}$
32. $\frac{7}{12} = \frac{}{96}$
33. $\frac{5}{14} = \frac{}{42}$
34. $\frac{2}{3} = \frac{}{42}$
35. $\frac{17}{24} = \frac{}{144}$
36. $\frac{5}{13} = \frac{}{169}$
37. $\frac{3}{8} = \frac{}{408}$
38. $\frac{9}{16} = \frac{}{272}$
39. $\frac{17}{40} = \frac{}{800}$
40. $\frac{9}{25} = \frac{}{1000}$

### *Objective B*

Write the fraction in simplest form.

41. $\frac{4}{12}$
42. $\frac{8}{22}$
43. $\frac{22}{44}$
44. $\frac{2}{14}$
45. $\frac{2}{12}$
46. $\frac{6}{25}$

**47.** $\frac{50}{75}$ **48.** $\frac{40}{36}$ **49.** $\frac{12}{8}$ **50.** $\frac{36}{9}$ **51.** $\frac{0}{30}$ **52.** $\frac{10}{10}$

**53.** $\frac{9}{22}$ **54.** $\frac{14}{35}$ **55.** $\frac{75}{25}$ **56.** $\frac{8}{60}$ **57.** $\frac{16}{84}$ **58.** $\frac{14}{45}$

**59.** $\frac{20}{44}$ **60.** $\frac{12}{35}$ **61.** $\frac{8}{36}$ **62.** $\frac{28}{44}$ **63.** $\frac{12}{16}$ **64.** $\frac{48}{35}$

**65.** $\frac{16}{12}$ **66.** $\frac{24}{18}$ **67.** $\frac{24}{40}$ **68.** $\frac{44}{60}$ **69.** $\frac{8}{88}$

**70.** $\frac{9}{90}$ **71.** $\frac{144}{36}$ **72.** $\frac{140}{297}$ **73.** $\frac{48}{144}$ **74.** $\frac{32}{120}$

**75.** $\frac{60}{100}$ **76.** $\frac{33}{110}$ **77.** $\frac{36}{16}$ **78.** $\frac{80}{45}$ **79.** $\frac{32}{160}$

## *APPLYING THE CONCEPTS*

**80.** Use the diagram to express the following ratios in lowest form.
a. The ratio of the dots in the triangle to the total number of dots.
b. The ratio of the number of dots in the triangle to the number of dots in the rectangle.
c. The ratio of the number of dots common to the triangle and rectangle to the dots in the rectangle.

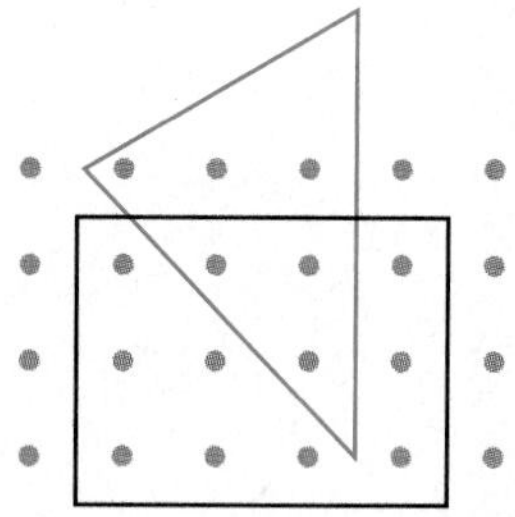

**81.** Show that $\frac{15}{24} = \frac{5}{8}$ by using a diagram.

**82.** Explain the procedure for finding equivalent fractions.
[W]

**83.** Explain the procedure for reducing fractions.
[W]

# 2.4 Addition of Fractions and Mixed Numbers

## *Objective A To add fractions with the same denominator*

Fractions with the same denominator are added by adding the numerators and placing the sum over the common denominator. After adding, write the sum in simplest form.

➡ Add: $\frac{2}{7} + \frac{4}{7}$

$$\begin{array}{r} \frac{2}{7} \\ + \frac{4}{7} \\ \hline \frac{6}{7} \end{array}$$

• **Add the numerators and place the sum over the common denominator.**

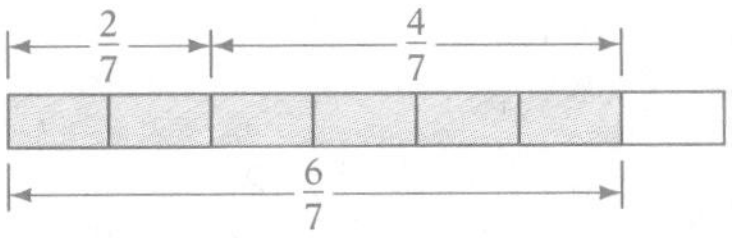


$$\frac{2}{7} + \frac{4}{7} = \frac{2+4}{7} = \frac{6}{7}$$

**Example 1** Add: $\frac{5}{12} + \frac{11}{12}$

**Solution**

$$\begin{array}{r} \frac{5}{12} \\ + \frac{11}{12} \\ \hline \frac{16}{12} = \frac{4}{3} = 1\frac{1}{3} \end{array}$$

**You Try It 1** Add: $\frac{3}{8} + \frac{7}{8}$

**Your solution**

*Solution on p. A13*

## *Objective B To add fractions with unlike denominators*

To add fractions with unlike denominators, first rewrite the fractions as equivalent fractions with a common denominator. The common denominator is the LCM of the denominators of the fractions.

➡ Find the total of $\frac{1}{2}$ and $\frac{1}{3}$.

The common denominator is the LCM of 2 and 3. LCM = 6. The LCM of denominators is sometimes called the **least common denominator** (LCD).

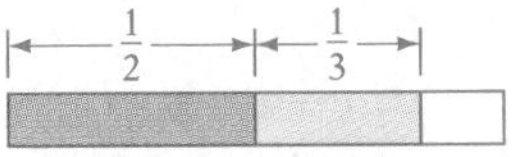


Write equivalent fractions using the LCM.

$$\begin{array}{r} \frac{1}{2} = \frac{3}{6} \\ + \frac{1}{3} = \frac{2}{6} \\ \hline \end{array}$$

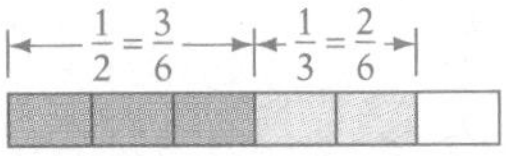


Add the fractions.

$$\begin{array}{r} \frac{1}{2} = \frac{3}{6} \\ + \frac{1}{3} = \frac{2}{6} \\ \hline \frac{5}{6} \end{array}$$

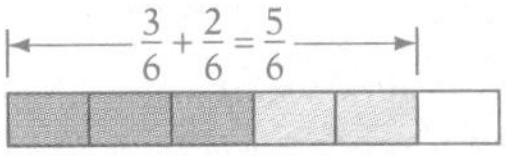

**Example 2** Find $\frac{7}{12}$ more than $\frac{3}{8}$.

**Solution**

$$\begin{array}{r} \frac{3}{8} = \frac{9}{24} \\ + \frac{7}{12} = \frac{14}{24} \\ \hline \frac{23}{24} \end{array}$$

• The LCM of 8 and 12 is 24.

**You Try It 2** Find the sum of $\frac{5}{12}$ and $\frac{9}{16}$.

**Your solution**

**Example 3** Add: $\frac{5}{8} + \frac{7}{9}$

**Solution**

$$\begin{array}{r} \frac{5}{8} = \frac{45}{72} \\ + \frac{7}{9} = \frac{56}{72} \\ \hline \frac{101}{72} = 1\frac{29}{72} \end{array}$$

• The LCM of 8 and 9 is 72.

**You Try It 3** Add: $\frac{7}{8} + \frac{11}{15}$

**Your solution**

**Example 4** Add: $\frac{2}{3} + \frac{3}{5} + \frac{5}{6}$

**Solution**

$$\begin{array}{r} \frac{2}{3} = \frac{20}{30} \\ \frac{3}{5} = \frac{18}{30} \\ + \frac{5}{6} = \frac{25}{30} \\ \hline \frac{63}{30} = 2\frac{3}{30} = 2\frac{1}{10} \end{array}$$

• The LCM of 3, 5, and 6 is 30.

**You Try It 4** Add: $\frac{3}{4} + \frac{4}{5} + \frac{5}{8}$

**Your solution**

*Solutions on p. A13*

### *Objective C To add whole numbers, mixed numbers, and fractions*

The sum of a whole number and a fraction is a mixed number.

➡ Add: $2 + \frac{2}{3}$

$$\boxed{2} + \frac{2}{3} = \boxed{\frac{6}{3}} + \frac{2}{3} = \frac{8}{3} = 2\frac{2}{3}$$

To add a whole number and a mixed number, write the fraction, then add the whole numbers.

➡ Add: $7\frac{2}{5} + 4$

Write the fraction.

$$\begin{array}{r} 7\frac{2}{5} \\ +\ 4 \\ \hline \frac{2}{5} \end{array}$$

Add the whole numbers.

$$\begin{array}{r} 7\frac{2}{5} \\ +\ 4 \\ \hline 11\frac{2}{5} \end{array}$$

To add two mixed numbers, add the fractional parts and then add the whole numbers. Remember to reduce the sum to simplest form.

➡ What is $5\frac{4}{9}$ added to $6\frac{14}{15}$?

The LCM of 9 and 15 is 48.

Add the fractional parts.

$$\begin{array}{r} 5\frac{4}{9} = 5\frac{20}{45} \\ +\ 6\frac{14}{15} = 6\frac{42}{45} \\ \hline \frac{62}{45} \end{array}$$

Add the whole numbers.

$$\begin{array}{r} 5\frac{4}{9} = 5\frac{20}{45} \\ +\ 6\frac{14}{15} = 6\frac{42}{45} \\ \hline 11\frac{62}{45} = 11 + 1\frac{17}{45} = 12\frac{17}{45} \end{array}$$

**Example 5** Add: $5 + \frac{3}{8}$

**Solution** $5 + \frac{3}{8} = 5\frac{3}{8}$

**You Try It 5** What is 7 added to $\frac{6}{11}$?

**Your solution**

**Example 6** Find 17 increased by $3\frac{3}{8}$.

**Solution**

$$\begin{array}{r} 17 \\ +\ 3\frac{3}{8} \\ \hline 20\frac{3}{8} \end{array}$$

**You Try It 6** Find the sum of 29 and $17\frac{5}{12}$.

**Your solution**

**Example 7** Add: $5\frac{2}{3} + 11\frac{5}{6} + 12\frac{7}{9}$

**Solution**

$$\begin{array}{r} 5\frac{2}{3} = 5\frac{12}{18} \\ 11\frac{5}{6} = 11\frac{15}{18} \\ +\ 12\frac{7}{9} = 12\frac{14}{18} \\ \hline 28\frac{41}{18} = 30\frac{5}{18} \end{array}$$

• LCM = 18.

**You Try It 7** Add: $7\frac{4}{5} + 6\frac{7}{10} + 13\frac{11}{15}$

**Your solution**

**Example 8** Add: $11\frac{5}{8} + 7\frac{5}{9} + 8\frac{7}{15}$

**Solution**

$$\begin{array}{r} 11\frac{5}{8} = 11\frac{225}{360} \\ 7\frac{5}{9} = 7\frac{200}{360} \\ +\ 8\frac{7}{15} = 8\frac{168}{360} \\ \hline 26\frac{593}{360} = 27\frac{233}{360} \end{array}$$

• LCM = 360.

**You Try It 8** Add: $9\frac{3}{8} + 17\frac{7}{12} + 10\frac{14}{15}$

**Your solution**

*Solutions on p. A13*

## *Objective D To solve application problems*

**Example 9**

A rain gauge collected $2\frac{1}{3}$ inches of rain in October, $5\frac{1}{2}$ inches in November, and $3\frac{3}{8}$ inches in December. Find the total rainfall for the 3 months.

**Strategy**
To find the total rainfall for the 3 months, add the 3 amounts of rainfall $\left(2\frac{1}{3}, 5\frac{1}{2}, \text{ and } 3\frac{3}{8}\right)$.

**Solution**

$$\begin{aligned} 2\frac{1}{3} &= 2\frac{8}{24} \\ 5\frac{1}{2} &= 5\frac{12}{24} \\ +\ 3\frac{3}{8} &= 3\frac{9}{24} \\ \hline & \ \ 10\frac{29}{24} = 11\frac{5}{24} \end{aligned}$$

The total rainfall for the 3 months was $11\frac{5}{24}$ inches.

**You Try It 9**

On Monday, you spent $4\frac{1}{2}$ hours in class, $3\frac{3}{4}$ hours studying, and $1\frac{1}{3}$ hours driving. Find the number of hours spent on these three activities.

**Your strategy**

**Your solution**

**Example 10**
Barbara Walsh worked 4 hours, $2\frac{1}{3}$ hours, and $5\frac{2}{3}$ hours this week on a part-time job. Barbara is paid \$6 an hour. How much did she earn this week?

**Strategy**
To find how much Barbara earned:
- Find the total number of hours worked.
- Multiply the total number of hours worked by the hourly wage (\$6).

**Solution**

$$\begin{array}{r} 4 \\ 2\frac{1}{3} \\ +\ 5\frac{2}{3} \\ \hline 11\frac{3}{3} \end{array} = 12 \text{ hours worked} \qquad \begin{array}{r} 12 \\ \times\ \$6 \\ \hline \$72 \end{array}$$

Barbara earned \$72 this week.

**You Try It 10**
Jeff Sapone, a carpenter, worked $1\frac{2}{3}$ hours of overtime on Monday, $3\frac{1}{3}$ hours of overtime on Tuesday, and 2 hours of overtime on Wednesday. At an overtime hourly rate of \$24, find Jeff's overtime pay for these 3 days.

**Your strategy**

**Your solution**

*Solutions on p. A14*

## 2.4 Exercises

### *Objective A*

Add.

**1.** $\frac{2}{7}+\frac{1}{7}$

**2.** $\frac{3}{11}+\frac{5}{11}$

**3.** $\frac{1}{2}+\frac{1}{2}$

**4.** $\frac{1}{3}+\frac{2}{3}$

**5.** $\frac{8}{11}+\frac{7}{11}$

**6.** $\frac{9}{13}+\frac{7}{13}$

**7.** $\frac{8}{5}+\frac{9}{5}$

**8.** $\frac{5}{3}+\frac{7}{3}$

**9.** $\frac{3}{5}+\frac{8}{5}+\frac{3}{5}$

**10.** $\frac{3}{8}+\frac{5}{8}+\frac{7}{8}$

**11.** $\frac{3}{4}+\frac{1}{4}+\frac{5}{4}$

**12.** $\frac{2}{7}+\frac{4}{7}+\frac{5}{7}$

**13.** $\frac{3}{8}+\frac{7}{8}+\frac{1}{8}$

**14.** $\frac{5}{12}+\frac{7}{12}+\frac{1}{12}$

**15.** $\frac{4}{15}+\frac{7}{15}+\frac{11}{15}$

**16.** $\frac{3}{4}+\frac{3}{4}+\frac{1}{4}$

**17.** $\frac{3}{16}+\frac{5}{16}+\frac{7}{16}$

**18.** $\frac{5}{18}+\frac{11}{18}+\frac{17}{18}$

**19.** $\frac{3}{11}+\frac{5}{11}+\frac{7}{11}$

**20.** $\frac{5}{7}+\frac{4}{7}+\frac{5}{7}$

**21.** Find the sum of $\frac{4}{9}$ and $\frac{5}{9}$.

**22.** Find the sum of $\frac{5}{12}$, $\frac{1}{12}$, and $\frac{11}{12}$.

**23.** Find the total of $\frac{5}{8}$, $\frac{3}{8}$, and $\frac{7}{8}$.

**24.** Find the total of $\frac{4}{13}$, $\frac{7}{13}$, and $\frac{11}{13}$.

### *Objective B*

Add.

**25.** $\frac{1}{2}+\frac{2}{3}$

**26.** $\frac{2}{3}+\frac{1}{4}$

**27.** $\frac{3}{14}+\frac{5}{7}$

**28.** $\frac{3}{5}+\frac{7}{10}$

**29.** $\frac{5}{9}+\frac{7}{15}$

**30.** $\frac{8}{15}+\frac{7}{20}$

**31.** $\frac{1}{6}+\frac{7}{9}$

**32.** $\frac{3}{8}+\frac{9}{14}$

**33.** $\frac{5}{12}+\frac{5}{16}$

**34.** $\frac{13}{40}+\frac{12}{25}$

**35.** $\frac{3}{20}+\frac{7}{30}$

**36.** $\frac{5}{12}+\frac{7}{30}$

**37.** $\frac{2}{3}+\frac{6}{19}$

**38.** $\frac{1}{2}+\frac{3}{29}$

**39.** $\frac{3}{7}+\frac{4}{21}$

Add.

**40.** $\frac{1}{3} + \frac{5}{6} + \frac{7}{9}$

**41.** $\frac{2}{3} + \frac{5}{6} + \frac{7}{12}$

**42.** $\frac{5}{6} + \frac{1}{12} + \frac{5}{16}$

**43.** $\frac{2}{9} + \frac{7}{15} + \frac{4}{21}$

**44.** $\frac{2}{3} + \frac{1}{5} + \frac{7}{12}$

**45.** $\frac{3}{4} + \frac{4}{5} + \frac{7}{12}$

**46.** $\frac{1}{4} + \frac{4}{5} + \frac{5}{9}$

**47.** $\frac{2}{3} + \frac{3}{5} + \frac{7}{8}$

**48.** $\frac{5}{16} + \frac{11}{18} + \frac{17}{24}$

**49.** $\frac{3}{10} + \frac{14}{15} + \frac{9}{25}$

**50.** $\frac{2}{3} + \frac{5}{8} + \frac{7}{9}$

**51.** $\frac{1}{3} + \frac{2}{9} + \frac{7}{8}$

**52.** Find $\frac{7}{8}$ added to $\frac{11}{12}$.

**53.** Find $\frac{13}{18}$ added to $\frac{5}{12}$.

**54.** What is $\frac{3}{8}$ added to $\frac{3}{5}$?

**55.** What is $\frac{5}{9}$ added to $\frac{7}{12}$?

**56.** Find the sum of $\frac{3}{8}$, $\frac{5}{6}$, and $\frac{7}{12}$.

**57.** Find the sum of $\frac{11}{12}$, $\frac{13}{24}$, and $\frac{4}{15}$.

**58.** Find the total of $\frac{1}{2}$, $\frac{5}{8}$, and $\frac{7}{9}$.

**59.** Find the total of $\frac{5}{14}$, $\frac{3}{7}$, and $\frac{5}{21}$.

***Objective C***

Add.

**60.** $1\frac{1}{2} + 2\frac{1}{6}$

**61.** $2\frac{2}{5} + 3\frac{3}{10}$

**62.** $4\frac{1}{2} + 5\frac{7}{12}$

**63.** $3\frac{3}{8} + 2\frac{5}{16}$

**64.** $2\frac{7}{9} + 3\frac{5}{12}$

**65.** $4\frac{7}{15} + 3\frac{11}{12}$

**66.** $4 + 5\frac{2}{7}$

**67.** $6\frac{8}{9} + 12$

**68.** $3\frac{5}{8} + 2\frac{11}{20}$

**69.** $4\frac{5}{12} + 6\frac{11}{18}$

**70.** $10\frac{2}{7} + 7\frac{21}{35}$

**71.** $16\frac{2}{3} + 8\frac{1}{4}$

**72.** $7\frac{5}{12} + 2\frac{9}{16}$

**73.** $9\frac{1}{2} + 3\frac{3}{11}$

**74.** $6\frac{1}{3} + 2\frac{3}{13}$

**75.** $8\frac{21}{40} + 6\frac{21}{32}$

**76.** $8\frac{29}{30} + 7\frac{11}{40}$

**77.** $17\frac{5}{16} + 3\frac{11}{24}$

**78.** $17\frac{3}{8} + 7\frac{7}{20}$

**79.** $14\frac{7}{12} + 29\frac{13}{21}$

**80.** $5\frac{7}{8} + 27\frac{5}{12}$

**81.** $7\frac{5}{6} + 3\frac{5}{9}$

**82.** $7\frac{5}{9} + 2\frac{7}{12}$

**83.** $3\frac{1}{2} + 2\frac{3}{4} + 1\frac{5}{6}$

**84.** $2\frac{1}{2} + 3\frac{2}{3} + 4\frac{1}{4}$

**85.** $3\frac{1}{3} + 7\frac{1}{5} + 2\frac{1}{7}$

**86.** $3\frac{1}{2} + 3\frac{1}{5} + 8\frac{1}{9}$

**87.** $6\frac{5}{9} + 6\frac{5}{12} + 2\frac{5}{18}$

**88.** $2\frac{3}{8} + 4\frac{7}{12} + 3\frac{5}{16}$

**89.** $2\frac{1}{8} + 4\frac{2}{9} + 5\frac{17}{18}$

**90.** $6\frac{5}{6} + 17\frac{2}{9} + 18\frac{5}{27}$

**91.** $4\frac{7}{20} + \frac{17}{80} + 25\frac{23}{60}$

**92.** Find the sum of $2\frac{4}{9}$ and $5\frac{7}{12}$.

**93.** Find the sum of $4\frac{7}{16}$ and $7\frac{11}{12}$.

**94.** Find $2\frac{3}{8}$ more than $5\frac{1}{3}$.

**95.** Find $5\frac{5}{6}$ more than $3\frac{3}{8}$.

**96.** What is $4\frac{3}{4}$ added to $9\frac{1}{3}$?

**97.** What is $4\frac{8}{9}$ added to $9\frac{1}{6}$?

**98.** Find the total of $2\frac{2}{3}$, $4\frac{5}{8}$, and $2\frac{2}{9}$.

**99.** Find the total of $1\frac{5}{8}$, $3\frac{5}{6}$, and $7\frac{7}{24}$.

### *Objective D* *Application Problems*

**100.** A family of four finds that $\frac{1}{3}$ of their income is spent on housing, $\frac{1}{8}$ is spent on transportation, and $\frac{1}{4}$ is spent on food. Find the total fractional amount of their income that is spent on these three items.

**101.** A table 36 inches high has a top that is $1\frac{1}{8}$ inches thick. Find the total thickness of the table top after a $\frac{3}{16}$-inch veneer is applied.

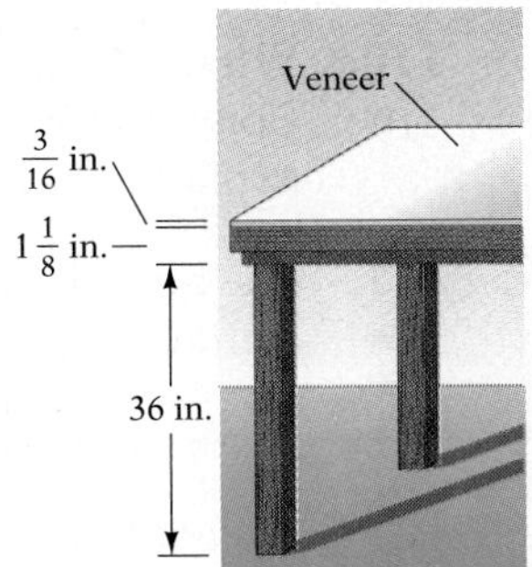


**102.** On March 29, the stock of AT&T was selling for $\$56\frac{3}{4}$. Then the price of the stock gained $\$7\frac{1}{8}$ per share by July 23. Find the price of the stock on July 23.

**103.** Find the length of the shaft.

**104.** Find the length of the shaft.

**105.** Assume that at the beginning of the month, American Airlines stock was selling at $\$158\frac{3}{8}$ per share. During the month, the stock gained $\$28\frac{3}{4}$ per share. Find the price of the stock at the end of the month.

**106.** You are working a part-time job that pays \$7 an hour. You worked 5, $3\frac{3}{4}$, $2\frac{1}{3}$, $1\frac{1}{4}$, and $7\frac{2}{3}$ hours during the last five days.
a. Find the total number of hours worked during the last five days.
b. Find your total wages for the five days.

**107.** Fred Thomson, a nurse, worked $2\frac{2}{3}$ hours of overtime on Monday, $1\frac{1}{4}$ hours on Wednesday, $1\frac{1}{3}$ on Friday, and $6\frac{3}{4}$ hours on Saturday.
a. Find the total number of overtime hours worked during the week.

b. At an overtime hourly wage of \$22 per hour, how much overtime pay does Fred receive?

**108.** Mt. Baldy had $5\frac{3}{4}$ inches of snow in December, $15\frac{1}{2}$ inches in January, and $9\frac{5}{8}$ inches in February. Find the total snowfall for the three months.

**109.** The price of a utility stock was $\$26\frac{3}{16}$ at the end of one month. The monthly gains for the next three months were $\$1\frac{1}{2}$, $\$\frac{5}{16}$, and $\$2\frac{5}{8}$ per share. Find the value of the stock at the end of the three months.

### *APPLYING THE CONCEPTS*

**110.** [W] What is a unit fraction? Find the sum of the three largest unit fractions. Is there a smallest unit fraction? If so, write it down. If not, explain why.

**111.** [W] Use a model to illustrate and explain the addition of fractions with unlike denominators.

**112.** [W] A survey was conducted to determine people's favorite color from among blue, green, red, purple, or other. The surveyor claims that $\frac{1}{3}$ of the people responded blue, $\frac{1}{6}$ responded green, $\frac{1}{8}$ responded red, $\frac{1}{12}$ responded purple, and $\frac{2}{5}$ responded some other color. Is this possible? Explain your answer.

**113.** The following are the average portions of each day that a person spends for each activity: sleeping, $\frac{1}{3}$; working, $\frac{1}{3}$; personal hygiene, $\frac{1}{24}$; eating, $\frac{1}{8}$; rest and relaxation, $\frac{1}{12}$. Do these five activities account for an entire day? Explain your answer.

# 2.5 Subtraction of Fractions and Mixed Numbers

## *Objective A To subtract fractions with the same denominator*

Fractions with the same denominator are subtracted by subtracting the numerators and placing the difference over the common denominator. After subtracting, write the fraction in simplest form.

➡ Subtract: $\frac{5}{7} - \frac{3}{7}$

$$\begin{array}{r} \frac{5}{7} \\ -\ \frac{3}{7} \\ \hline \frac{2}{7} \end{array}$$

• Subtract the numerators and place the difference over the common denominator

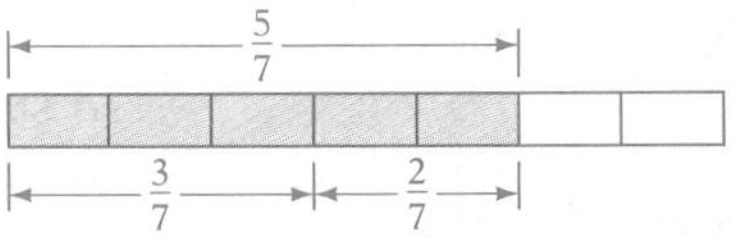


$$\frac{5}{7} - \frac{3}{7} = \frac{5-3}{7} = \frac{2}{7}$$

**Example 1** Find $\frac{17}{30}$ less $\frac{11}{30}$.

**Solution**

$$\begin{array}{r} \frac{17}{30} \\ -\ \frac{11}{30} \\ \hline \frac{6}{30} = \frac{1}{5} \end{array}$$

**You Try It 1** Subtract: $\frac{16}{27} - \frac{7}{27}$

**Your solution**

*Solution on p. A14*

## *Objective B To subtract fractions with unlike denominators*

To subtract fractions with unlike denominators, first rewrite the fractions as equivalent fractions with a common denominator. As with adding fractions, the common denominator is the LCM of the denominators of the fractions.

➡ Subtract: $\frac{5}{6} - \frac{1}{4}$

The common denominator is the LCM of 6 and 4. LCM = 12.

Build equivalent fractions using the LCM.

$$\begin{array}{r} \frac{5}{6} = \frac{10}{12} \\ -\ \frac{1}{4} = \frac{3}{12} \\ \hline \end{array}$$

Subtract the fractions.

$$\begin{array}{r} \frac{5}{6} = \frac{10}{12} \\ -\ \frac{1}{4} = \frac{3}{12} \\ \hline \frac{7}{12} \end{array}$$

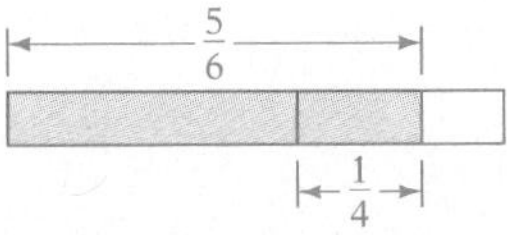


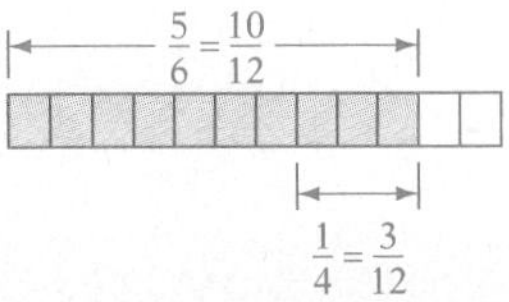


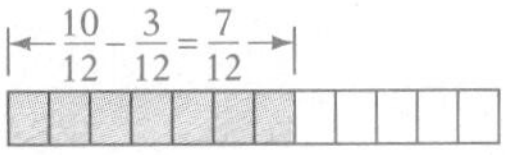

**Example 2** Subtract: $\frac{11}{16} - \frac{5}{12}$

**Solution**

$$\begin{array}{r} \frac{11}{16} = \frac{33}{48} \\ -\frac{5}{12} = \frac{20}{48} \\ \hline \frac{13}{48} \end{array}$$

• LCM = 48

**You Try It 2** Subtract: $\frac{13}{18} - \frac{7}{24}$

**Your solution**

*Solution on p. A14*

### *Objective C To subtract whole numbers, mixed numbers, and fractions*

To subtract mixed numbers without borrowing, subtract the fractional parts and then subtract the whole numbers.

➡ Subtract: $5\frac{5}{6} - 2\frac{3}{4}$

Subtract the fractional parts.

$$\begin{array}{r} 5\frac{5}{6} = 5\frac{10}{12} \\ -2\frac{3}{4} = 2\frac{9}{12} \\ \hline \frac{1}{12} \end{array}$$

• The LCM of 6 and 4 is 12.

Subtract the whole numbers.

$$\begin{array}{r} 5\frac{5}{6} = 5\frac{10}{12} \\ -2\frac{3}{4} = 2\frac{9}{12} \\ \hline 3\frac{1}{12} \end{array}$$

Subtraction of mixed numbers sometimes involves borrowing.

➡ Subtract: $5 - 2\frac{5}{8}$

Borrow 1 from 5.

$$\begin{array}{r} 5 = \overset{4}{\cancel{5}}1 \\ -2\frac{5}{8} = 2\frac{5}{8} \\ \hline \end{array}$$

Write 1 as a fraction so that the fractions have the same denominators.

$$\begin{array}{r} 5 = 4\frac{8}{8} \\ -2\frac{5}{8} = 2\frac{5}{8} \\ \hline \end{array}$$

Subtract the mixed numbers.

$$\begin{array}{r} 5 = 4\frac{8}{8} \\ -2\frac{5}{8} = 2\frac{5}{8} \\ \hline 2\frac{3}{8} \end{array}$$

➡ Subtract: $7\frac{1}{6} - 2\frac{5}{8}$

Write equivalent fractions using the LCM.

$$\begin{array}{r} 7\frac{1}{6} = 7\frac{4}{24} \\ -2\frac{5}{8} = 2\frac{15}{24} \\ \hline \end{array}$$

Borrow 1 from 7. Add the 1 to $\frac{4}{24}$. Write $1\frac{4}{24}$ as $\frac{28}{24}$.

$$\begin{array}{r} 7\frac{1}{6} = \overset{6}{\cancel{7}}1\frac{4}{24} = 6\frac{28}{24} \\ -2\frac{5}{8} = 2\frac{15}{24} = 2\frac{15}{24} \\ \hline \end{array}$$

Subtract the mixed numbers.

$$\begin{array}{r} 7\frac{1}{6} = 6\frac{28}{24} \\ -2\frac{5}{8} = 2\frac{15}{24} \\ \hline 4\frac{13}{24} \end{array}$$

**Example 3** Subtract: $15\frac{7}{8} - 12\frac{2}{3}$

**Solution** LCM = 24.

$$\begin{array}{r} 15\frac{7}{8} = 15\frac{21}{24} \\ -\ 12\frac{2}{3} = 12\frac{16}{24} \\ \hline 3\frac{5}{24} \end{array}$$

**You Try It 3** Subtract: $17\frac{5}{9} - 11\frac{5}{12}$

**Your solution**

**Example 4** Subtract: $9 - 4\frac{3}{11}$

**Solution** LCM = 11.

$$\begin{array}{r} 9 \quad = 8\frac{11}{11} \\ -\ 4\frac{3}{11} = 4\frac{3}{11} \\ \hline 4\frac{8}{11} \end{array}$$

**You Try It 4** Subtract: $8 - 2\frac{4}{13}$

**Your solution**

**Example 5** Find $11\frac{5}{12}$ decreased by $2\frac{11}{16}$.

**Solution** LCM = 48.

$$\begin{array}{r} 11\frac{5}{12} = 11\frac{20}{48} = 10\frac{68}{48} \\ -\ 2\frac{11}{16} = 2\frac{33}{48} = 2\frac{33}{48} \\ \hline 8\frac{35}{48} \end{array}$$

**You Try It 5** Subtract: $21\frac{7}{9}$ and $7\frac{11}{12}$.

**Your solution**

*Solutions on p. A14*

## *Objective D* *To solve application problems*

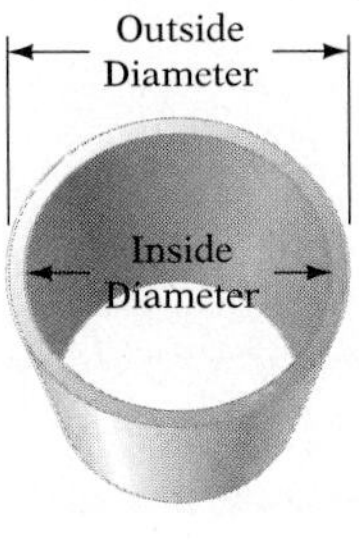


Find the inside diameter of the bushing with the outside diameter of $3\frac{3}{8}$ in. and the wall thickness is $\frac{1}{4}$ in.

$$\begin{array}{r} \frac{1}{4} \\ +\ \frac{1}{4} \\ \hline \frac{2}{4} = \frac{1}{2} \end{array}$$

- **Add $\frac{1}{4}$ to $\frac{1}{4}$ to find the total thickness of the two walls.**

$$\begin{array}{r} 3\frac{3}{8} = 3\frac{3}{8} = 2\frac{11}{8} \\ -\ \frac{1}{2} = \frac{4}{8} = \frac{4}{8} \\ \hline 2\frac{7}{8} \end{array}$$

- **Subtract the total thickness of the two walls to find the inside diameter.**

The inside diameter of the busing is $2\frac{7}{8}$ inches.

**Example 6**

A $2\frac{2}{3}$-inch piece is cut from a $6\frac{5}{8}$-inch board. How much of the board is left?

**Strategy**

To find the length remaining, subtract the length of the piece cut from the total length of the board.

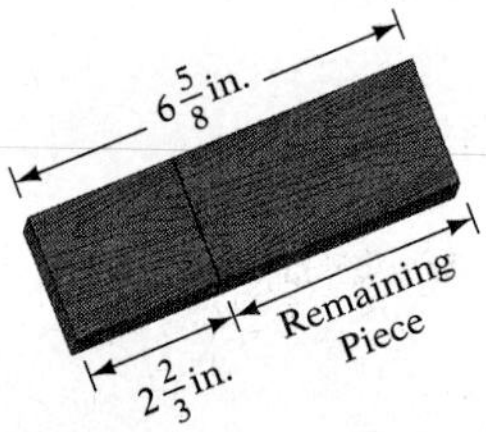


**Solution**

$$\begin{array}{r} 6\frac{5}{8} = 6\frac{15}{24} = 5\frac{39}{24} \\ -\ 2\frac{2}{3} = 2\frac{16}{24} = 2\frac{16}{24} \\ \hline 3\frac{23}{24} \end{array}$$

$3\frac{23}{24}$ inches of the board are left.

**You Try It 6**

A flight from New York to Los Angeles takes $5\frac{1}{2}$ hours. After the plane is in the air for $2\frac{3}{4}$ hours, how much flight time remains?

**Your strategy**

**Your solution**

**Example 7**

Two painters are staining a house. In 1 day one painter stained $\frac{1}{3}$ of the house, and the other stained $\frac{1}{4}$ of the house. How much of the job remains to be done?

**Strategy**

To find how much of the job remains:

- Find the total amount of the house already stained $\left(\frac{1}{3} + \frac{1}{4}\right)$.
- Subtract the amount already stained from 1, which represents the complete job.

**Solution**

$$\begin{array}{r} \frac{1}{3} = \frac{4}{12} \\ +\ \frac{1}{4} = \frac{3}{12} \\ \hline \frac{7}{12} \end{array} \qquad \begin{array}{r} 1 = \frac{12}{12} \\ -\ \frac{7}{12} = \frac{7}{12} \\ \hline \frac{5}{12} \end{array}$$

$\frac{5}{12}$ of the house remains to be stained.

**You Try It 7**

A patient is put on a diet to lose 24 pounds in 3 months. The patient lost $7\frac{1}{2}$ pounds the first month and $5\frac{3}{4}$ pounds the second month. How much weight must be lost the third month to achieve the goal?

**Your strategy**

**Your solution**

*Solutions on p. A15*

## 2.5 Exercises

### Objective A

Subtract.

1. $\frac{9}{17} - \frac{7}{17}$

2. $\frac{11}{15} - \frac{3}{15}$

3. $\frac{11}{12} - \frac{7}{12}$

4. $\frac{13}{15} - \frac{4}{15}$

5. $\frac{9}{20} - \frac{7}{20}$

6. $\frac{27}{40} - \frac{13}{40}$

7. $\frac{48}{55} - \frac{13}{55}$

8. $\frac{42}{65} - \frac{17}{65}$

9. $\frac{11}{24} - \frac{5}{24}$

10. $\frac{23}{30} - \frac{13}{30}$

11. $\frac{17}{42} - \frac{5}{42}$

12. $\frac{29}{48} - \frac{13}{48}$

13. What is $\frac{5}{14}$ less than $\frac{13}{14}$?

14. What is $\frac{7}{19}$ less than $\frac{17}{19}$?

15. Find the difference between $\frac{7}{8}$ and $\frac{5}{8}$.

16. Find the difference between $\frac{7}{12}$ and $\frac{5}{12}$.

17. What is $\frac{18}{23}$ minus $\frac{9}{23}$?

18. What is $\frac{7}{9}$ minus $\frac{3}{9}$?

19. Find $\frac{17}{24}$ decreased by $\frac{11}{24}$.

20. Find $\frac{19}{30}$ decreased by $\frac{11}{30}$.

### Objective B

Subtract.

21. $\frac{2}{3} - \frac{1}{6}$

22. $\frac{7}{8} - \frac{5}{16}$

23. $\frac{5}{8} - \frac{2}{7}$

24. $\frac{5}{6} - \frac{3}{7}$

25. $\frac{5}{7} - \frac{3}{14}$

26. $\frac{7}{10} - \frac{3}{5}$

27. $\frac{5}{9} - \frac{7}{15}$

28. $\frac{8}{15} - \frac{7}{20}$

29. $\frac{7}{9} - \frac{1}{6}$

30. $\frac{9}{14} - \frac{3}{8}$

31. $\frac{5}{12} - \frac{5}{16}$

32. $\frac{12}{25} - \frac{13}{40}$

Subtract.

33. $\begin{array}{r} \frac{46}{51} \\ -\frac{3}{17} \\ \hline \end{array}$

34. $\begin{array}{r} \frac{9}{16} \\ -\frac{17}{32} \\ \hline \end{array}$

35. $\begin{array}{r} \frac{21}{35} \\ -\frac{5}{14} \\ \hline \end{array}$

36. $\begin{array}{r} \frac{19}{40} \\ -\frac{3}{16} \\ \hline \end{array}$

37. $\begin{array}{r} \frac{29}{60} \\ -\frac{3}{40} \\ \hline \end{array}$

38. What is $\frac{3}{5}$ less than $\frac{11}{12}$?

39. What is $\frac{5}{9}$ less than $\frac{11}{15}$?

40. Find the difference between $\frac{11}{24}$ and $\frac{7}{18}$.

41. Find the difference between $\frac{9}{14}$ and $\frac{5}{42}$.

42. Find $\frac{11}{12}$ decreased by $\frac{11}{15}$.

43. Find $\frac{17}{20}$ decreased by $\frac{7}{15}$.

44. What is $\frac{13}{20}$ minus $\frac{1}{6}$?

45. What is $\frac{5}{6}$ minus $\frac{7}{9}$?

*Objective C*

Subtract.

46. $\begin{array}{r} 5\frac{7}{12} \\ -\ 2\frac{5}{12} \\ \hline \end{array}$

47. $\begin{array}{r} 16\frac{11}{15} \\ -\ 11\frac{8}{15} \\ \hline \end{array}$

48. $\begin{array}{r} 72\frac{21}{23} \\ -\ 16\frac{17}{23} \\ \hline \end{array}$

49. $\begin{array}{r} 19\frac{16}{17} \\ -\ 9\frac{7}{17} \\ \hline \end{array}$

50. $\begin{array}{r} 6\frac{1}{3} \\ -\ 2 \\ \hline \end{array}$

51. $\begin{array}{r} 5\frac{7}{8} \\ -\ 1 \\ \hline \end{array}$

52. $\begin{array}{r} 10 \\ -\ 6\frac{1}{3} \\ \hline \end{array}$

53. $\begin{array}{r} 3 \\ -\ 2\frac{5}{21} \\ \hline \end{array}$

54. $\begin{array}{r} 6\frac{2}{5} \\ -\ 4\frac{4}{5} \\ \hline \end{array}$

55. $\begin{array}{r} 16\frac{3}{8} \\ -\ 10\frac{7}{8} \\ \hline \end{array}$

56. $\begin{array}{r} 25\frac{4}{9} \\ -\ 16\frac{7}{9} \\ \hline \end{array}$

57. $\begin{array}{r} 8\frac{3}{7} \\ -\ 2\frac{6}{7} \\ \hline \end{array}$

58. $\begin{array}{r} 16\frac{2}{5} \\ -\ 8\frac{4}{9} \\ \hline \end{array}$

59. $\begin{array}{r} 23\frac{7}{8} \\ -\ 16\frac{2}{3} \\ \hline \end{array}$

60. $\begin{array}{r} 6 \\ -\ 4\frac{3}{5} \\ \hline \end{array}$

61. $\begin{array}{r} 65\frac{8}{35} \\ -\ 16\frac{11}{14} \\ \hline \end{array}$

62. $\begin{array}{r} 82\frac{4}{33} \\ -\ 16\frac{5}{22} \\ \hline \end{array}$

63. $\begin{array}{r} 101\frac{2}{9} \\ -\ 16 \\ \hline \end{array}$

64. $\begin{array}{r} 77\frac{5}{18} \\ -\ 61 \\ \hline \end{array}$

65. $\begin{array}{r} 17 \\ -\ 7\frac{8}{13} \\ \hline \end{array}$

**66.** What is $5\frac{3}{8}$ less than $8\frac{1}{9}$?

**67.** What is $7\frac{3}{5}$ less than $23\frac{3}{20}$?

**68.** Find the difference between $9\frac{2}{7}$ and $3\frac{1}{4}$.

**69.** Find the difference between $12\frac{3}{8}$ and $7\frac{5}{12}$.

**70.** What is $5\frac{3}{5}$ minus $2\frac{5}{6}$?

**71.** What is $10\frac{5}{9}$ minus $5\frac{11}{15}$?

**72.** Find $6\frac{1}{3}$ decreased by $3\frac{3}{5}$.

**73.** Find $8\frac{5}{8}$ decreased $6\frac{5}{6}$.

***Objective D*** *Application Problems*

**74.** Find the missing dimension.

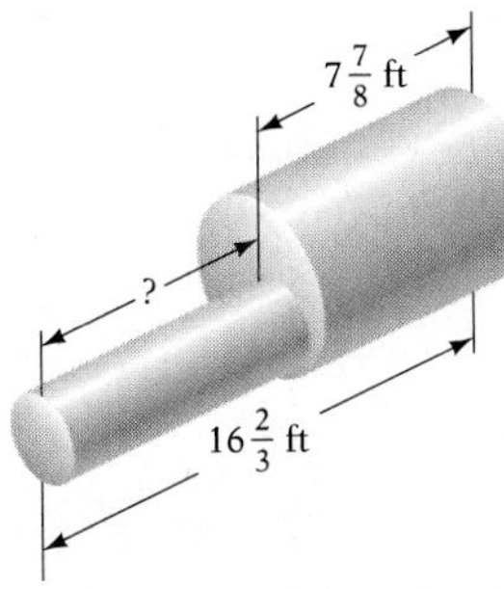


**75.** Find the missing dimension.

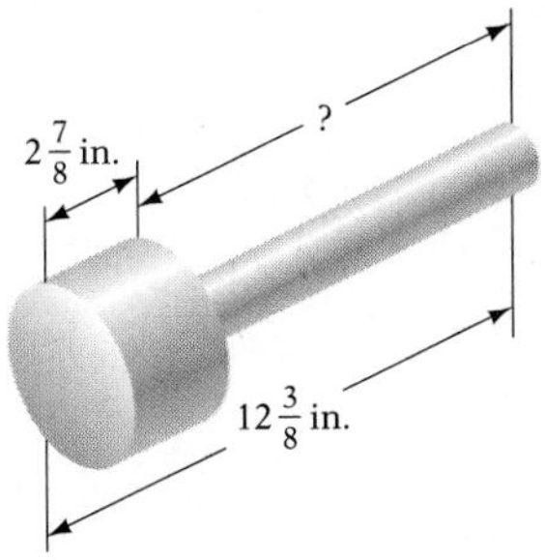


**76.** The horses in the Kentucky Derby run $1\frac{1}{4}$ mile. In the Belmont Stakes they run $1\frac{1}{2}$ mile, and in the Preakness Stakes they run $1\frac{3}{16}$ mile. How much farther do the horses run in the Kentucky Derby than in the Preakness Stakes? How much farther do they run in the Belmont Stakes than in the Preakness Stakes?

**77.** In the running high jump in the 1948 Summer Olympic Games, Alice Coachman's distance was $66\frac{1}{8}$ in. In the same event in the 1956 Summer Olympics, Mildred McDaniel jumped $69\frac{1}{4}$ in., and in the 1988 Olympic Games Louise Ritter jumped 80 in. Find the difference between Ritter's distance and Coachman's distance. Find the difference between Ritter's distance and McDaniel's distance.

**78.** Two hikers plan a 3-day $27\frac{1}{2}$-mile backpack trip carrying a total of 80 pounds. The hikers plan to travel $7\frac{3}{8}$ miles the first day and $10\frac{1}{3}$ miles the second day.
a. How many miles do the hikers plan to travel the first two days?
b. How many miles will be left to travel on the third day?

**79.** A 12-mile walkathon has three checkpoints. The first is $3\frac{3}{8}$ miles from the starting point. The second checkpoint is $4\frac{1}{3}$ miles from the first.

a. How many miles is it from the starting point to the second checkpoint?

b. How many miles is it from the second checkpoint to the finish line?

**80.** A patient with high blood pressure is put on a diet to lose 38 pounds in 3 months. The patient loses $12\frac{1}{2}$ pounds the first month and $16\frac{3}{8}$ pounds the second month. How much weight must be lost the third month for the goal to be achieved?

**81.** A wrestler is entered in the 172-pound weight class in the conference finals coming up in 3 weeks. The wrestler needs to lose $12\frac{3}{4}$ pounds. If $5\frac{1}{4}$ pounds are lost the first week and $4\frac{1}{4}$ the second week, how many pounds must be lost the third week for the desired weight to be reached?

### *APPLYING THE CONCEPTS*

The figure at the right shows a selected portion of the New York Stock Exchange stock prices. The figure shows the closing Friday price of the stocks, the change in price for the week, and the Friday change in the price of the stocks. To find the price of "AsiaPc" at the beginning of the week, subtract $2\frac{1}{2}$ from 23. Use the figure for Exercises 82–85.

| | Div. | PE | Wk. vol (100's) | Rel Stance | Fri. Close | Fri. Chg. | Wk. Chg. |
|---|---|---|---|---|---|---|---|
| AquilaGas n | ... | 13 | 2486 | 41 | 13 | +1/8 | +11/4 |
| Aracrz n | ... | ... | 5750 | 86 | 12 | +3/8 | -1/4 |
| Acadn n | 2.42 | 33 | 1600 | 71 | 223/8 | ... | -1/8 |
| ArchDn b | .10 | 17 | 68078 | 42 | 233/4 | +1/4 | +13/8 |
| ArcCh | 2.50 | 22 | 673 | 53 | 431/2 | +1/2 | +11/4 |
| Argent n | 1.72 | 9 | 2566 | 83 | 221/8 | -1/4 | -1/8 |
| ArgFd n q | ... | ... | 2537 | 86 | 141/4 | ... | -3/8 |
| AriP pfW | 1.23 | ... | 1401 | 42 | 243/4 | ... | ... |
| Arkla | .28 | 22 | 18140 | 14 | 73/4 | ... | +1/4 |
| Armco | ... | ... | 23472 | 39 | 61/4 | +1/4 | +1/4 |
| ArmWI | 1.20 | 33 | 9380 | 98 | 49 | +5/8 | +21/2 |
| ArrowE | ... | 16 | 6323 | 69 | 38 | +1/4 | +1/8 |
| Artra | ... | ... | 745 | 69 | 63/4 | +1/4 | -1/8 |
| Arvin | .76 | 17 | 4222 | 41 | 303/8 | +3/8 | +23/4 |
| ARX | ... | 13 | 2073 | 88 | 35/8 | ... | +1/8 |
| ASA q | 2.00 | ... | 3972 | 86 | 485/8 | -5/8 | -5/8 |
| Asarco | .40 | ... | 7814 | 48 | 221/4 | +3/4 | -1/2 |
| AshOil | 1.00 | 14 | 7429 | 68 | 32 | +3/8 | -11/8 |
| AshOil pf | 3.12 | 26 | 757 | 61 | 581/2 | +1/8 | -11/2 |
| AsiaPc e q | 1.74 | ... | 8926 | 97 | 23 | +5/8 | +21/2 |

**82.** Find the price of one share of ArcCh stock at the beginning of the week.

**83.** Find the price of one share of Arvin stock at the beginning of the week.

**84.** Find the price of one share of Arvin stock at the beginning of Friday's trading.

**85.** Find the price of one share of ArmWl at the beginning of the week.

**86.** Fill in the square to produce a true statement: $5\frac{1}{3} - \square = 2\frac{1}{2}$.

**87.** Fill in the square to produce a true statement: $\square - 4\frac{1}{2} = 1\frac{5}{8}$.

**88.** Fill-in the blank squares at the right so that the sum of the numbers along any row, column, or diagonal is the same. The resulting square is called a magic square.

| | | |
|---|---|---|
| | | $\frac{3}{4}$ |
| | $\frac{5}{8}$ | |
| $\frac{1}{2}$ | | $\frac{7}{8}$ |

**89.** If $\frac{4}{15}$ of an electrician's income is spent for housing, what fraction of the electrician's income is not spent for housing?

# 2.6 Multiplication of Fractions and Mixed Numbers

## *Objective A* *To multiply fractions*

The product of two fractions is the product of the numerators over the product of the denominators.

➡ Multiply: $\frac{2}{3} \times \frac{4}{5}$

$$\frac{2}{3} \times \frac{4}{5} = \frac{2 \cdot 4}{3 \cdot 5} = \frac{8}{15}$$

- Multiply the numerators.
- Multiply the denominators.

The product $\frac{2}{3} \times \frac{4}{5}$ can be read "$\frac{2}{3}$ times $\frac{4}{5}$" or "$\frac{2}{3}$ of $\frac{4}{5}$."

Reading the times sign as "of" is useful in application problems.

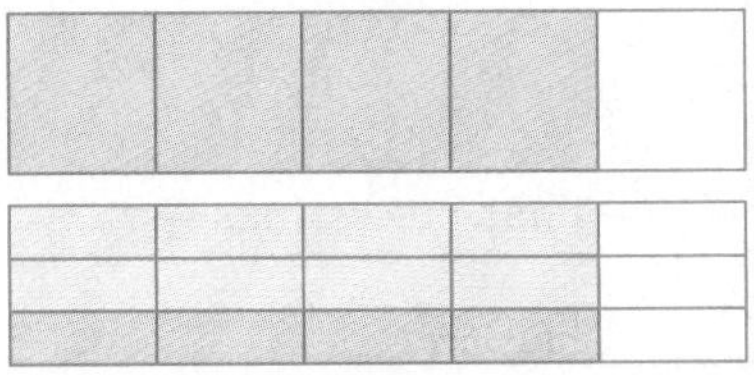

$\frac{4}{5}$ of the bar is shaded.

Shade $\frac{2}{3}$ of the $\frac{4}{5}$ already shaded.

$\frac{8}{15}$ of the bar is then shaded.

$\frac{2}{3}$ of $\frac{4}{5} = \frac{2}{3} \times \frac{4}{5} = \frac{8}{15}$

After multiplying two fractions, write the product in simplest form.

➡ Multiply: $\frac{3}{4} \times \frac{14}{15}$

$$\frac{3}{4} \times \frac{14}{15} = \frac{3 \cdot 14}{4 \cdot 15}$$

- Multiply numerators.
- Multiply denominators.

$$= \frac{3 \cdot 2 \cdot 7}{2 \cdot 2 \cdot 3 \cdot 5}$$

- Write the prime factorization of each number.

$$= \frac{\overset{1}{\cancel{3}} \cdot \overset{1}{\cancel{2}} \cdot 7}{\underset{1}{\cancel{2}} \cdot 2 \cdot \underset{1}{\cancel{3}} \cdot 5} = \frac{7}{10}$$

- Eliminate the common factors. Then multiply the factors of the numerator and denominator.

This example could also be worked by using the GCF.

$$\frac{3}{4} \times \frac{14}{15} = \frac{42}{60}$$

- Multiply numerators.
- Multiply denominators.

$$= \frac{6 \cdot 7}{6 \cdot 10}$$

- The GCF of 42 and 60 is 6. Factor 6 from 42 and 60.

$$= \frac{\overset{1}{\cancel{6}} \cdot 7}{\underset{1}{\cancel{6}} \cdot 10} = \frac{7}{10}$$

- Eliminate the GCF.

**Example 1**

Multiply: $\frac{4}{15}$ and $\frac{5}{28}$

**Solution**

$$\frac{4}{15} \times \frac{5}{28} = \frac{4 \cdot 5}{15 \cdot 28} = \frac{\overset{1}{\cancel{2}} \cdot \overset{1}{\cancel{2}} \cdot \overset{1}{\cancel{5}}}{3 \cdot \underset{1}{\cancel{5}} \cdot \underset{1}{\cancel{2}} \cdot \underset{1}{\cancel{2}} \cdot 7} = \frac{1}{21}$$

**You Try It 1**

Multiply: $\frac{4}{21}$ and $\frac{7}{44}$

**Your solution**

**Example 2**

Find the product of $\frac{9}{20}$ and $\frac{33}{35}$.

**Solution**

$$\frac{9}{20} \times \frac{33}{35} = \frac{9 \cdot 33}{20 \cdot 35} = \frac{3 \cdot 3 \cdot 3 \cdot 11}{2 \cdot 2 \cdot 5 \cdot 5 \cdot 7} = \frac{297}{700}$$

**You Try It 2**

Find the product of $\frac{2}{21}$ and $\frac{10}{33}$.

**Your solution**

**Example 3**

What is $\frac{14}{9}$ times $\frac{12}{7}$?

**Solution**

$$\frac{14}{9} \times \frac{12}{7} = \frac{14 \cdot 12}{9 \cdot 7} = \frac{2 \cdot \overset{1}{\cancel{7}} \cdot 2 \cdot 2 \cdot \overset{1}{\cancel{3}}}{3 \cdot \underset{1}{\cancel{3}} \cdot \underset{1}{\cancel{7}}} = \frac{8}{3} = 2\frac{2}{3}$$

**You Try It 3**

What is $\frac{16}{5}$ times $\frac{15}{24}$?

**Your solution**

*Solutions on p. A15*

## *Objective B To multiply whole numbers, mixed numbers, and fractions*

To multiply a whole number by a fraction or mixed number, first write the whole number as a fraction with a denominator of 1.

➡ Multiply: $4 \times \frac{3}{7}$

$$4 \times \frac{3}{7} = \frac{4}{1} \times \frac{3}{7} = \frac{4 \cdot 3}{1 \cdot 7} = \frac{2 \cdot 2 \cdot 3}{7} = \frac{12}{7} = 1\frac{5}{7}$$

• Write 4 with a denominator of 1; then multiply the fractions.

When one or more of the factors in a product is a mixed number, write the mixed number as an improper fraction before multiplying.

➡ Multiply: $2\frac{1}{3} \times \frac{3}{14}$

$$2\frac{1}{3} \times \frac{3}{14} = \frac{7}{3} \times \frac{3}{14} = \frac{7 \cdot 3}{3 \cdot 14} = \frac{\overset{1}{\cancel{7}} \cdot \overset{1}{\cancel{3}}}{\underset{1}{\cancel{3}} \cdot 2 \cdot \underset{1}{\cancel{7}}} = \frac{1}{2}$$

• Write $2\frac{1}{3}$ as an improper fraction; then multiply the fractions.

**Example 4**

Multiply: $4\frac{5}{6} \times \frac{12}{13}$

**Solution**

$$4\frac{5}{6} \times \frac{12}{13} = \frac{29}{6} \times \frac{12}{13} = \frac{29 \cdot 12}{6 \cdot 13}$$

$$= \frac{29 \cdot \overset{1}{\cancel{2}} \cdot 2 \cdot \overset{1}{\cancel{3}}}{\underset{1}{\cancel{2}} \cdot \underset{1}{\cancel{3}} \cdot 13} = \frac{58}{13} = 4\frac{6}{13}$$

**You Try It 4**

Multiply: $5\frac{2}{5} \times \frac{5}{9}$

**Your solution**

**Example 5**

Find $5\frac{2}{3}$ times $4\frac{1}{2}$.

**Solution**

$$5\frac{2}{3} \times 4\frac{1}{2} = \frac{17}{3} \times \frac{9}{2} = \frac{17 \cdot 9}{3 \cdot 2}$$

$$= \frac{17 \cdot \overset{1}{\cancel{3}} \cdot 3}{\underset{1}{\cancel{3}} \cdot 2} = \frac{51}{2} = 25\frac{1}{2}$$

**You Try It 5**

Multiply: $3\frac{2}{5} \times 6\frac{1}{4}$

**Your solution**

**Example 6**

Multiply: $4\frac{2}{5} \times 7$

**Solution**

$$4\frac{2}{5} \times 7 = \frac{22}{5} \times \frac{7}{1} = \frac{22 \cdot 7}{5 \cdot 1}$$

$$= \frac{2 \cdot 11 \cdot 7}{5} = \frac{154}{5} = 30\frac{4}{5}$$

**You Try It 6**

Multiply: $3\frac{2}{7} \times 6$

**Your solution**

*Solutions on p. A15 and A16*

## *Objective C To solve application problems*

| *Length (ft)* | *Weight (lb/ft)* |
|---|---|
| $6\frac{1}{2}$ | $\frac{3}{8}$ |
| $8\frac{5}{8}$ | $1\frac{1}{4}$ |
| $10\frac{3}{4}$ | $2\frac{1}{2}$ |
| $12\frac{7}{12}$ | $4\frac{1}{3}$ |

The table at the left lists the length of steel rods and the weight per foot. The weight per foot is measured in pounds for each foot of rod and is abbreviated as (lb/ft).

➡ Find the weight of the steel bar that is $10\frac{3}{4}$ feet long.

**Strategy**

To find the weight of a steel bar, multiply its length by the weight per foot.

**Solution** $10\frac{3}{4} \times 2\frac{1}{2} = \frac{43}{4} \times \frac{5}{2} = \frac{43 \cdot 5}{4 \cdot 2} = \frac{215}{8} = 26\frac{7}{8}$

The weight of the $10\frac{3}{4}$-ft rod is $26\frac{7}{8}$ lb.

**Example 7**

An electrician earns $150 for each day worked. What are the electrician's earnings for working $4\frac{1}{2}$ days?

**Strategy**

To find the electrician's total earnings, multiply the daily earnings ($150) by the number of days worked $\left(4\frac{1}{2}\right)$.

**Solution**

$$150 \times 4\frac{1}{2} = \frac{150}{1} \times \frac{9}{2} = \frac{150 \cdot 9}{1 \cdot 2} = 675$$

The electrician's earnings are $675.

**You Try It 7**

Over the last 10 years, a house increased in value by $3\frac{1}{2}$ times. The price of the house 10 years ago was $30,000. What is the value of the house today?

**Your strategy**

**Your solution**

**Example 8**

The value of a small office building and the land on which it is built is $90,000. The value of the land is $\frac{1}{4}$ the total value. What is the value of the building (in dollars)?

**Strategy**

To find the value of the building:

- Find the value of the land $\left(\frac{1}{4} \times 90{,}000\right)$.
- Subtract the value of the land from the total value.

**Solution**

$$90{,}000 \times \frac{1}{4} = \frac{90{,}000}{4} = 22{,}500 \quad \text{value of the land}$$

$$\begin{array}{r} 90{,}000 \\ -\,22{,}500 \\ \hline 67{,}500 \end{array}$$

The value of the building is $67,500.

**You Try It 8**

A paint company bought a drying chamber and an air compressor for spray painting. The total cost of the two items was $60,000. The drying chamber's cost was $\frac{4}{5}$ of the total cost. What was the cost of the air compressor?

**Your strategy**

**Your solution**

*Solutions on p. A16*

## 2.6 Exercises

### *Objective A*

Multiply.

1. $\frac{2}{3} \times \frac{7}{8}$
2. $\frac{1}{2} \times \frac{2}{3}$
3. $\frac{5}{16} \times \frac{7}{15}$
4. $\frac{3}{8} \times \frac{6}{7}$
5. $\frac{1}{2} \times \frac{5}{6}$
6. $\frac{1}{6} \times \frac{1}{8}$
7. $\frac{2}{5} \times \frac{5}{6}$
8. $\frac{11}{12} \times \frac{6}{7}$
9. $\frac{11}{12} \times \frac{3}{5}$
10. $\frac{2}{5} \times \frac{4}{9}$
11. $\frac{1}{6} \times \frac{6}{7}$
12. $\frac{3}{5} \times \frac{10}{11}$
13. $\frac{1}{5} \times \frac{5}{8}$
14. $\frac{6}{7} \times \frac{14}{15}$
15. $\frac{4}{9} \times \frac{15}{16}$
16. $\frac{8}{9} \times \frac{27}{4}$
17. $\frac{3}{5} \times \frac{3}{10}$
18. $\frac{5}{6} \times \frac{1}{2}$
19. $\frac{3}{8} \times \frac{5}{12}$
20. $\frac{3}{10} \times \frac{6}{7}$
21. $\frac{16}{9} \times \frac{27}{8}$
22. $\frac{5}{8} \times \frac{16}{15}$
23. $\frac{3}{2} \times \frac{4}{9}$
24. $\frac{5}{3} \times \frac{3}{7}$
25. $\frac{5}{16} \times \frac{12}{11}$
26. $\frac{7}{8} \times \frac{3}{14}$
27. $\frac{2}{9} \times \frac{1}{5}$
28. $\frac{1}{10} \times \frac{3}{8}$
29. $\frac{5}{12} \times \frac{6}{7}$
30. $\frac{1}{3} \times \frac{9}{2}$
31. $\frac{15}{8} \times \frac{16}{3}$
32. $\frac{5}{6} \times \frac{4}{15}$
33. $\frac{1}{2} \times \frac{2}{15}$
34. $\frac{3}{8} \times \frac{5}{16}$
35. $\frac{2}{9} \times \frac{15}{17}$
36. $\frac{5}{7} \times \frac{14}{15}$
37. $\frac{3}{8} \times \frac{15}{41}$
38. $\frac{5}{12} \times \frac{42}{65}$
39. $\frac{16}{33} \times \frac{55}{72}$
40. $\frac{8}{3} \times \frac{21}{32}$
41. $\frac{12}{5} \times \frac{5}{3}$
42. $\frac{17}{9} \times \frac{81}{17}$
43. $\frac{16}{85} \times \frac{125}{84}$
44. $\frac{19}{64} \times \frac{48}{95}$
45. $\frac{18}{25} \times \frac{15}{40}$

46. Multiply $\frac{7}{12}$ and $\frac{15}{42}$.

47. Multiply $\frac{32}{9}$ and $\frac{3}{8}$.

48. Find the product of $\frac{5}{9}$ and $\frac{3}{20}$.

49. Find the product of $\frac{7}{3}$ and $\frac{15}{14}$.

50. What is $\frac{1}{2}$ times $\frac{8}{15}$?

51. What is $\frac{3}{8}$ times $\frac{12}{17}$?

*Objective B*

Multiply.

52. $4 \times \frac{3}{8}$

53. $14 \times \frac{5}{7}$

54. $\frac{2}{3} \times 6$

55. $\frac{5}{12} \times 40$

56. $\frac{1}{3} \times 1\frac{1}{3}$

57. $\frac{2}{5} \times 2\frac{1}{2}$

58. $1\frac{7}{8} \times \frac{4}{15}$

59. $2\frac{1}{5} \times \frac{5}{22}$

60. $55 \times \frac{3}{10}$

61. $\frac{5}{14} \times 49$

62. $4 \times 2\frac{1}{2}$

63. $9 \times 3\frac{1}{3}$

64. $2\frac{1}{7} \times 3$

65. $5\frac{1}{4} \times 8$

66. $3\frac{2}{3} \times 5$

67. $4\frac{2}{9} \times 3$

68. $\frac{1}{2} \times 3\frac{3}{7}$

69. $\frac{3}{8} \times 4\frac{4}{5}$

70. $6\frac{1}{8} \times \frac{4}{7}$

71. $5\frac{1}{3} \times \frac{5}{16}$

72. $5\frac{1}{8} \times 5$

73. $6\frac{1}{9} \times 2$

74. $\frac{3}{8} \times 4\frac{1}{2}$

75. $\frac{5}{7} \times 2\frac{1}{3}$

76. $6 \times 2\frac{2}{3}$

77. $6\frac{1}{8} \times 0$

78. $1\frac{1}{3} \times 2\frac{1}{4}$

79. $2\frac{5}{8} \times \frac{3}{23}$

80. $2\frac{5}{8} \times 3\frac{2}{5}$

81. $5\frac{3}{16} \times 5\frac{1}{3}$

82. $3\frac{1}{7} \times 2\frac{1}{8}$

83. $16\frac{5}{8} \times 1\frac{1}{16}$

**84.** $2\frac{2}{5} \times 3\frac{1}{12}$

**85.** $2\frac{2}{3} \times \frac{3}{20}$

**86.** $5\frac{1}{5} \times 3\frac{1}{13}$

**87.** $3\frac{3}{4} \times 2\frac{3}{20}$

**88.** $10\frac{1}{4} \times 3\frac{1}{5}$

**89.** $12\frac{3}{5} \times 1\frac{3}{7}$

**90.** $5\frac{3}{7} \times 5\frac{1}{4}$

**91.** $6\frac{1}{2} \times 1\frac{3}{13}$

**92.** Multiply $2\frac{1}{2}$ and $3\frac{3}{5}$.

**93.** Multiply $4\frac{3}{8}$ and $3\frac{3}{5}$.

**94.** Find the product of $2\frac{1}{8}$ and $\frac{5}{17}$.

**95.** Find the product of $12\frac{2}{5}$ and $3\frac{7}{31}$.

**96.** What is $1\frac{3}{8}$ times $2\frac{1}{5}$?

**97.** What is $3\frac{1}{8}$ times $2\frac{4}{7}$?

## *Objective C* *Application Problems*

**98.** A car can travel $23\frac{1}{2}$ miles on 1 gallon of gas. How far can the car travel on $\frac{1}{2}$ gallon of gas?

**99.** Maria Rivera can walk $3\frac{1}{2}$ miles in 1 hour. At this rate, how far can Maria walk in $\frac{1}{3}$ hour?

**100.** A Honda Civic travels 38 miles on each gallon of gasoline. How many miles can the car travel on $9\frac{1}{2}$ gallons of gasoline?

**101.** A board is $5\frac{3}{4}$ feet long. One-third of the board is cut off. What is the length of the piece cut off?

**102.** A cook is using a recipe that calls for $3\frac{1}{2}$ cups of flour. The cook wants to double the recipe. How much flour should the cook use?

**103.** The F-1 engine in the first stage of the Saturn 5 rocket burns 214,000 gallons of propellant in 1 minute. The first stage burns $2\frac{1}{2}$ minutes before burnout. How much propellant is used before burnout?

**104.** A family budgets $\frac{2}{5}$ of its monthly income of $3200 per month for housing and utilities.
a. What amount is budgeted for housing and utilities?
b. What amount remains for purposes other than housing and utilities?

**105.** Ann Rodman read $\frac{2}{3}$ of a book containing 432 pages.
a. How many pages did Ann read?
b. How many pages remain to be read?

**106.** The parents of the Newton Junior High School Choir members are making robes for the choir. Each robe requires $2\frac{5}{8}$ yards of material at a cost of $8 per yard. Find the total cost of 24 choir robes.

**107.** A college spends $\frac{5}{8}$ of its monthly income on employee salaries. During one month the college had an income of \$712,000. How much of the monthly income remained after the employee's salaries were paid?

The table at the right shows the length of steel rods and their weight per foot. Use this table for Exercises 108–110.

| *Length (ft)* | *Weight (lb/ft)* |
|---|---|
| $6\frac{1}{2}$ | $\frac{3}{8}$ |
| $8\frac{5}{8}$ | $1\frac{1}{4}$ |
| $10\frac{3}{4}$ | $2\frac{1}{2}$ |
| $12\frac{7}{12}$ | $4\frac{1}{3}$ |

**108.** Find the weight of $6\frac{1}{2}$ feet of steel rod.

**109.** Find the weight of $12\frac{7}{12}$ feet of steel rod.

**110.** Find the total weight of $8\frac{5}{8}$ feet and $10\frac{3}{4}$ feet of steel rod.

## APPLYING THE CONCEPTS

**111.** The product of 1 and a number is $\frac{1}{2}$. Find the number.

**112.** Our calendar is based on the solar year, which is $365\frac{1}{4}$ days. Use this fact to explain leap years.

**113.** If two positive fractions, each less than 1, are multiplied, is the product less than 1?

**114.** Is the product of two positive fractions always greater than either one of the two numbers? If so, explain why. If not, give an example.

**115.** Determine the square of $\frac{2}{3}$.

**116.** A state park has reserved $\frac{4}{5}$ of its total acreage for a wildlife preserve. Three-fourths of the wildlife preserve is heavily wooded. What fraction of the state park is heavily wooded?

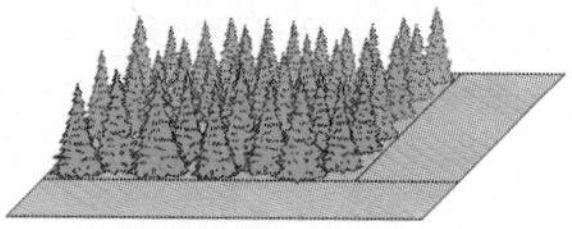

State Park

**117.** The manager of a mutual fund has one-half of the portfolio invested in bonds. Of the amount invested in bonds, $\frac{3}{8}$ is invested in corporate bonds. What fraction of the total portfolio is invested in corporate bonds?

**118.** Which of the labeled points on the number line at the right could be the graph of the product of $B$ and $C$?

**119.** Fill-in the circles on the square at the right with the fractions $\frac{1}{6}$, $\frac{5}{18}$, $\frac{4}{9}$, $\frac{5}{9}$, $\frac{2}{3}$, $\frac{3}{4}$, $1\frac{1}{9}$, $1\frac{1}{2}$, and $2\frac{1}{4}$ so that the product of any three connected circles is equal to $\frac{5}{18}$. (*Note*: There is more than one answer.)

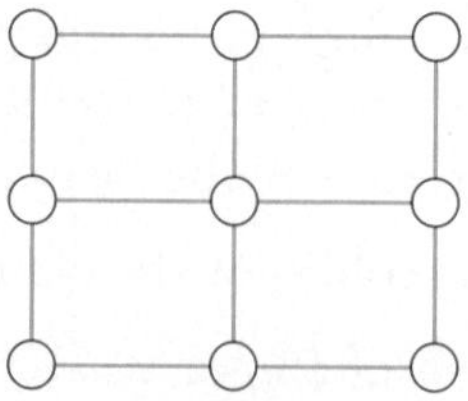

# 2.7 Division of Fractions and Mixed Numbers

## Objective A To divide fractions

The **reciprocal** of a fraction is the fraction with the numerator and denominator interchanged. For instance, the reciprocal of $\frac{2}{3}$ is $\frac{3}{2}$. The process of interchanging the numerator and denominator of a fraction is called **inverting** the fraction.

To find the reciprocal of a whole number, first write the whole number as a fraction with a denominator of 1; then find the reciprocal of that fraction.

The reciprocal of 5 is $\frac{1}{5}$. $\left(\text{Think } 5 = \frac{5}{1}\right)$

Reciprocals are used to rewrite division problems as related multiplication problems. Look at the following two problems:

$8 \div 2 = 4$ $\qquad$ $8 \times \frac{1}{2} = 4$

8 divided by 2 is 4. $\qquad$ 8 times the reciprocal of 2 is 4.

"Divided by" means the same as "times the reciprocal of." Thus "÷ 2" can be replaced with "$\times \frac{1}{2}$," and the answer will be the same. Fractions are divided by making this replacement.

➡ Divide: $\frac{2}{3} \div \frac{3}{4}$ $\qquad$ $\frac{2}{3} \div \frac{3}{4} = \frac{2}{3} \times \frac{4}{3} = \frac{2 \cdot 4}{3 \cdot 3} = \frac{2 \cdot 2 \cdot 2}{3 \cdot 3} = \frac{8}{9}$

**Example 1** Divide: $\frac{5}{8} \div \frac{4}{9}$

**Solution** $\frac{5}{8} \div \frac{4}{9} = \frac{5}{8} \times \frac{9}{4} = \frac{5 \cdot 9}{8 \cdot 4}$

$= \frac{5 \cdot 3 \cdot 3}{2 \cdot 2 \cdot 2 \cdot 2 \cdot 2} = \frac{45}{32} = 1\frac{13}{32}$

**You Try It 1** Divide: $\frac{3}{7} \div \frac{2}{3}$

**Your solution**

**Example 2** Find the quotient of $\frac{3}{5}$ and $\frac{12}{25}$.

**Solution** $\frac{3}{5} \div \frac{12}{25} = \frac{3}{5} \times \frac{25}{12} = \frac{3 \cdot 25}{5 \cdot 12}$

$= \frac{\overset{1}{\cancel{3}} \cdot \overset{1}{\cancel{5}} \cdot 5}{\underset{1}{\cancel{5}} \cdot 2 \cdot 2 \cdot \underset{1}{\cancel{3}}} = \frac{5}{4} = 1\frac{1}{4}$

**You Try It 2** Divide: $\frac{3}{4} \div \frac{9}{10}$

**Your solution**

*Solutions on p. A16*

## Objective B To divide whole numbers, mixed numbers, and fractions

To divide a fraction and a whole number, first write the whole number as a fraction with a denominator of 1.

➡ Divide: $\frac{3}{7} \div 5$

$\frac{3}{7} \div \boxed{5} = \frac{3}{7} \div \boxed{\frac{5}{1}} = \frac{3}{7} \times \frac{1}{5} = \frac{3 \cdot 1}{7 \cdot 5} = \frac{3}{35}$

• Write 5 with a denominator of 1; then divide the fractions.

When one of the numbers in a quotient is a mixed number, write the mixed number as an improper fraction before dividing.

➡ Divide: $4\frac{2}{3} \div \frac{8}{15}$

Write $4\frac{2}{3}$ as an improper fraction; then divide the fractions.

$$4\frac{2}{3} \div \frac{8}{15} = \frac{14}{3} \div \frac{8}{15} = \frac{14}{3} \times \frac{15}{8} = \frac{14 \cdot 15}{3 \cdot 8} = \frac{\overset{1}{\not 2} \cdot 7 \cdot \overset{1}{\not 3} \cdot 5}{\underset{1}{\not 3} \cdot 2 \cdot \underset{1}{\not 2} \cdot 2} = \frac{35}{4} = 8\frac{3}{4}$$

➡ Divide: $1\frac{13}{15} \div 4\frac{4}{5}$

Write the mixed numbers as improper fractions. Then divide the fractions.

$$1\frac{13}{15} \div 4\frac{4}{5} = \frac{28}{15} \div \frac{24}{5} = \frac{28}{15} \times \frac{5}{24} = \frac{28 \cdot 5}{15 \cdot 24} = \frac{\overset{1}{\not 2} \cdot \overset{1}{\not 2} \cdot 7 \cdot \overset{1}{\not 5}}{3 \cdot \underset{1}{\not 5} \cdot \underset{1}{\not 2} \cdot \underset{1}{\not 2} \cdot 2 \cdot 3} = \frac{7}{18}$$

**Example 3** Divide $\frac{4}{9}$ by 5.

**Solution**

$$\frac{4}{9} \div 5 = \frac{4}{9} \div \frac{5}{1} = \frac{4}{9} \times \frac{1}{5}$$
$$= \frac{4 \cdot 1}{9 \cdot 5} = \frac{2 \cdot 2}{3 \cdot 3 \cdot 5} = \frac{4}{45}$$

**You Try It 3** Divide $\frac{5}{7}$ by 6.

**Your solution**

**Example 4**

Find the quotient of $\frac{3}{8}$ and $2\frac{1}{10}$.

**Solution**

$$\frac{3}{8} \div 2\frac{1}{10} = \frac{3}{8} \div \frac{21}{10} = \frac{3}{8} \times \frac{10}{21}$$
$$= \frac{3 \cdot 10}{8 \cdot 21} = \frac{\overset{1}{\not 3} \cdot \overset{1}{\not 2} \cdot 5}{\underset{1}{\not 2} \cdot 2 \cdot 2 \cdot \underset{1}{\not 3} \cdot 7} = \frac{5}{28}$$

**You Try It 4**

Find the quotient of $12\frac{3}{5}$ and 7.

**Your solution**

**Example 5** Divide: $2\frac{3}{4} \div 1\frac{5}{7}$

**Solution**

$$2\frac{3}{4} \div 1\frac{5}{7} = \frac{11}{4} \div \frac{12}{7} = \frac{11}{4} \times \frac{7}{12} = \frac{11 \cdot 7}{4 \cdot 12}$$
$$= \frac{11 \cdot 7}{2 \cdot 2 \cdot 2 \cdot 2 \cdot 3} = \frac{77}{48} = 1\frac{29}{48}$$

**You Try It 5** Divide: $3\frac{2}{3} \div 2\frac{2}{5}$

**Your solution**

*Solutions on pp. A16 and A17*

**Example 6** Divide: $1\frac{13}{15} \div 4\frac{1}{5}$

**Solution**

$$1\frac{13}{15} \div 4\frac{1}{5} = \frac{28}{15} \div \frac{21}{5} = \frac{28}{15} \times \frac{5}{21} = \frac{28 \cdot 5}{15 \cdot 21}$$

$$= \frac{2 \cdot 2 \cdot \overset{1}{\cancel{7}} \cdot \overset{1}{\cancel{5}}}{3 \cdot \underset{1}{\cancel{5}} \cdot 3 \cdot \underset{1}{\cancel{7}}} = \frac{4}{9}$$

**You Try It 6** Divide: $2\frac{5}{6} \div 8\frac{1}{2}$

**Your solution**

**Example 7** Divide: $4\frac{3}{8} \div 7$

**Solution**

$$4\frac{3}{8} \div 7 = \frac{35}{8} \div \frac{7}{1} = \frac{35}{8} \times \frac{1}{7}$$

$$= \frac{35 \cdot 1}{8 \cdot 7} = \frac{5 \cdot \overset{1}{\cancel{7}}}{2 \cdot 2 \cdot 2 \cdot \underset{1}{\cancel{7}}} = \frac{5}{8}$$

**You Try It 7** Divide: $6\frac{2}{5} \div 4$

**Your solution**

*Solutions on p. A17*

## *Objective C To solve application problems*

**Example 8**

A car used $15\frac{1}{2}$ gallons of gasoline on a 310-mile trip. How many miles can this car travel on 1 gallon of gasoline?

**Strategy**
To find the number of miles, divide the number of miles traveled by the number of gallons of gasoline used.

**Solution**

$$310 \div 15\frac{1}{2} = \frac{310}{1} \div \frac{31}{2}$$

$$= \frac{310}{1} \times \frac{2}{31}$$

$$= \frac{310 \cdot 2}{1 \cdot 31}$$

$$= 20$$

The car travels 20 miles on 1 gallon of gasoline.

**You Try It 8**

Leon Dern purchased a $\frac{1}{2}$-ounce gold coin for \$195. What was the price of 1 ounce?

**Your strategy**

**Your solution**

*Solution on p. A17*

**Example 9**

A 12-foot board is cut into pieces $2\frac{1}{4}$ feet long for use as bookshelves. What is the length of the remaining piece after as many shelves as possible are cut?

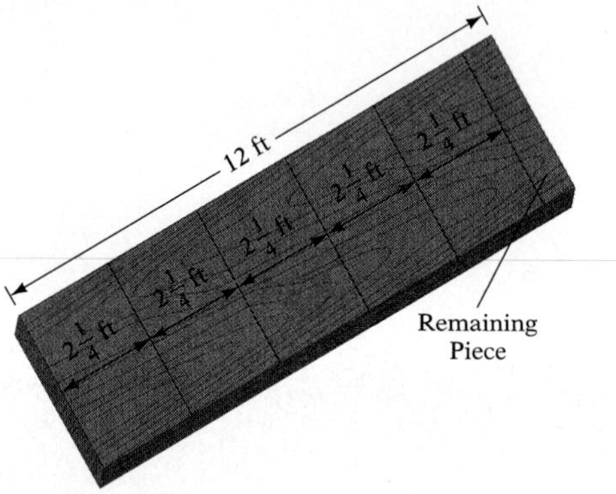


**Strategy**

To find the length of the remaining piece:

- Divide the total length by the length of each shelf $\left(12 \div 2\frac{1}{4}\right)$. This will give you the number of shelves cut, with a certain fraction of a shelf left over.
- Multiply the fraction left over by the length of a shelf to determine the actual length of the piece remaining.

**Solution**

$$12 \div 2\frac{1}{4} = \frac{12}{1} \div \frac{9}{4} = \frac{12}{1} \times \frac{4}{9}$$
$$= \frac{12 \cdot 4}{1 \cdot 9} = \frac{16}{3} = 5\frac{1}{3}$$

5 pieces $2\frac{1}{4}$ feet long

1 piece $\frac{1}{3}$ of $2\frac{1}{4}$ feet long

$$\frac{1}{3} \times 2\frac{1}{4} = \frac{1}{3} \times \frac{9}{4} = \frac{1 \cdot 9}{3 \cdot 4} = \frac{3}{4}$$

The length of the piece remaining is $\frac{3}{4}$ foot.

**You Try It 9**

A 16-foot board is cut into pieces $3\frac{1}{3}$ feet long for shelves for a bookcase. What is the length of the remaining piece after as many shelves as possible are cut?

**Your strategy**

**Your solution**

*Solution on p. A17*

## 2.7 Exercises

### *Objective A*

Divide.

1. $\frac{1}{3} \div \frac{2}{5}$

2. $\frac{3}{7} \div \frac{3}{2}$

3. $\frac{31}{40} \div \frac{11}{15}$

4. $\frac{3}{7} \div \frac{3}{7}$

5. $0 \div \frac{1}{2}$

6. $0 \div \frac{3}{4}$

7. $\frac{16}{33} \div \frac{4}{11}$

8. $\frac{5}{24} \div \frac{15}{36}$

9. $\frac{11}{15} \div \frac{1}{12}$

10. $\frac{2}{9} \div \frac{16}{19}$

11. $\frac{15}{16} \div \frac{16}{39}$

12. $\frac{2}{15} \div \frac{3}{5}$

13. $\frac{8}{9} \div \frac{4}{5}$

14. $\frac{11}{15} \div \frac{5}{22}$

15. $\frac{12}{13} \div \frac{4}{9}$

16. $\frac{1}{9} \div \frac{2}{3}$

17. $\frac{10}{21} \div \frac{5}{7}$

18. $\frac{2}{5} \div \frac{4}{7}$

19. $\frac{3}{8} \div \frac{5}{12}$

20. $\frac{5}{9} \div \frac{15}{32}$

21. $\frac{1}{2} \div \frac{1}{4}$

22. $\frac{1}{3} \div \frac{1}{9}$

23. $\frac{1}{5} \div \frac{1}{10}$

24. $\frac{4}{15} \div \frac{2}{5}$

25. $\frac{7}{15} \div \frac{14}{5}$

26. $\frac{5}{8} \div \frac{15}{2}$

27. $\frac{14}{3} \div \frac{7}{9}$

28. $\frac{7}{4} \div \frac{9}{2}$

29. $\frac{5}{9} \div \frac{25}{3}$

30. $\frac{5}{16} \div \frac{3}{8}$

31. $\frac{2}{3} \div \frac{1}{3}$

32. $\frac{4}{9} \div \frac{1}{9}$

33. $\frac{5}{7} \div \frac{2}{7}$

34. $\frac{5}{6} \div \frac{1}{9}$

35. $\frac{2}{3} \div \frac{2}{9}$

36. $\frac{5}{12} \div \frac{5}{6}$

37. $4 \div \frac{2}{3}$

38. $\frac{2}{3} \div 4$

39. $\frac{3}{2} \div 3$

40. $3 \div \frac{3}{2}$

41. Divide $\frac{7}{8}$ by $\frac{3}{4}$.

42. Divide $\frac{7}{12}$ by $\frac{3}{4}$.

43. Find the quotient of $\frac{5}{7}$ and $\frac{3}{14}$.

44. Find the quotient of $\frac{6}{11}$ and $\frac{9}{32}$.

***Objective B***

Divide.

45. $\frac{5}{6} \div 25$

46. $22 \div \frac{3}{11}$

47. $6 \div 3\frac{1}{3}$

48. $5\frac{1}{2} \div 11$

49. $3\frac{1}{3} \div \frac{3}{8}$

50. $6\frac{1}{2} \div \frac{1}{2}$

51. $\frac{3}{8} \div 2\frac{1}{4}$

52. $\frac{5}{12} \div 4\frac{4}{5}$

53. $1\frac{1}{2} \div 1\frac{3}{8}$

54. $2\frac{1}{4} \div 1\frac{3}{8}$

55. $8\frac{1}{4} \div 2\frac{3}{4}$

56. $3\frac{5}{9} \div 32$

57. $4\frac{1}{5} \div 21$

58. $6\frac{8}{9} \div \frac{31}{36}$

59. $5\frac{3}{5} \div \frac{7}{10}$

60. $\frac{11}{12} \div 2\frac{1}{3}$

61. $\frac{7}{8} \div 3\frac{1}{4}$

62. $\frac{5}{16} \div 5\frac{3}{8}$

63. $\frac{9}{14} \div 3\frac{1}{7}$

64. $27 \div \frac{3}{8}$

65. $35 \div \frac{7}{24}$

66. $\frac{3}{8} \div 2\frac{3}{4}$

67. $\frac{11}{18} \div 2\frac{2}{9}$

68. $\frac{21}{40} \div 3\frac{3}{10}$

69. $\frac{6}{25} \div 4\frac{1}{5}$

70. $2\frac{1}{16} \div 2\frac{1}{2}$

71. $7\frac{3}{5} \div 1\frac{7}{12}$

72. $1\frac{2}{3} \div \frac{3}{8}$

73. $16 \div \frac{2}{3}$

74. $2\frac{1}{2} \div 3\frac{5}{14}$

75. $1\frac{5}{8} \div 4$

76. $13\frac{3}{8} \div \frac{1}{4}$

77. $16 \div 1\frac{1}{2}$

78. $9 \div \frac{7}{8}$

79. $15\frac{1}{3} \div 2\frac{2}{9}$

80. $16\frac{5}{8} \div 1\frac{2}{3}$

81. $24\frac{4}{5} \div 2\frac{3}{5}$

82. $1\frac{1}{3} \div 5\frac{8}{9}$

83. $13\frac{2}{3} \div 0$

**84.** $82\frac{3}{5} \div 19\frac{1}{10}$ **85.** $45\frac{3}{5} \div 15$ **86.** $102 \div 1\frac{1}{2}$ **87.** $44\frac{5}{12} \div 32\frac{5}{6}$

**88.** $0 \div 3\frac{1}{2}$ **89.** $8\frac{2}{7} \div 1$ **90.** $6\frac{9}{16} \div 1\frac{3}{32}$ **91.** $5\frac{5}{14} \div 3\frac{4}{7}$

**92.** $8\frac{8}{9} \div 2\frac{13}{18}$ **93.** $10\frac{1}{5} \div 1\frac{7}{10}$ **94.** $7\frac{3}{8} \div 1\frac{27}{32}$ **95.** $4\frac{5}{11} \div 2\frac{1}{3}$

**96.** Divide $7\frac{7}{9}$ by $5\frac{5}{6}$.

**97.** Divide $2\frac{3}{4}$ by $1\frac{23}{32}$.

**98.** Find the quotient of $8\frac{1}{4}$ and $1\frac{5}{11}$.

**99.** Find the quotient of $\frac{14}{17}$ and $3\frac{1}{9}$.

## *Objective C* Application Problems

**100.** Individual cereal boxes contain $\frac{3}{4}$ ounce of cereal. How many boxes can be filled with 600 ounces of cereal?

**101.** A box of bran flakes contains 20 ounces of cereal. How many $1\frac{1}{4}$-ounce portions can be served from this box?

**102.** A $\frac{5}{8}$-carat diamond was purchased for \$1200. What would a similar diamond weighing 1 carat cost?

**103.** The Inverness Investor Group bought $8\frac{1}{3}$ acres of land for \$200,000. What was the cost of each acre?

**104.** KU Energy stock is offered for $31\frac{5}{8}$ per share. How many shares can you buy for \$1265?

**105.** The Tuesday attendance at the dinner theater was 280. The theater was $\frac{2}{3}$ full. What is the capacity of the theater?

**106.** The Hammond Company purchased $9\frac{3}{4}$ acres for a housing project. One and one-half acres were set aside for a park.

a. How many acres are available for housing?

b. How many $\frac{1}{4}$-acre parcels of land can be sold after the land for the park is set aside?

**107.** A chef purchased a roast that weighed $10\frac{3}{4}$ pounds. After the fat was trimmed and the bone removed, the roast weighed $9\frac{1}{3}$ pounds.

a. What was the total weight of the fat and bone?

b. How many $\frac{1}{3}$-pound servings can be cut from the roast?

**108.** One-tenth of a shipment of $18\frac{1}{3}$ pounds of grapes was spoiled. How many $\frac{1}{2}$-pound packages of unspoiled grapes can be packaged from this shipment?

**109.** A 15-foot board is cut into pieces $3\frac{1}{2}$ feet long for a bookcase. What is the length of the piece remaining after as many shelves as possible have been cut?

**110.** A scale of $\frac{1}{2}$ inch to 1 foot is used to draw the plans for a house. The scale measurements for three walls are given in the table at the right. Complete the table to determine the actual wall lengths for the three walls a, b, and c.

| *Wall* | *Scale* | *Actual Wall Length* |
|---|---|---|
| a | $6\frac{1}{4}$ in. | ? |
| b | 9 in. | ? |
| c | $7\frac{7}{8}$ in. | ? |

## APPLYING THE CONCEPTS

**111.** On a map, two cities are $4\frac{5}{8}$ inches apart. If $\frac{3}{8}$ inch on the map represents 60 miles, what is the number of miles between the two cities?

**112.** [W] Is the quotient always less than the dividend in a division problem? Explain.

**113.** Fill in the box to make a true statement.

a. $\frac{3}{4} \cdot \square = \frac{1}{2}$    b. $\frac{2}{3} \cdot \square = 1\frac{3}{4}$

**114.** A page of type in a certain textbook is $7\frac{1}{2}$ inches wide. If the page is divided into three equal columns, with $\frac{3}{8}$ inch between columns, how wide is each column?

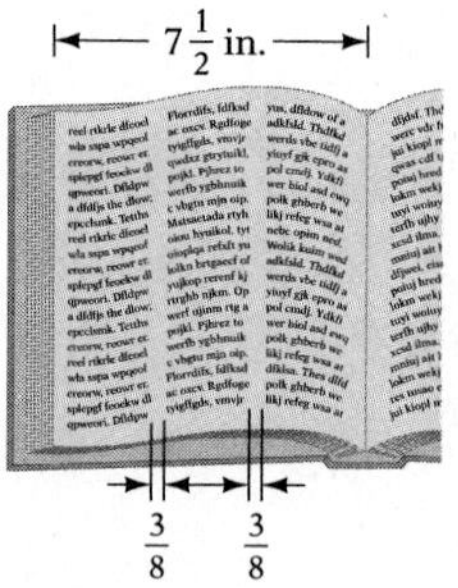


**115.** A whole number is both multiplied and divided by the same proper fraction. Which is greater, the product or the quotient?

**116.** Fractions are multiplied by multiplying the numerators and multiplying the denominators. Consider dividing fractions in a similar manner. Divide the numerators and divide the denominators.

$$\frac{4}{15} \div \frac{2}{5} = \frac{4 \div 2}{15 \div 5} = \frac{2}{3}$$

This gives the same answer as the traditional method.

$$\frac{4}{15} \div \frac{2}{5} = \frac{4}{15} \cdot \frac{5}{2} = \frac{20}{30} = \frac{2}{3}$$

Try this method for the following division problems.

a. $\frac{5}{6} \div \frac{1}{3}$    b. $\frac{4}{21} \div \frac{2}{7}$    c. $\frac{5}{9} \div \frac{5}{3}$    d. $\frac{15}{16} \div \frac{3}{8}$

**117.** Does the method of dividing fractions shown above always work? Give an example of when the traditional method would be a better choice.

# 2.8 Order, Exponents, and the Order of Operations Agreement

## Objective A To identify the order relation between two fractions

**POINT OF INTEREST**

Leonardo of Pisa, who was also called Fibonacci (c. 1175–1250), is credited with bringing the Hindu-Arabic number system to the western world and promoting its use instead of the cumbersome Roman numeral system.

He was also influential in promoting the idea of the fraction bar. His notation, however, was very different from what we have today. For instance, he wrote $\frac{3\ 5}{4\ 7}$ to mean $\frac{5}{7}+\frac{3}{7\cdot 4}$.

Recall that whole numbers can be graphed as points on the number line. Fractions can also be graphed as points on the number line.

The graph of $\frac{3}{4}$ on the number line

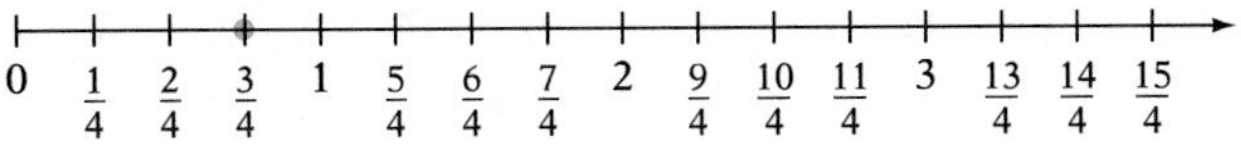

The number line can be used to determine the order relation between two fractions. A fraction that appears to the left of a given fraction is less than the given fraction. A fraction that appears to the right of a given fraction is greater than the given fraction.

$\frac{1}{8}<\frac{3}{8}$ $\frac{6}{8}>\frac{3}{8}$

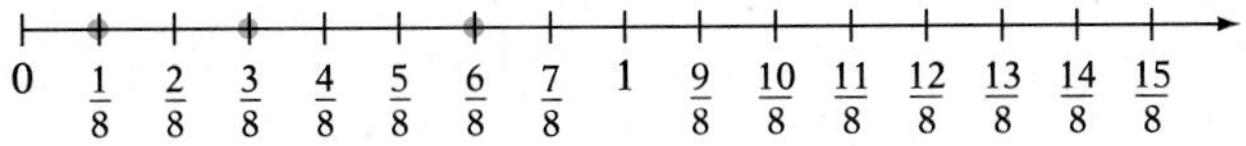

To find the order of relation between two fractions with the same denominator, compare numerators. The fraction that has the smaller numerator is the smaller fraction. When the denominators are different, begin by writing equivalent fractions with a common denominator; then compare numerators.

➡ Find the order relation between $\frac{11}{18}$ and $\frac{5}{8}$.

The LCM of 18 and 8 is 72.

$\frac{11}{18}=\frac{44}{72}$ ← Smaller numerator

$\frac{5}{8}=\frac{45}{72}$ ← Larger numerator

$\frac{11}{18}<\frac{5}{8}$ or $\frac{5}{8}>\frac{11}{18}$

**Example 1** Place the correct symbol, < or >, between the numbers.

$\frac{5}{12}$ $\frac{7}{18}$

**Solution** $\frac{5}{12}=\frac{15}{36}$ $\frac{7}{18}=\frac{14}{36}$

$\frac{5}{12}>\frac{7}{18}$

**You Try It 1** Place the correct symbol, < or >, between the numbers.

$\frac{9}{14}$ $\frac{13}{21}$

**Your solution**

*Solution on p. A18*



## Objective B To simplify expressions containing exponents

Repeated multiplication of the same fraction can be written in two ways:

$$\frac{1}{2}\cdot\frac{1}{2}\cdot\frac{1}{2}\cdot\frac{1}{2} \text{ or } \left(\frac{1}{2}\right)^4 \leftarrow \text{Exponent}$$

The exponent indicates how many times the fraction occurs as a factor in the multiplication. The expression $\left(\frac{1}{2}\right)^4$ is in exponential notation.

**Example 2** Simplify: $\left(\frac{5}{6}\right)^3 \cdot \left(\frac{3}{5}\right)^2$

**Solution**

$$\left(\frac{5}{6}\right)^8 \cdot \left(\frac{3}{5}\right)^2 = \left(\frac{5}{6} \cdot \frac{5}{6} \cdot \frac{5}{6}\right) \cdot \left(\frac{3}{5} \cdot \frac{3}{5}\right)$$

$$= \frac{\overset{1}{\cancel{5}} \cdot \overset{1}{\cancel{5}} \cdot 5 \cdot \overset{1}{\cancel{3}} \cdot \overset{1}{\cancel{3}}}{2 \cdot \underset{1}{\cancel{3}} \cdot 2 \cdot \underset{1}{\cancel{3}} \cdot 2 \cdot 3 \cdot \underset{1}{\cancel{5}} \cdot \underset{1}{\cancel{5}}} = \frac{5}{24}$$

**You Try It 2** Simplify: $\left(\frac{7}{11}\right)^2 \cdot \left(\frac{2}{7}\right)$

**Your solution**

*Solution on p. A18*

## *Objective C To use the Order of Operations Agreement to simplify expressions*

The Order of Operations Agreement is used for fractions as well as whole numbers.

**Step 1** Do all operations inside parentheses.
**Step 2** Simplify any number expressions containing exponents.
**Step 3** Do multiplications and divisions as they occur from left to right.
**Step 4** Do additions and subtractions as they occur from left to right.

➡ Simplify $\frac{14}{15} - \left(\frac{1}{2}\right)^2 \times \left(\frac{2}{3} + \frac{4}{5}\right)$ by using the Order of Operations Agreement.

$\frac{14}{15} - \left(\frac{1}{2}\right)^2 \times \left(\frac{2}{3} + \frac{4}{5}\right)$ **1.** Perform operations in parentheses.

$\frac{14}{15} - \left(\frac{1}{2}\right)^2 \times \frac{22}{15}$ **2.** Simplify expressions with exponents.

$\frac{14}{15} - \frac{1}{4} \times \frac{22}{15}$ **3.** Do multiplications and divisions as they occur from left to right.

$\frac{14}{15} - \frac{11}{30}$ **4.** Do additions and subtractions as they occur from left to right.

$\frac{17}{30}$

One or more of the above steps may not be needed to simplify an expression. In that case, proceed to the next step in the Order of Operations Agreement.

**Example 3** Simplify: $\left(\frac{3}{4}\right)^2 \div \left(\frac{3}{8} - \frac{1}{12}\right)$

**Solution**

$$\left(\frac{3}{4}\right)^2 \div \left(\frac{3}{8} - \frac{1}{12}\right)$$

$$\left(\frac{3}{4}\right)^2 \div \left(\frac{7}{24}\right)$$

$$\frac{9}{16} \div \frac{7}{24} = \frac{27}{14} = 1\frac{13}{14}$$

**You Try It 3** Simplify:
$\left(\frac{1}{13}\right)^2 \cdot \left(\frac{1}{4} + \frac{1}{6}\right) \div \frac{5}{13}$

**Your solution**

*Solution on p. A18*

## 2.8 Exercises

### *Objective A*

Place the correct symbol, < or >, between the two numbers.

1. $\frac{11}{40}\quad\frac{19}{40}$

2. $\frac{92}{103}\quad\frac{19}{103}$

3. $\frac{2}{3}\quad\frac{5}{7}$

4. $\frac{2}{5}\quad\frac{3}{8}$

5. $\frac{5}{8}\quad\frac{7}{12}$

6. $\frac{11}{16}\quad\frac{17}{24}$

7. $\frac{7}{9}\quad\frac{11}{12}$

8. $\frac{5}{12}\quad\frac{7}{15}$

9. $\frac{13}{14}\quad\frac{19}{21}$

10. $\frac{13}{18}\quad\frac{7}{12}$

11. $\frac{7}{24}\quad\frac{11}{30}$

12. $\frac{13}{36}\quad\frac{19}{48}$

### *Objective B*

Simplify.

13. $\left(\frac{3}{8}\right)^2$

14. $\left(\frac{5}{12}\right)^2$

15. $\left(\frac{2}{9}\right)^3$

16. $\left(\frac{5}{7}\right)^3$

17. $\left(\frac{1}{2}\right)\cdot\left(\frac{2}{3}\right)^2$

18. $\left(\frac{2}{3}\right)\cdot\left(\frac{1}{2}\right)^4$

19. $\left(\frac{1}{3}\right)^2\cdot\left(\frac{3}{5}\right)^3$

20. $\left(\frac{2}{5}\right)^2\cdot\left(\frac{5}{8}\right)^3$

21. $\left(\frac{2}{5}\right)^3\cdot\left(\frac{5}{7}\right)^2$

22. $\left(\frac{5}{9}\right)^3\cdot\left(\frac{18}{25}\right)^2$

23. $\left(\frac{1}{3}\right)^4\cdot\left(\frac{9}{11}\right)^2$

24. $\left(\frac{8}{15}\right)^3\cdot\left(\frac{5}{8}\right)^4$

25. $\left(\frac{1}{2}\right)^6\cdot\left(\frac{32}{35}\right)^2$

26. $\left(\frac{2}{3}\right)^4\cdot\left(\frac{81}{100}\right)^2$

27. $\left(\frac{1}{6}\right)\cdot\left(\frac{6}{7}\right)^2\cdot\left(\frac{2}{3}\right)$

28. $\left(\frac{2}{7}\right)\cdot\left(\frac{7}{8}\right)^2\cdot\left(\frac{8}{9}\right)$

29. $3\cdot\left(\frac{3}{5}\right)^3\cdot\left(\frac{1}{3}\right)^2$

30. $4\cdot\left(\frac{3}{4}\right)^3\cdot\left(\frac{4}{7}\right)^2$

31. $11\cdot\left(\frac{3}{8}\right)^3\cdot\left(\frac{8}{11}\right)^2$

32. $5\cdot\left(\frac{3}{5}\right)^3\cdot\left(\frac{2}{3}\right)^4$

33. $\left(\frac{2}{7}\right)^2\cdot\left(\frac{7}{9}\right)^2\cdot\left(\frac{9}{11}\right)^2$

### Objective C

Simplify using the Order of Operations Agreement.

**34.** $\frac{1}{2} - \frac{1}{3} + \frac{2}{3}$

**35.** $\frac{2}{5} + \frac{3}{10} - \frac{2}{3}$

**36.** $\frac{1}{3} \div \frac{1}{2} + \frac{3}{4}$

**37.** $\frac{3}{5} \div \frac{6}{7} + \frac{4}{5}$

**38.** $\frac{4}{5} + \frac{3}{7} \cdot \frac{14}{15}$

**39.** $\frac{2}{3} + \frac{5}{8} \cdot \frac{16}{35}$

**40.** $\left(\frac{3}{4}\right)^2 - \frac{5}{12}$

**41.** $\left(\frac{3}{5}\right)^3 - \frac{3}{25}$

**42.** $\frac{5}{6} \cdot \left(\frac{2}{3} - \frac{1}{6}\right) + \frac{7}{18}$

**43.** $\frac{3}{4} \cdot \left(\frac{11}{12} - \frac{7}{8}\right) + \frac{5}{16}$

**44.** $\frac{7}{12} - \left(\frac{2}{3}\right)^2 + \frac{5}{8}$

**45.** $\frac{11}{16} - \left(\frac{3}{4}\right)^2 + \frac{7}{12}$

**46.** $\frac{3}{4} \cdot \left(\frac{4}{9}\right)^2 + \frac{1}{2}$

**47.** $\frac{9}{10} \cdot \left(\frac{2}{3}\right)^3 + \frac{2}{3}$

**48.** $\left(\frac{1}{2} + \frac{3}{4}\right) \div \frac{5}{8}$

**49.** $\left(\frac{2}{3} + \frac{5}{6}\right) \div \frac{5}{9}$

**50.** $\frac{3}{8} \div \left(\frac{5}{12} + \frac{3}{8}\right)$

**51.** $\frac{7}{12} \div \left(\frac{2}{3} + \frac{5}{9}\right)$

**52.** $\left(\frac{3}{8}\right)^2 \div \left(\frac{3}{7} + \frac{3}{14}\right)$

**53.** $\left(\frac{5}{6}\right)^2 \div \left(\frac{5}{12} + \frac{2}{3}\right)$

**54.** $\frac{2}{5} \div \frac{3}{8} \cdot \frac{4}{5}$

### APPLYING THE CONCEPTS

**55.** $\frac{2}{3} < \frac{3}{4}$. Is $\frac{2+3}{3+4}$ less than $\frac{2}{3}$, greater than $\frac{2}{3}$, or between $\frac{2}{3}$ and $\frac{3}{4}$?

**56.** [W] A farmer died and left 17 horses to be divided among 3 children. The first child was to receive $\frac{1}{2}$ of the horses, the second child $\frac{1}{3}$ of the horses, and the third child $\frac{1}{9}$ of the horses. The executor for the family's estate realized that 17 horses could not be divided by halves, thirds, or ninths and so added a neighbor's horse to the farmer's. With 18 horses, the executor gave 9 horses to the first child, 6 horses to the second child, and 2 horses to the third child. This accounted for the 17 horses, so the executor returned the borrowed horse to the neighbor. Explain why this worked.

# Projects in Mathematics

**Music** In musical notation, notes are printed on a staff, which is a set of five horizontal lines and the spaces between them. The notes of a musical composition are grouped into measures, or bars. Vertical lines separate measures on a staff. The shape of a note indicates how long it should be held. The whole note has the longest time value of any note. Each time value is divided by 2 in order to find the next smallest note value.

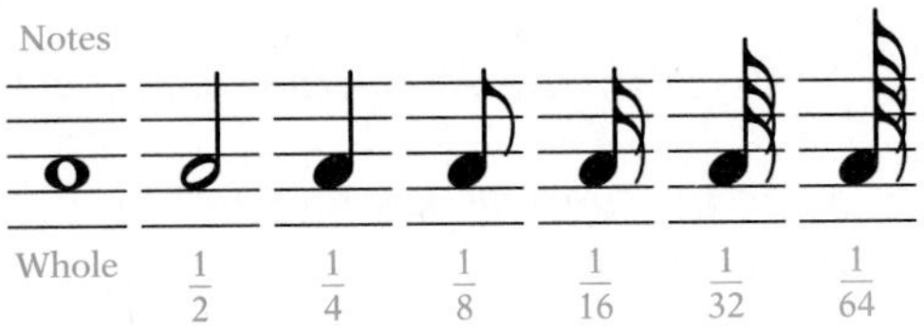

The time signature is a fraction that appears at the beginning of a piece of music. The numerator of the fraction indicates the number of beats in a measure. The denominator indicates what kind of note receives one beat. For example, music written in $\frac{2}{4}$ time has 2 beats to a measure, and a quarter note receives one beat. One measure in $\frac{2}{4}$ time may have 1 half note, 2 quarter notes, and 4 eighth notes or any other combination of notes totaling 2 beats. Other common time signatures include $\frac{4}{4}$, $\frac{3}{4}$, and $\frac{6}{8}$.

***Exercises:***

1. Explain the meaning of the 6 and the 8 in the time signature $\frac{6}{8}$. Give some possible combinations of notes in one measure of a piece written in $\frac{4}{4}$ time.
2. What does a dot at the right of a note indicate? What is the effect of a dot at the right of a half note? A quarter note? An eighth note?
3. Symbols called rests are used to indicate periods of silence in a piece of music. What symbols are used to indicate the different time values of rests?
4. Find some examples of musical compositions written in different time signatures. Use a few measures from each to show that the sum of the time values of the notes and rests in each measure equals the numerator of the time signature.

**Construction** Suppose you are involved in building your own home. Design a stairway from the first floor of the house to the second floor. Some of the questions you will need to answer follow.

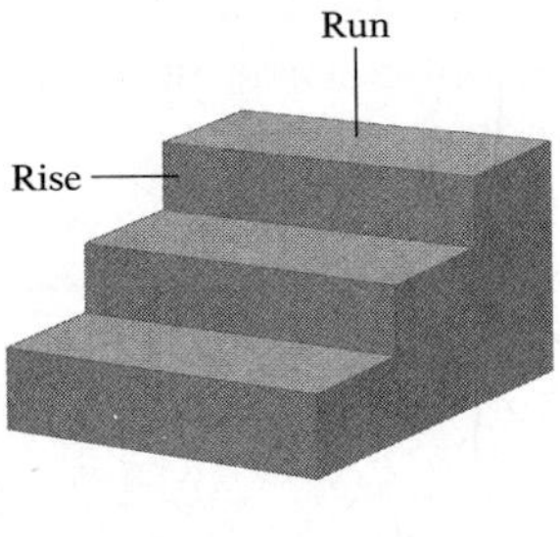


What is the distance from the floor of the first story to the floor of the second story?

Typically, what is the number of steps in a stairway?

What is a reasonable length for the run of each step?

What width wood is being used to build the staircase?

In designing the stairway, remember that each riser should be the same height and each run should be the same length. And the width of the wood used for the steps will have to be incorporated in the calculation.

# Chapter Summary

**Key Words** The *least common multiple* (LCM) is the smallest common multiple of two or more numbers.

The *greatest common factor* (GCF) is the largest common factor of two or more numbers.

A *fraction* can represent the number of equal parts of a whole.

A *proper fraction* is a fraction less than 1.

A *mixed number* is a number greater than 1 with a whole number part and a fractional part.

An *improper fraction* is a fraction greater than or equal to 1.

Equal fractions with different denominators are called *equivalent fractions*.

A fraction is in *simplest form* when there are no common factors in the numerator and the denominator.

The *reciprocal* of a fraction is the fraction with the numerator and denominator interchanged.

*Inverting* is the process of finding the reciprocal of a fraction.

**Essential Rules** ***Addition of Fractions with Like Denominators***

To add fractions with like denominators, add the numerators and place the sum over the common denominator.

***Addition of Fractions with Unlike Denominators***

To add fractions with unlike denominators, first rewrite the fractions as equivalent fractions with the same denominator. Then add the numerators and place the sum over the common denominator.

***Subtraction of Fractions with Like Denominators***

To subtract fractions with like denominators, subtract the numerators and place the difference over the common denominator.

***Subtraction of Fractions with Unlike Denominators***

To subtract fractions with unlike denominators, rewrite the fractions as equivalent fractions with the same denominator. Then subtract the numerators and place the difference over the common denominator.

***Multiplication of Fractions***

To multiply two fractions, multiply the numerators and place the product over the product of the denominators.

***Division of Fractions***

To divide two fractions, multiply by the reciprocal of the divisor.

# Chapter Review

## SECTION 2.1

1. Find the LCM of 18 and 12.

2. Find the LCM of 18 and 27.

3. Find the GCF of 20 and 48.

4. Find the GCF of 15 and 25.

## SECTION 2.2

5. Express the shaded portion of the circles as an improper fraction.

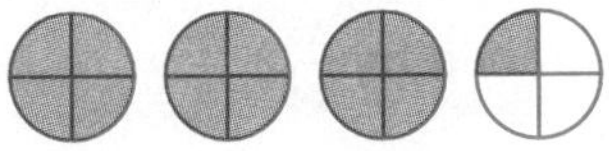

6. Express the shaded portion of the circles as a mixed number.

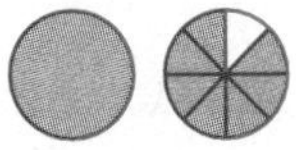

7. Write $\frac{17}{5}$ as a mixed number.

8. Write $2\frac{5}{7}$ as an improper fraction.

## SECTION 2.3

9. Write an equivalent fraction with the given denominator.

$\frac{2}{3} = \frac{}{36}$

10. Write an equivalent fraction with the given denominator.

$\frac{8}{11} = \frac{}{44}$

11. Write $\frac{30}{45}$ in simplest form.

12. Write $\frac{16}{44}$ in simplest form.

## SECTION 2.4

13. Add: $\frac{3}{8} + \frac{5}{8} + \frac{1}{8}$

14. Find the total of $\frac{2}{3}$, $\frac{5}{6}$, and $\frac{2}{9}$.

15. Add: $\frac{3}{8} + 1\frac{2}{3} + 3\frac{5}{6}$

16. Add: $4\frac{4}{9} + 2\frac{1}{6} + 11\frac{17}{27}$

17. During three months of the rainy season, $5\frac{7}{8}$, $6\frac{2}{3}$, and $8\frac{3}{4}$ inches of rain fell. Find the total rainfall for the three months.

## SECTION 2.5

18. Subtract: $\frac{11}{18} - \frac{5}{18}$

19. Find $\frac{17}{24}$ decreased by $\frac{3}{16}$.

20. Subtract: $\begin{array}{r} 18\frac{1}{6} \\ -\ 3\frac{5}{7} \\ \hline \end{array}$

21. Subtract: $\begin{array}{r} 16 \\ -\ 5\frac{7}{8} \\ \hline \end{array}$

22. A 15-mile race has three checkpoints. The first checkpoint is $4\frac{1}{2}$ miles from the starting point. The second checkpoint is $5\frac{3}{4}$ miles from the first checkpoint. How many miles is the second checkpoint from the finish line?

## SECTION 2.6

23. Multiply: $\frac{5}{12} \times \frac{4}{25}$

24. What is $\frac{11}{50}$ multiplied by $\frac{25}{44}$?

25. Multiply: $2\frac{1}{3} \times 3\frac{7}{8}$

26. Multiply: $2\frac{1}{4} \times 7\frac{1}{3}$

27. A compact car gets 36 miles on each gallon of gasoline. How many miles can the car travel on $6\frac{3}{4}$ gallons of gasoline?

## SECTION 2.7

28. Divide: $\frac{5}{6} \div \frac{5}{12}$

29. What is $\frac{15}{28}$ divided by $\frac{5}{7}$?

30. Divide: $1\frac{1}{3} \div \frac{2}{3}$

31. Divide: $8\frac{2}{3} \div 2\frac{3}{5}$

32. A home building contractor bought $4\frac{2}{3}$ acres for \$168,000. What was the cost of each acre?

## SECTION 2.8

33. Place the correct symbol, $<$ or $>$, between the two numbers.
$\frac{11}{18} \quad \frac{17}{24}$

34. Simplify: $\left(\frac{3}{4}\right)^3 \cdot \frac{20}{27}$

35. Simplify: $\frac{2}{7}\left(\frac{5}{8} - \frac{1}{3}\right) \div \frac{3}{5}$

36. Simplify: $\left(\frac{4}{5} - \frac{2}{3}\right)^2 \div \frac{4}{15}$

# Chapter Test

1. Find the LCM of 24 and 40.

2. Find the GCF of 24 and 80.

3. Express the shaded portion of the circles as an improper fraction.

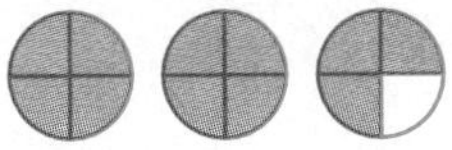

4. Write $\frac{18}{5}$ as a mixed number.

5. Write $9\frac{4}{5}$ as an improper fraction.

6. Write an equivalent fraction with the given denominator.

$$\frac{5}{8} = \frac{\phantom{0}}{72}$$

7. Write $\frac{40}{64}$ in simplest form.

8. Add: $\frac{7}{12} + \frac{11}{12} + \frac{5}{12}$

9. Add:

$$\begin{array}{r} \frac{5}{6} \\ \frac{7}{9} \\ + \frac{1}{15} \\ \hline \end{array}$$

10. What is $12\frac{5}{12}$ more than $9\frac{17}{20}$?

11. The rainfall for a 3-month period was $11\frac{1}{2}$ inches, $7\frac{5}{8}$ inches, and $2\frac{1}{3}$ inches. Find the total rainfall for the 3 months.

12. Subtract: $\frac{17}{24} - \frac{11}{24}$

13. What is $\frac{9}{16}$ minus $\frac{5}{12}$?

14. Subtract:

$$\begin{array}{r} 23\frac{1}{8} \\ - \; 9\frac{9}{44} \\ \hline \end{array}$$

**15.** Chris Aguilar bought 100 shares of a utility stock at $\$24\frac{1}{2}$ per share. The stock gained $\$5\frac{5}{8}$ during the first month of ownership and lost $\$2\frac{1}{4}$ during the second month. Find the value of 1 share of the utility stock at the end of the second month.

**16.** Multiply: $\frac{9}{11} \times \frac{44}{81}$

**17.** What is $5\frac{2}{3}$ multiplied by $1\frac{7}{17}$?

**18.** An electrician earns $120 for each day worked. What is the total of the electrician's earnings for working $3\frac{1}{2}$ days?

**19.** Divide: $\frac{5}{9} \div \frac{7}{18}$

**20.** Find the quotient of $6\frac{2}{3}$ and $3\frac{1}{6}$?

**21.** Grant Miura bought $7\frac{1}{4}$ acres of land for a housing project. One and three-fourths acres were set aside for a park, and the remaining land was developed into $\frac{1}{2}$-acre lots. How many lots were available for sale?

**22.** Place the correct symbol, $<$ or $>$, between the two numbers.

$\frac{3}{8} \quad \frac{5}{12}$

**23.** Simplify: $\left(\frac{2}{3}\right)^4 \cdot \frac{27}{32}$

**24.** Simplify: $\left(\frac{3}{4}\right)^2 \div \left(\frac{2}{3} + \frac{5}{6}\right) - \frac{1}{12}$

**25.** Simplify: $\left(\frac{1}{4}\right)^3 \div \left(\frac{1}{8}\right)^2 - \frac{1}{6}$

# Cumulative Review

1. Round 290,496 to the nearest thousand.

2. Subtract: $\begin{array}{r} 390{,}047 \\ -\ 98{,}769 \\ \hline \end{array}$

3. Find the product of 926 and 79.

4. Divide: $57\overline{)30{,}792}$

5. Simplify: $4 \cdot (6 - 3) \div 6 - 1$

6. Find the prime factorization of 44.

7. Find the LCM of 30 and 42.

8. Find the GCF of 60 and 80.

9. Write $7\frac{2}{3}$ as an improper fraction.

10. Write $\frac{25}{4}$ as a mixed number.

11. Write an equivalent fraction with the given denominator.
$\frac{5}{16} = \frac{\phantom{0}}{48}$

12. Write $\frac{24}{60}$ in simplest form.

13. What is $\frac{9}{16}$ more than $\frac{7}{12}$?

14. Add: $\begin{array}{r} 3\frac{7}{8} \\ 7\frac{5}{12} \\ +\ 2\frac{15}{16} \\ \hline \end{array}$

15. Find $\frac{3}{8}$ less than $\frac{11}{12}$.

16. Subtract: $\begin{array}{r} 5\frac{1}{6} \\ -\ 3\frac{7}{18} \\ \hline \end{array}$

**17.** Multiply: $\frac{3}{8} \times \frac{14}{15}$

**18.** Multiply: $3\frac{1}{8} \times 2\frac{2}{5}$

**19.** Divide: $\frac{7}{16} \div \frac{5}{12}$

**20.** Find the quotient of $6\frac{1}{8}$ and $2\frac{1}{3}$.

**21.** Simplify: $\left(\frac{1}{2}\right)^3 \cdot \frac{8}{9}$

**22.** Simplify: $\left(\frac{1}{2} + \frac{1}{3}\right) \div \left(\frac{2}{5}\right)^2$

**23.** Molly O'Brien had $1359 in a checking account. During the week, Molly wrote checks of $128, $54, and $315. Find the amount in the checking account at the end of the week.

**24.** The tickets for a movie were $5 for an adult and $2 for a student. Find the total income from the sale of 87 adult tickets and 135 student tickets.

**25.** Find the total weight of three packages that weigh $1\frac{1}{2}$ pounds, $7\frac{7}{8}$ pounds, and $2\frac{2}{3}$ pounds.

**26.** A board $2\frac{5}{8}$ feet long is cut from a board $7\frac{1}{3}$ feet long. What is the length of the remaining piece?

**27.** A car travels 27 miles on each gallon of gasoline. How many miles can the car travel on $8\frac{1}{3}$ gallons of gasoline?

**28.** Jimmy Santos purchased $10\frac{1}{3}$ acres of land to build a housing development. Jimmy donated 2 acres for a park. How many $\frac{1}{3}$-acre parcels can be sold from the remaining land?

Chapter 3

# Decimals

## Objectives

**Section 3.1**

To write decimals in standard form and in words

To round a decimal to a given place value

**Section 3.2**

To add decimals

To solve application problems

**Section 3.3**

To subtract decimals

To solve application problems

**Section 3.4**

To multiply decimals

To solve application problems

**Section 3.5**

To divide decimals

To solve application problems

**Section 3.6**

To convert fractions to decimals

To convert decimals to fractions

To identify the order relation between two decimals or between a decimal and a fraction

# Decimal Fractions

How would you like to add $\frac{37,544}{23,465} + \frac{5184}{3456}$? These two fractions are very cumbersome, and it would take even a mathematician some time to get the answer.

Well, around 1550, help with such problems arrived with the publication of a book called *La Disme* (*The Tenth*), which urged the use of decimal fractions. A decimal fraction is one in which the denominator is 10, 100, 1000, 10,000, and so on.

This book suggested that all whole numbers were "units" and when written would end with the symbol ⓪. For example, the number 294⓪ would be the number two hundred ninety-four. This is very much like the way numbers are currently written (except for the ⓪).

For a fraction between 0 and 1, a unit was broken down into parts called "primes." The fraction three-tenths would be written

$\frac{3}{10}$ = 3①     The ① was used to mean the end of the primes, or what are now called tenths.

Each prime was further broken down into "seconds," and each second was broken down into "thirds," and so on. The primes ended with ①, the seconds ended with ②, and the thirds ended with ③.

Examples of these numbers in the old notation and our modern fraction notation are shown below.

$\frac{37}{100}$ = 3①7②     $\frac{257}{1000}$ = 2①5②7③

After completing this chapter, you might come back to the problem in the first line. Use decimals instead of fractions to find the answer. The answer is 3.1.

# 3.1 Introduction to Decimals

## Objective A To write decimals in standard form and in words

**POINT OF INTEREST**

Leonardo of Pisa (Fibonacci) anticipated the decimal system. He wrote $\frac{2\;\;7}{10\;10}5$ to mean $5 + \frac{7}{10} + \frac{2}{10 \cdot 10}$.

The smallest human bone is found in the middle ear and measures 0.135 inch in length. The number 0.135 is in **decimal notation.**

Note the relationship between fractions and numbers written in decimal notation.

| Three-tenths | Three-hundredths | Three-thousandths |
|---|---|---|
| $\frac{3}{10} = 0.\underline{3}$ | $\frac{3}{100} = 0.\underline{03}$ | $\frac{3}{1000} = 0.\underline{003}$ |
| 1 zero 1 decimal place | 2 zeros 2 decimal places | 3 zeros 3 decimal places |

A number written in decimal notation has three parts.

| 351 | . | 7089 |
|---|---|---|
| **Whole-number part** | **Decimal point** | **Decimal part** |

A number written in decimal notation is often called simply a **decimal.** The position of a digit in a decimal determines the digit's place value.

In the decimal 351.7089, the position of the digit 9 determines that its place value is ten-thousandths.

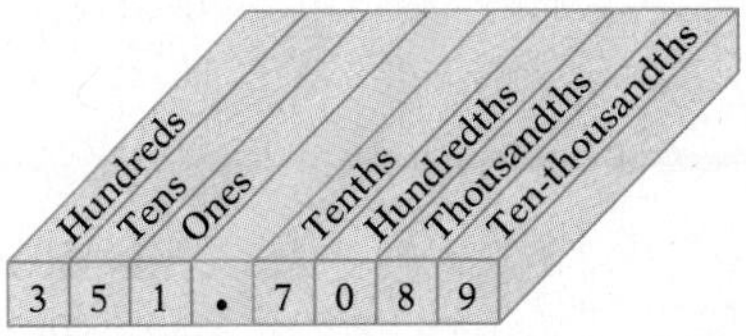


When writing a decimal in words, write the decimal part as if it were a whole number; then name the place value of the last digit.

0.6481 Six thousand four hundred eighty-one ten-thousandths

549.238 Five hundred forty-nine and two hundred thirty-eight thousandths (The decimal point is read as "and.")

To write a decimal in standard form, zeros may have to be inserted after the decimal point so that the last digit is in the given place-value position.

Five and thirty-eight <u>hundredths</u>

8 is in the hundredths' place. 5.3<u>8</u>

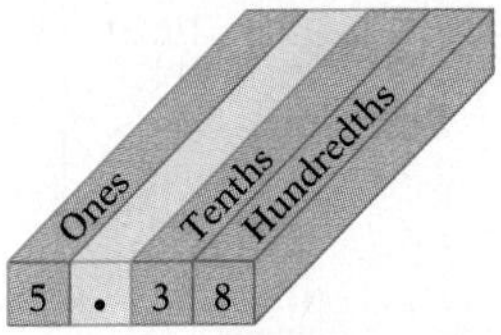


Nineteen and four <u>thousandths</u>

4 is in the thousandths' place. Insert two zeros so that 4 is in the thousandths place. 19.00<u>4</u>

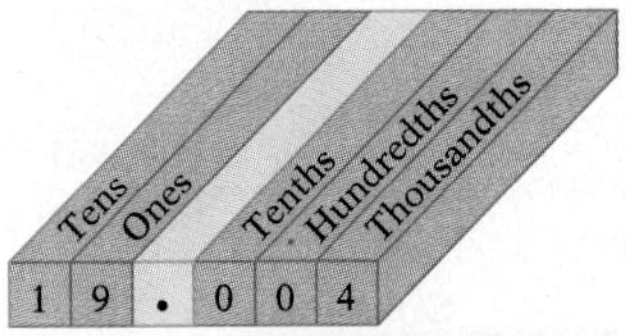


Seventy-one <u>ten-thousandths</u>

1 is in the ten-thousandths' place. Insert two zeros so that 1 is in the ten-thousandths place. 0.007<u>1</u>

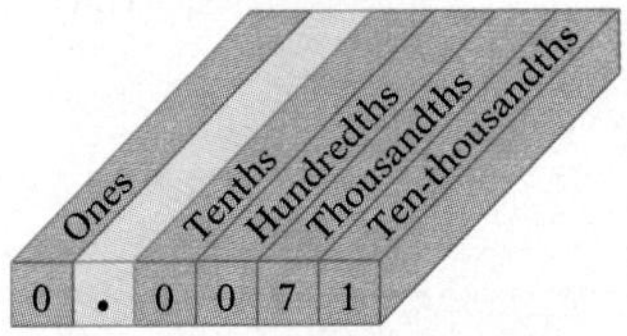

**Example 1** Write 307.4027 in words.

**Solution** Three hundred seven and four thousand twenty-seven ten-thousandths

**You Try It 1** Write 209.05838 in words.

**Your solution**

**Example 2** Write six hundred seven and seven hundred eight hundred-thousandths in standard form.

**Solution** 607.00708

**You Try It 2** Write forty-two thousand and two hundred seven millionths in standard form.

**Your solution**

*Solutions on p. A18*

## *Objective B To round a decimal to a given place value*

Rounding decimals is similar to rounding whole numbers except that the digits to the right of the given place value are dropped instead of being replaced by zeros.

If the digit to the right of the given place value is less than 5, that digit and all digits to the right are dropped. If the digit to the right of the given place value is greater than or equal to 5, increase the given place value by 1 and drop all digits to its right.

➡ Round 26.3799 to the nearest hundredth.

Given place value (7)
26.3799
9 > 5 Increase 7 by 1 and drop all digits to the right of 7.

26.3799 rounded to the nearest hundredth is 26.38.

**Example 3** Round 0.39275 to the nearest ten-thousandth.

**Solution**

Given place value (7)
0.39275
5 = 5

0.3928

**You Try It 3** Round 4.349254 to the nearest hundredth.

**Your solution**

**Example 4** Round 42.0237412 to the nearest hundred-thousandth.

**Solution**

Given place value (4)
42.0237412
1 < 5

42.02374

**You Try It 4** Round 3.290532 to the nearest hundred-thousandth.

**Your solution**

*Solutions on p. A18*

## 3.1 Exercises

### *Objective A*

Write each decimal in words.

**1.** 0.27

**2.** 0.92

**3.** 1.005

**4.** 3.067

**5.** 36.4

**6.** 59.7

**7.** 0.00035

**8.** 0.00092

**9.** 10.007

**10.** 20.009

**11.** 52.00095

**12.** 64.00037

**13.** 0.0293

**14.** 0.0717

**15.** 6.324

**16.** 8.916

**17.** 276.3297

**18.** 418.3115

**19.** 216.0729

**20.** 976.0317

**21.** 4625.0379

**22.** 2986.0925

**23.** 1.00001

**24.** 3.00003

Write each decimal in standard form.

**25.** Seven hundred sixty-two thousandths

**26.** Two hundred ninety-five thousandths

**27.** Sixty-two millionths

**28.** Forty-one millionths

**29.** Eight and three hundred four ten-thousandths

**30.** Four and nine hundred seven ten-thousandths

**31.** Three hundred four and seven hundredths

**32.** Eight hundred ninety-six and four hundred seven thousandths

Write each decimal in standard form.

**33.** Three hundred sixty-two and forty-eight thousandths

**34.** Seven hundred eighty-four and eighty-four thousandths

**35.** Three thousand forty-eight and two thousand two ten-thousandths

**36.** Seven thousand sixty-one and nine thousand one ten-thousandths

### *Objective B*

Round each decimal to the given place value.

**37.** 7.359 Tenths

**38.** 6.405 Tenths

**39.** 23.009 Tenths

**40.** 89.19204 Tenths

**41.** 22.68259 Hundredths

**42.** 16.30963 Hundredths

**43.** 480.325 Hundredths

**44.** 670.974 Hundredths

**45.** 1.03925 Thousandths

**46.** 7.072854 Thousandths

**47.** 1946.3745 Thousandths

**48.** 62.009435 Thousandths

**49.** 0.029876 Ten-thousandths

**50.** 0.012346 Ten-thousandths

**51.** 1.702596 Nearest whole number

**52.** 2.079239 Hundred-thousandths

**53.** 0.0102903 Millionths

**54.** 0.1009754 Millionths

### *APPLYING THE CONCEPTS*

**55.** [W] To what decimal place value are timed events in the Olympics recorded? Provide some specific examples of events and the winning times in each.

**56.** [W] Provide an example of a situation in which a decimal is always rounded up, even if the digit to the right is less than 5. Provide an example of a situation in which a decimal is always rounded down, even if the digit to the right is 5 or greater than 5. (*Hint*: Think about situations in which money changes hands.)

**57.** Indicate which zeros of the number, if any, need not be entered on a calculator.
a. 23.500 b. 0.000235 c. 300.0005 d. 0.004050

**58.** A decimal number was rounded to 6. Between what two numbers, to the nearest tenth, was the number?

**59.** A decimal number was rounded to 10.2. Between what two numbers to the nearest hundredth, was the number?

# 3.2 Addition of Decimals

## Objective A To add decimals

**POINT OF INTEREST**
Simon Stevin (1548–1620) was the first to name decimal numbers. He wrote the number 2.345 as 2 0 3 1 4 2 5 3. He called the whole number part the *commencement,* the tens digit was *prime,* the hundredths digit was *second,* the thousandths digit was *thirds* and so on.

To add decimals, write the numbers so that the decimal points are on a vertical line. Add as for whole numbers, and write the decimal point in the sum directly below the decimal points in the addends.

➡ Add: 0.237 + 4.9 + 27.32

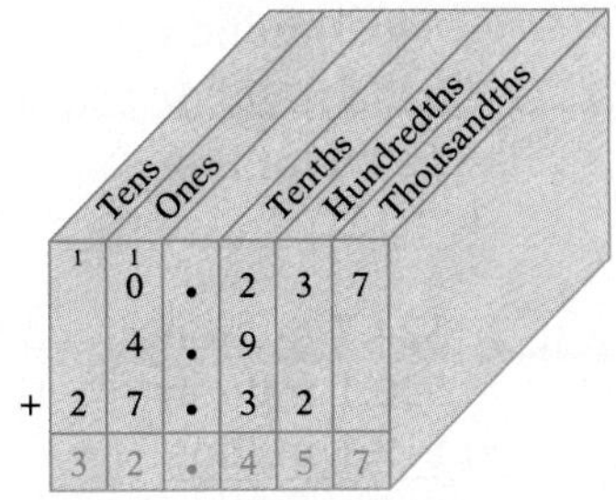


Note that by placing the decimal points on a vertical line, we make sure that digits of the same place value are added.

**Example 1** Find the sum of 42.3, 162.903, and 65.0729.

**Solution**

```
   111
    42.3
   162.903
+   65.0729
   270.2759
```

**You Try It 1** Add: 4.62 + 27.9 + 0.62054

**Your solution**

**Example 2** Add: 0.83 + 7.942 + 15

**Solution**

```
   11
    0.83
    7.942
+  15.
   23.772
```

**You Try It 2** Add: 6.05 + 12 + 0.374

**Your solution**

*Solutions on p. A18*

**ESTIMATION**

**Estimating the Sum of Two or More Decimals**

Estimate and then use your calculator to find 23.037 + 16.7892.

To estimate the sum of two or more numbers, round each number to the same place value. In this case, we will round to the nearest whole number. Then add. The estimated answer is 40.

$$\begin{array}{rcr} 23.037 & \approx & 23 \\ +\ 16.7892 & \approx & +\ 17 \\ \hline & & 40 \end{array}$$

Now use your calculator to find the exact result. The exact answer is 39.8262.

23.037 [+] 16.7892 [=] 39.8262

## *Objective B To solve application problems*

The table at the the right lists the team payrolls, in millions of dollars, of the professional football teams.

**Comparing Team Payrolls**

Payrolls computed to nearest $100,000

| Team | 1993 | 1992 | Change |
|---|---|---|---|
| San Francisco 49ers | $41.4 mil. | $34.3 mil. | +$ 7.1 |
| Washington Redskins | $40.3 | $31.7 | +$ 8.6 |
| Buffalo Bills | $39.1 | $29.9 | +$ 9.2 |
| New York Giants | $38.5 | $31.5 | +$ 7.0 |
| New Orleans Saints | $36.3 | $29.5 | +$ 6.8 |
| Green Bay Packers | $35.3 | $28.3 | +$ 7.0 |
| Phoenix Cardinals | $35.2 | $22.3 | +$12.9 |
| Kansas City Chiefs | $35.1 | $24.1 | +$11.0 |
| New York Jets | $34.9 | $24.8 | +$10.1 |
| Cleveland Browns | $34.4 | $27.2 | +$ 7.2 |
| Miami Dolphins | $34.2 | $33.9 | +$ 0.3 |
| Atlanta Falcons | $33.9 | $31.6 | +$ 2.3 |
| Los Angeles Raiders | $33.3 | $27.9 | +$ 5.4 |
| Indianapolis Colts | $33.2 | $32.8 | +$ 0.4 |
| Chicago Bears | $31.9 | $23.9 | +$ 8.0 |
| Detroit Lions | $30.7 | $24.9 | +$ 5.8 |
| Denver Broncos | $30.4 | $23.9 | +$ 6.5 |
| Houston Oilers | $29.6 | $26.7 | +$ 2.9 |
| Los Angeles Rams | $28.4 | $26.1 | +$ 2.3 |
| Tampa Bay Buccaneers | $28.2 | $19.7 | +$ 8.5 |
| Minnesota Vikings | $26.7 | $26.7 | even |
| San Diego Chargers | $26.2 | $25.1 | +$ 1.1 |
| Philadelphia Eagles | $26.2 | $27.1 | $–0.9 |
| Pittsburgh Steelers | $26.1 | $17.6 | +$ 8.5 |
| Seattle Seahawks | $25.7 | $24.7 | +$ 1.0 |
| Dallas Cowboys | $24.7 | $25.2 | $–0.5 |
| New England Patriots | $20.1 | $21.9 | $–1.8 |
| Cincinnati Bengals | $19.1 | $28.2 | $–9.1 |

Find the combined payroll for the three highest-payroll teams and the three lowest-payroll teams in 1993.

$$\begin{array}{r} \$41.4 \text{ million} \\ \$40.3 \text{ million} \\ +\ \$39.1 \text{ million} \\ \hline \$120.8 \text{ million} \end{array}$$

Add the payrolls of the San Francisco 49ers, the Washington Redskins, and the Buffalo Bills. The combined payroll for the three highest payroll teams was $120.8 million.

$$\begin{array}{r} \$24.7 \text{ million} \\ \$20.1 \text{ million} \\ +\ \$19.1 \text{ million} \\ \hline \$63.9 \text{ million} \end{array}$$

Add the payrolls of the Dallas Cowboys, the New England Patriots, and the Cincinnati Bengals. The combined payroll for the three lowest-payroll teams was $63.9 million.

**Example 3**

Dan Burhoe earned a salary of $138.50 for working 5 days this week as a food server. He also received $22.92, $15.80, $19.65, $39.20, and $27.70 in tips during the 5 days. Find the total income for the week.

**Strategy**

To find the total income, add the tips ($22.92, $15.80, $19.65, $39.20, and $27.70) to the salary ($138.50).

**Solution**

$$\begin{array}{r} \$138.50 \\ 22.92 \\ 15.80 \\ 19.65 \\ 39.20 \\ +\quad 27.70 \\ \hline \$263.77 \end{array}$$

Dan's total income for the week was $263.77.

**You Try It 3**

Anita Khavari, an insurance executive, earns a salary of $425 every four weeks. During the past 4-week period, she received commissions of $485.60, $599.46, $326.75, and $725.42. Find her total income for the past 4-week period.

**Your strategy**

**Your solution**

*Solution on p. A19*

## 3.2 Exercises

### *Objective A*

Add.

**1.** 16.008 + 2.0385 + 132.06

**2.** 17.32 + 1.0579 + 16.5

**3.** 1.792 + 67 + 27.0526

**4.** 8.772 + 1.09 + 26.5027

**5.** 3.02 + 62.7 + 3.924

**6.** 9.06 + 4.976 + 59.6

**7.** 82.006 + 9.95 + 0.927

**8.** 0.826 + 8.76 + 79.005

**9.** 4.307 + 99.82 + 9.078

**10.** 0.3 + 0.07

**11.** 0.29 + 0.4

**12.** 1.007 + 2.1

**13.** 7.3 + 9.005

**14.** 4.9257 + 27.05 + 9.0063

**15.** 8.72 + 99.073 + 2.9736

**16.** 62.4 + 9.827 + 692.44

**17.** 8 + 89.43 + 7.0659

Estimate by rounding to the nearest whole number. Then use your calculator to add.

**18.** 342.42 + 89.625 + 176.2

**19.** 219.9 + 0.872 + 13.42

**20.** 823.9 + 82.65 + 46.923

**21.** 678.92 + 97.6 + 5.423

### *Objective B* Application Problems

**22.** A family receives an electric bill of $70.42, a gas bill for $54.55, and a garbage collection bill of $23.14. Find the total bill for the three services.

**23.** Find the length of the shaft.

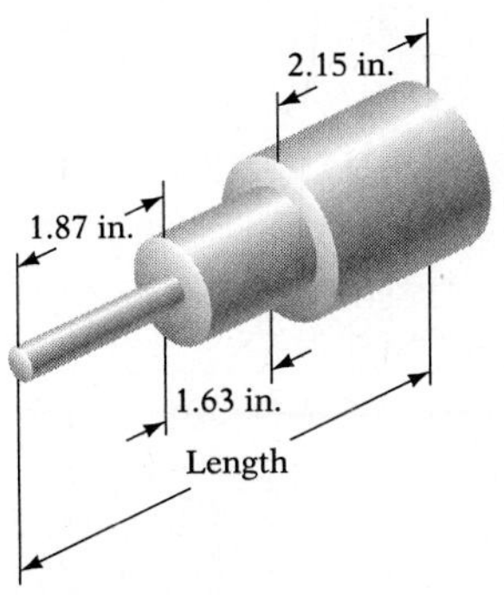


**24.** Find the length of the shaft.

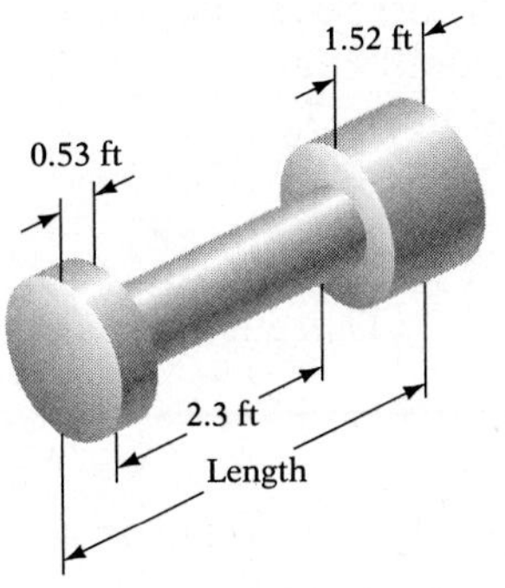

**25.** Commuting, Mae Chan used 12.4 gallons of gas the first week, 9.8 gallons the second week, 15.2 gallons the third week, and 10.4 gallons the fourth week. Find the total amount of gas she used during the 4 weeks.

**26.** The odometer on a family's car reads 24,835.9 miles. The car was driven 8.2 miles on Friday, 82.6 miles on Saturday, and 133.6 miles on Sunday.
a. How many miles was the car driven during the three days?
b. Find the odometer reading at the end of the three days.

**27.** You have $2143.57 in your checking account. You make deposits of $210.98, $45.32, $1236.34, and $27.99. Find the amount in your checking account after you have made the deposits if no money has been withdrawn.

The figure at the right shows the number of viewers who watch television each night of the week. Use this figure for Exercises 28–30.

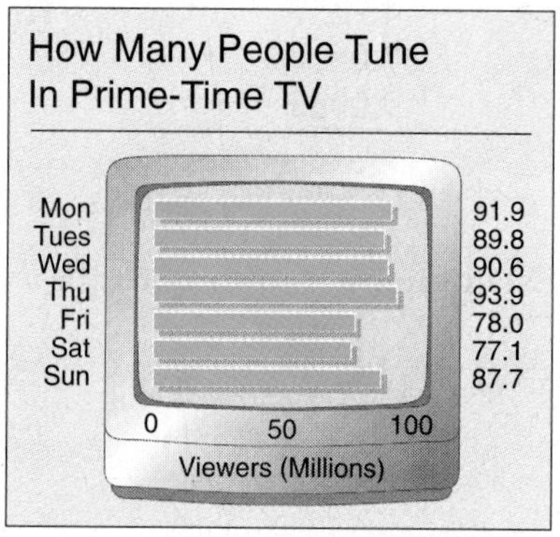


**28.** Find the total number of television viewers for Friday, Saturday, and Sunday nights.

**29.** Find the total number of television viewers for Monday, Tuesday, Wednesday, and Thursday nights.

**30.** Find the total number of television viewers for the week.

## *APPLYING THE CONCEPTS*

The table at the right gives the prices for selected products in a grocery store. Use this table for Exercises 31 and 32.

| *Product* | *Cost* |
|---|---|
| Raisin Bran | 3.45 |
| Butter | 2.69 |
| Bread | 1.23 |
| Popcorn | .89 |
| Potatoes | 1.09 |
| Cola (6-pack) | .98 |
| Mayonnaise | 2.25 |
| Lunch meat | 3.31 |
| Milk | 2.18 |
| Toothpaste | 2.45 |

**31.** Does a customer with $10 have enough money to purchase raisin bran, bread, milk, lunch meat, and butter?

**32.** Name three items that would cost more than $6 but less than $7. (There is more than one answer.)

**33.** Can a piece of rope 4 feet long be wrapped around the box shown at the right?

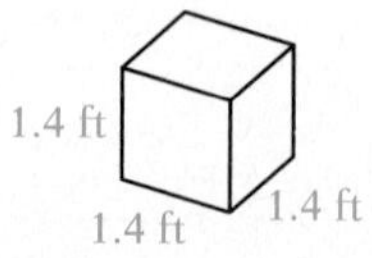

# 3.3 Subtraction of Decimals

## *Objective A* *To subtract decimals*

To subtract decimals, write the numbers so that the decimal points are on a vertical line. Subtract as for whole numbers, and write the decimal point in the difference directly below the decimal point in the subtrahend.

➡ Subtract 21.532 − 9.875 and check.

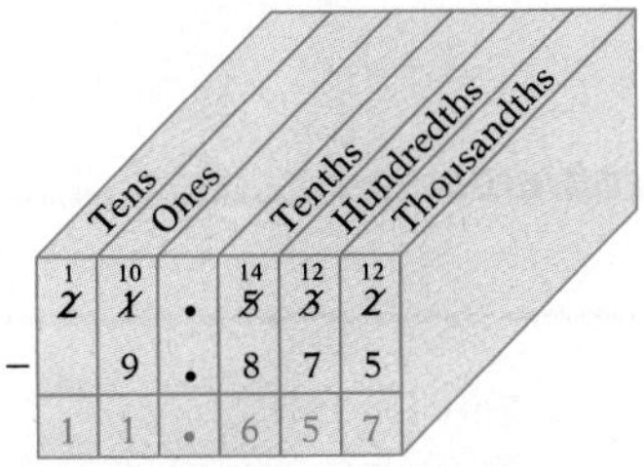


Placing the decimal points on a vertical line ensures that digits of the same place value are subtracted.

| | | |
|---|---|---|
| Check: | Subtrahend | 9.875 |
| | + Difference | + 11.657 |
| | = Minuend | 21.532 |

➡ Subtract 4.3 − 1.7942 and check.

$$\begin{array}{r} \overset{3}{\not4}.\overset{12}{\not3}\overset{9}{\not0}\overset{9}{\not0}\overset{10}{\not0} \\ -\,1.7942 \\ \hline 2.5058 \end{array}$$

If necessary, insert zeros in the minuend before subtracting.

Check:

$$\begin{array}{r} 1.7942 \\ +\,2.5058 \\ \hline 4.3000 \end{array}$$

**Example 1** Subtract 39.047 − 7.96 and check.

**Solution**

$$\begin{array}{r} 3\overset{8}{\not9}.\overset{9}{\not0}\overset{14}{\not4}7 \\ -\;\;7.96\phantom{0} \\ \hline 31.087 \end{array}$$

Check:

$$\begin{array}{r} 7.96 \\ +\,31.087 \\ \hline 39.047 \end{array}$$

**You Try It 1** Subtract 72.039 − 8.47 and check.

**Your solution**

**Example 2** Find 9.23 less than 29 and check.

**Solution**

$$\begin{array}{r} \overset{1}{\not2}\overset{18}{\not9}.\overset{9}{\not0}\overset{10}{\not0} \\ -\;\;9.23 \\ \hline 19.77 \end{array}$$

Check:

$$\begin{array}{r} 9.23 \\ +\,19.77 \\ \hline 29.00 \end{array}$$

**You Try It 2** Subtract 35 − 9.67 and check.

**Your solution**

**Example 3** Subtract 1.2 − 0.8235 and check.

**Solution**

$$\begin{array}{r} \overset{0}{\not1}.\overset{11}{\not2}\overset{9}{\not0}\overset{9}{\not0}\overset{10}{\not0} \\ -\,0.8235 \\ \hline 0.3765 \end{array}$$

Check:

$$\begin{array}{r} 0.8235 \\ +\,0.3765 \\ \hline 1.2000 \end{array}$$

**You Try It 3** Subtract 3.7 − 1.9715 and check.

**Your solution**

*Solutions on p. A19*

## ESTIMATION

### Estimating the Difference Between Two Decimals

Estimate and then use your calculator to find 820.2306 − 475.74815.

To estimate the difference between two numbers, round each number to the same place value. In this case we will round to the nearest ten. Then subtract. The estimated answer is 340.

$$\begin{array}{rcr} 820.2306 & \approx & 820 \\ -\ 475.74815 & \approx & -\ 480 \\ \hline & & 340 \end{array}$$

Now use your calculator to find the exact result. The exact answer is 344.48245.

820.2306 [−] 475.74815 [=] 344.48245

## Objective B To solve application problems

**Example 4**
You bought a book for $15.87. How much change did you receive from a $20.00 bill?

**Strategy**
To find the amount of change, subtract the cost of the book ($15.87) from $20.00.

**Solution**

$$\begin{array}{r} \$20.00 \\ -\ 15.87 \\ \hline \$\ 4.13 \end{array}$$

You receive $4.13 in change.

**You Try It 4**
Your breakfast cost $3.85. How much change did you receive from a $5.00 bill?

**Your strategy**

**Your solution**

**Example 5**
You had a balance of $62.41 in your checking account. You then bought a cassette for $8.95, film for $3.17, and a skateboard for $39.77. After paying for these items with a check, how much do you have left in your checking account?

**Strategy**
To find the new balance:

- Find the total cost of the three items ($8.95 + $3.17 + $39.77).
- Subtract the total cost from the old balance ($62.41).

**Solution**

$$\begin{array}{r} \$\ 8.95 \\ 3.17 \\ +\ 39.77 \\ \hline \$51.89 \end{array} \text{ total cost} \qquad \begin{array}{r} \$62.41 \\ -\ 51.89 \\ \hline \$10.52 \end{array}$$

The new balance is $10.52

**You Try It 5**
You had a balance of $2472.69 in your checking account. You then wrote checks for $1025.60, $79.85, and $162.47. Find the new balance in your checking account.

**Your strategy**

**Your solution**

*Solutions on p. A19*

## 3.3 Exercises

*Objective A*

Subtract and check.

1. 24.037 − 18.41

2. 26.029 − 19.31

3. 123.07 − 9.4273

4. 214 − 7.143

5. 16.5 − 9.7902

6. 13.2 − 8.6205

7. 235.79 − 20.093

8. 463.27 − 40.095

9. 63.005 − 9.1274

10. 23.004 − 7.2175

11. 92 − 19.2909

12. 41.2405 − 25.2709

13. 7.01 − 2.325

14. 8.07 − 5.392

15. 19.0035 − 8.967

16. 0.32 − 0.0058

17. 0.78 − 0.0073

18. 3.005 − 1.982

19. 6.007 − 2.734

20. 352.16 − 90.994

21. 872 − 80.753

22. 724.32 − 69

23. 625.46 − 77.509

24. 362.394 − 19.4672

25. 421.385 − 17.5293

26. 19 − 10.372

27. 23.4 − 0.921

Estimate by rounding to the nearest ten. Then use your calculator to subtract.

28. 620.59 − 132.79

29. 835.07 − 244.82

30. 67.3 − 19.793

31. 84.1 − 48.906

Estimate by rounding to the nearest whole number (nearest one). Then use your calculator to subtract.

32. 93.079256 − 66.09249

33. 3.7529 − 1.00784

34. 76.53902 − 45.73005

35. 9.07325 − 1.924

**Objective B** *Application Problems*

**36.** A patient has a fever of 102.3°F. Normal temperature is 98.6°F. How many degrees above normal is the patient's temperature?

**37.** The manager of the Edgewater Cafe takes a reading of the cash register tape each hour. At 1:00 P.M. the tape read $967.54; at 2:00 P.M. the tape read $1437.15. Find the amount of sales between 1:00 P.M. and 2:00 P.M.

**38.** Find the missing dimension.

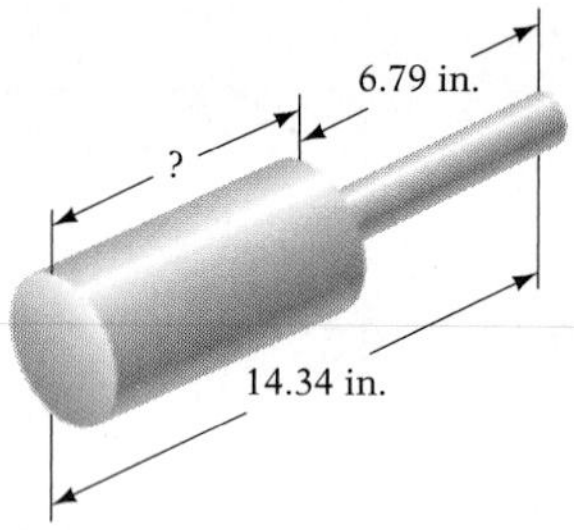


**39.** Find the missing dimension.

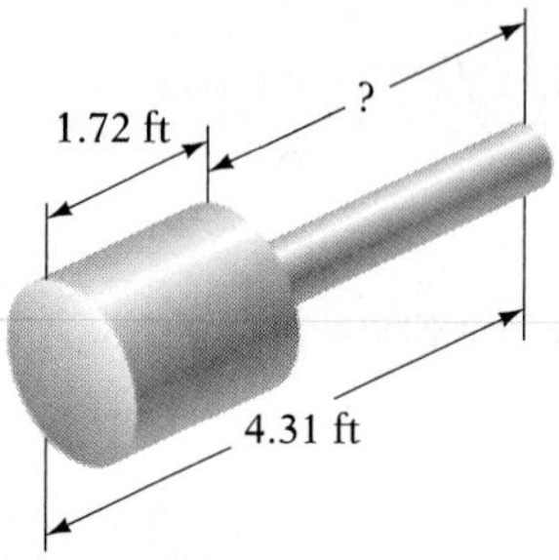


**40.** You had a balance of $1,029.74 in your checking account. You then wrote checks for $67.92, $43.10, and $496.34.
a. Find the total amount of the checks written.
b. Find the new balance in your checking account.

**41.** The price of gasoline is $1.22 per gallon after the price rose $.07 one month and $.12 the second month. Find the price of gasoline before these increases in price.

**42.** Rainfall for the last 3 months of the year was 1.42 inches, 5.39 inches, and 3.55 inches. The normal rainfall for the last 3 months of the year is 11.22 inches. How many inches below normal was the rainfall?

The table at the right shows the amount that a wage earner would have to earn to have the buying power of a $15,000 salary in 1977. Use the table for Exercises 43 and 44.

**Keeping Up With Inflation**

How much a person would have to earn each year to have the same buying power that a $15,000 salary had in June 1977.

| | | | |
|---|---|---|---|
| 1978 | $16,729.81 | 1986 | $27,306.25 |
| 1979 | $18,953.85 | 1987 | $28,517.09 |
| 1980 | $21,326.16 | 1988 | $29,777.38 |
| 1981 | $23,228.94 | 1989 | $31,161.22 |
| 1982 | $24,118.54 | 1990 | $33,063.99 |
| 1983 | $25,032.85 | 1991 | $34,077.14 |
| 1984 | $26,021.29 | 1992 | $35,085.58 |
| 1985 | $27,009.74 | 1993[1] | $35,683.30 |

1–first six months

**43.** How much more earnings would a wage earner need in 1992 to equal the buying power of a $15,000 wage in 1977?

**44.** How much more earnings would a wage earner need in 1980 to equal the buying power of a $15,000 wage in 1977?

*APPLYING THE CONCEPTS*

**45.** Find the largest amount by which the estimate of the sum of two decimals with tenths, hundredths, and thousandths places could differ from the exact sum.

**46.** Grace Herrera owned 453.472 shares of a mutual fund on January 1, 1994. On December 31, 1994, she had 617.005 shares. What was the increase in the number of shares for the year?

**47.** The average speed for the winner of the 1991 Indianapolis 500 was 176.460 mph. The average speed for the winner in 1992 was 134.477 mph. What is the difference between the average speeds for those two years?

# 3.4 Multiplication of Decimals

## Objective A To multiply decimals

**POINT OF INTEREST**
Benjamin Banneker (1731–1806) was the first African American to earn distinction as a mathematician and scientist. He was on the survey team that determined the boundaries of Washington, D.C. The mathematics of surveying requires extensive use of decimals.

Decimals are multiplied as if they were whole numbers; then the decimal point is placed in the product. Writing the decimals as fractions shows where to write the decimal point in the product.

$$0.\underline{3} \times 5 = \frac{3}{10} \times \frac{5}{1} = \frac{15}{10} = 1.\underline{5}$$

1 decimal place — 1 decimal place

$$0.\underline{3} \times 0.\underline{5} = \frac{3}{10} \times \frac{5}{10} = \frac{15}{100} = 0.\underline{15}$$

1 decimal place — 1 decimal place — 2 decimal places

$$0.\underline{3} \times 0.\underline{05} = \frac{3}{10} \times \frac{5}{100} = \frac{15}{1000} = 0.\underline{015}$$

1 decimal place — 2 decimal places — 3 decimal places

To multiply decimals, multiply the numbers as in whole numbers. Write the decimal point in the product so that the number of decimal places in the product is the sum of the decimal places in the factors.

➡ Multiply: 21.4 × 0.36

| | |
|---|---|
| 21.4 | 1 decimal place |
| × 0.36 | 2 decimal places |
| 1284 | |
| 642 | |
| 7.704 | 3 decimal places |

➡ Multiply: 0.037 × 0.08

| | |
|---|---|
| 0.037 | 3 decimal places |
| × 0.08 | 2 decimal places |
| 0.00296 | 5 decimal places |

• Two zeros must be inserted between the 2 and the decimal point so that there are 5 decimal places in the product.

To multiply a decimal by a power of 10 (10, 100, 1000,...), move the decimal point to the right the same number of places as there are zeros in the power of 10.

| | |
|---|---|
| 3.8925 × 10 | = 38.925 |
| 1 zero | 1 decimal place |
| 3.8925 × 100 | = 389.25 |
| 2 zeros | 2 decimal places |
| 3.8925 × 1000 | = 3892.5 |
| 3 zeros | 3 decimal places |
| 3.8925 × 10,000 | = 38,925. |
| 4 zeros | 4 decimal places |
| 3.8925 × 100,000 | = 389,250. |
| 5 zeros | 5 decimal places |

Note that a zero must be inserted before the decimal point.

Note that if the power of 10 is written in exponential notation, the exponent indicates how many places to move the decimal point.

$3.8925 \times 10^1 = 38.925$
1 decimal place

$3.8925 \times 10^2 = 389.25$
2 decimal places

$3.8925 \times 10^3 = 3892.5$
3 decimal places

$3.8925 \times 10^4 = 38{,}925.$
4 decimal places

$3.8925 \times 10^5 = 389{,}250.$
5 decimal places

**Example 1** Multiply: $920 \times 3.7$

**Solution**

```
  920
× 3.7
 644 0
2760
3404.0
```

**You Try It 1** Multiply: $870 \times 4.6$

**Your solution**

**Example 2** Find 0.00079 multiplied by 0.025.

**Solution**

```
   0.00079
×    0.025
       395
      158
0.00001975
```

**You Try It 2** Find 0.000086 multiplied by 0.057.

**Your solution**

**Example 3** Find the product of 3.69 and 2.07.

**Solution**

```
  3.69
× 2.07
  2583
 7380
7.6383
```

**You Try It 3** Find the product of 4.68 and 6.03.

**Your solution**

**Example 4** Multiply: $42.07 \times 10{,}000$

**Solution** $42.07 \times 10{,}000 = 420{,}700$

**You Try It 4** Multiply: $6.9 \times 1000$

**Your solution**

**Example 5** Find 3.01 times $10^3$.

**Solution** $3.01 \times 10^3 = 3010$

**You Try It 5** Find 4.0273 times $10^2$.

**Your solution**

*Solutions on p. A20*

**ESTIMATION**

### Estimating the Product of Two Decimals

Estimate and then use your calculator to find 28.259 × 0.029.

To estimate a product, round each number so that there is one non-zero digit. Then multiply.

$$\begin{array}{rcr} 28.259 & \approx & 30 \\ \times\ 0.029 & \approx & \times\ 0.03 \\ \hline & & 0.90 \end{array}$$

The estimated answer is 0.90.

Now use your calculator to find the exact answer.

28.259 [×] 0.029 [=] 0.819511

The exact answer is 0.819511.

## *Objective B* *To solve application problems*

The table below lists water rates and meter fees for a city. This table is used for Example 6 and You Try It 6.

| *Water Charges* | |
|---|---|
| Commercial | $1.39/1,000 gal |
| Comm Restaurant | $1.39/1,000 gal |
| Industrial | $1.39/1,000 gal |
| Institutional | $1.39/1,000 gal |
| Res—No Sewer | |
| Residential—SF | |
| >0 <200 gal. per day | $1.15/1,000 gal |
| >200 <1,500 gal. per day | $1.39/1,000 gal |
| >1,500 gal. per day | $1.54/1,000 gal |

| *Meter Charges* | |
|---|---|
| *Meter* | *Meter Fee* |
| 5/8″ & 3/4″ | $13.50 |
| 1″ | $21.80 |
| 1-1/2″ | $42.50 |
| 2″ | $67.20 |
| 3″ | $133.70 |
| 4″ | $208.20 |
| 6″ | $415.10 |
| 8″ | $663.70 |

**Example 6**
Find the total bill for an industrial water user with a 6-inch meter that uses 152,000 gallons of water for July and August.

**Strategy**
To find the cost of water:
- Find the cost of water by multiplying the cost per 1000 gallons (1.39) by the number of 1000-gallon units.
- Add the cost of the water to the meter fee ($415.10).

**Solution**

$$\text{Cost of water} = \frac{152{,}000}{1{,}000} \cdot 1.39 = 211.28$$

$$\text{Total cost} = 211.28 + 415.10 = 626.38$$

The total cost is $626.38.

**You Try It 6**
Find the total bill for a commercial user that used 5000 gallons of water per day for July and August. The user has a 3-inch meter.

**Your strategy**

**Your solution**

*Solution on p. A20*

**Example 7**

It costs \$.036 an hour to operate an electric motor. How much does it cost to operate the motor for 120 hours?

**Strategy**

To find the cost of running the motor for 120 hours, multiply the hourly cost (\$.036) by the number of hours the motor is run (120).

**Solution**

$$\begin{array}{r} \$.036 \\ \times\ 120 \\ \hline 720 \\ 36\phantom{0} \\ \hline \$4.320 \end{array}$$

The cost of running the motor for 120 hours is \$4.32.

**You Try It 7**

The cost of electricity to run a freezer for 1 hour is \$.035. This month the freezer has run for 210 hours. Find the total cost of running the freezer this month.

**Your strategy**

**Your solution**

**Example 8**

Jason Ng earns a salary of \$280 for a 40-hour work week. This week he worked 12 hours of overtime at a rate of \$10.50 for each hour of overtime worked. Find his total income for the week.

**Strategy**

To find Jason's total income for the week:

- Find the overtime pay by multiplying the hourly overtime rate (\$10.50) by the number of hours of overtime worked (12).
- Add the overtime pay to the weekly salary (\$280).

**Solution**

$$\begin{array}{r} \$10.50 \\ \times\quad 12 \\ \hline 21\ 00 \\ 105\ 0\phantom{0} \\ \hline \$126.00 \end{array} \text{ overtime pay} \qquad \begin{array}{r} \$280.00 \\ +\ 126.00 \\ \hline \$406.00 \end{array}$$

Jason's total income for this week is \$406.00.

**You Try It 8**

You make a down payment of \$175 on a stereo and agree to make payments of \$37.18 a month for the next 18 months to repay the remaining balance. Find the total cost of the stereo.

**Your strategy**

**Your solution**

*Solutions on p. A21*

## 3.4 Exercises

### *Objective A*

Multiply.

| | | | | |
|---|---|---|---|---|
| **1.** $0.9 \times 0.4$ | **2.** $0.7 \times 0.9$ | **3.** $0.5 \times 0.6$ | **4.** $0.3 \times 0.7$ | **5.** $0.5 \times 0.5$ |
| **6.** $0.7 \times 0.7$ | **7.** $0.9 \times 0.5$ | **8.** $0.2 \times 0.6$ | **9.** $7.7 \times 0.9$ | **10.** $3.4 \times 0.4$ |
| **11.** $9.2 \times 0.2$ | **12.** $2.6 \times 0.7$ | **13.** $7.2 \times 0.6$ | **14.** $6.8 \times 0.4$ | **15.** $7.4 \times 0.1$ |
| **16.** $3.8 \times 0.1$ | **17.** $7.9 \times 5$ | **18.** $9.3 \times 7$ | **19.** $0.68 \times 4$ | **20.** $0.83 \times 9$ |
| **21.** $0.67 \times 0.9$ | **22.** $0.84 \times 0.3$ | **23.** $0.16 \times 0.6$ | **24.** $0.47 \times 0.8$ | **25.** $2.5 \times 5.4$ |
| **26.** $3.9 \times 1.9$ | **27.** $8.4 \times 9.5$ | **28.** $7.6 \times 5.8$ | **29.** $0.83 \times 5.2$ | **30.** $0.24 \times 2.7$ |
| **31.** $0.46 \times 3.9$ | **32.** $0.78 \times 6.8$ | **33.** $0.2 \times 0.3$ | **34.** $0.3 \times 0.3$ | **35.** $0.24 \times 0.3$ |
| **36.** $0.17 \times 0.5$ | **37.** $1.47 \times .09$ | **38.** $6.37 \times 0.05$ | **39.** $8.92 \times 0.004$ | **40.** $6.75 \times 0.007$ |

Multiply.

**41.** $0.49 \times 0.16$

**42.** $0.38 \times 0.21$

**43.** $7.6 \times 0.01$

**44.** $5.1 \times 0.01$

**45.** $8.62 \times 4$

**46.** $5.83 \times 7$

**47.** $64.5 \times 9$

**48.** $37.8 \times 8$

**49.** $2.19 \times 9.2$

**50.** $1.25 \times 5.6$

**51.** $1.85 \times 0.023$

**52.** $37.8 \times 0.052$

**53.** $0.478 \times 0.37$

**54.** $0.526 \times 0.22$

**55.** $48.3 \times 0.0041$

**56.** $67.2 \times 0.0086$

**57.** $2.437 \times 6.1$

**58.** $4.237 \times 0.54$

**59.** $0.413 \times 0.0016$

**60.** $0.517 \times 0.0029$

**61.** $94.73 \times 0.57$

**62.** $89.23 \times 0.62$

**63.** $8.005 \times 0.067$

**64.** $9.032 \times 0.019$

**65.** $4.29 \times 0.1$

**66.** $6.78 \times 0.1$

**67.** $5.29 \times 0.4$

**68.** $6.78 \times 0.5$

**69.** $0.68 \times 0.7$

**70.** $0.56 \times 0.9$

**71.** $1.4 \times 0.73$

**72.** $6.3 \times 0.37$

**73.** $5.2 \times 7.3$

**74.** $7.4 \times 2.9$

**75.** $3.8 \times 0.61$

**76.** $7.2 \times 0.72$

**77.** $0.32 \times 10$

**78.** $6.93 \times 10$

**79.** $0.065 \times 100$

**80.** $0.039 \times 100$

**81.** $6.2856 \times 1000$

Multiply.

**82.** $3.2954 \times 1000$

**83.** $3.2 \times 1000$

**84.** $0.006 \times 10{,}000$

**85.** $3.57 \times 10{,}000$

**86.** $8.52 \times 10^1$

**87.** $0.63 \times 10^1$

**88.** $82.9 \times 10^2$

**89.** $0.039 \times 10^2$

**90.** $6.8 \times 10^3$

**91.** $4.9 \times 10^4$

**92.** $6.83 \times 10^4$

**93.** $0.067 \times 10^2$

**94.** $0.052 \times 10^2$

**95.** Find the product of 0.0035 and 3.45.

**96.** Find the product of 237 and 0.34.

**97.** Multiply 3.005 by 0.00392.

**98.** Multiply 20.34 by 1.008.

**99.** Multiply 1.348 by 0.23.

**100.** Multiply 0.000358 by 3.56.

**101.** Find the product of 23.67 and 0.0035.

**102.** Find the product of 0.00346 and 23.1.

**103.** Find the product of 5, 0.45, and 2.3.

**104.** Find the product of 0.03, 23, and 9.45.

Estimate and then use your calculator to multiply.

105. $28.5 \times 3.2$

106. $86.3 \times 4.4$

107. $2.38 \times 0.44$

108. $9.82 \times 0.77$

109. $0.866 \times 4.5$

110. $0.239 \times 8.2$

111. $4.34 \times 2.59$

112. $6.87 \times 9.98$

113. $8.434 \times 0.044$

114. $7.037 \times 0.094$

115. $28.44 \times 1.12$

116. $86.57 \times 7.33$

117. $49.6854 \times 39.0672$

118. $2.00547 \times 9.672$

119. $0.00456 \times 0.009542$

120. $7.00637 \times .0128$

***Objective B*** *Application Problems*

**121.** It costs $.18 per mile to rent a car. Find the cost to rent a car that is driven 114 miles.

**122.** An electric motor costing $315.45 has an operating cost of $.027 for 1 hour of operation. Find the cost to run the motor for 56 hours (Round to the nearest cent.)

**123.** Four hundred empty soft drink cans weigh 18.75 pounds. A recycling center pays $.75 per pound for the cans. Find the amount received for the 400 cans (round to the nearest cent).

**124.** A recycling center pays $.035 per pound for newspapers. Find the amount received from recycling 420 pounds of newspapers.

**125.** A broker's fee for buying stock is 0.045 times the price of the stock. An investor bought 100 shares of stock at $38.50 per share. Find the broker's fee.

**126.** A broker's fee for buying stock is 0.052 times the price of the stock. You bought 100 shares of stock at $62.75 per share. Find the broker's fee (round to the nearest cent).

**127.** You bought a car for $2000 down and made payments of $127.50 each month for 36 months.
a. Find the amount of the payments over the 36 months.
b. Find the total cost of the car.

**128.** As a nurse, Rob Martinez earns a salary of $344 for a 40-hour work week. This week he worked 15 hours of overtime at a rate of $12.90 for each hour of overtime worked.
a. Find the amount of overtime pay.
b. Find Rob's total income for the week.

**129.** Bay Area Rental Cars charges $12 a day and $.12 per mile for renting a car. You rented a car for 3 days and drove 235 miles. Find the total cost of renting the car.

**130.** A taxi costs $1.50 and $0.20 for each $\frac{1}{8}$ mile driven. Find the cost of hiring a taxi to get from the airport to the hotel—a distance of 5.5 miles.

## APPLYING THE CONCEPTS

**131.** Anthony Schmidt owns 236.147 shares of a mutual fund. The current value of each share is \$8.67. What is the value of the shares?

**132.** Show how the decimal is placed in the product of 1.3 × 2.31 by first writing each number as a fraction and then multiplying. Now change back to decimal notation.

The table at the right lists three pieces of steel required for a repair project. Use this table for Exercises 133 and 134.

| *Grade of Steel* | *Weight (Pounds per Foot)* | *Required Number of Feet* | *Cost per Pound* |
|---|---|---|---|
| 1 | 2.2 | 8 | \$1.20 |
| 2 | 3.4 | 6.5 | 1.35 |
| 3 | 6.75 | 15.4 | 1.94 |

**133.** Find the total cost of each of the grades of steel.

**134.** Find the total cost of the three pieces of steel.

A confectioner ships holiday packs of candy and nuts anywhere in the United States. Following is a price list of nuts and candy and also a table of shipping charges to zones in the United States. For any fraction of a pound, use the next higher weight. Use these tables for Exercise 135.

| *Code* | *Description* | *Price* |
|---|---|---|
| 112 | Almonds 16 oz | 4.75 |
| 116 | Cashews 8 oz | 2.90 |
| 117 | Cashews 16 oz | 5.50 |
| 130 | Macadamias 7 oz | 5.25 |
| 131 | Macadamias 16 oz | 9.95 |
| 149 | Pecan halves 8 oz | 6.25 |
| 155 | Mixed nuts 10 oz | 4.80 |
| 160 | Cashew brittle 8 oz | 1.95 |
| 182 | Pecan roll 8 oz | 3.70 |
| 199 | Chocolate peanuts 8 oz | 1.90 |

| *Pounds* | *Zone 1* | *Zone 2* | *Zone 3* | *Zone 4* |
|---|---|---|---|---|
| 1–3 | 6.55 | 6.85 | 7.25 | 7.75 |
| 4–6 | 7.10 | 7.40 | 7.80 | 8.30 |
| 7–9 | 7.50 | 7.80 | 8.20 | 8.70 |
| 10–12 | 7.90 | 8.20 | 8.60 | 9.10 |

**135.** Find the cost of sending the following orders to the given mail zone.

a.

| *Code* | *Quantity* |
|---|---|
| 116 | 2 |
| 130 | 1 |
| 149 | 3 |
| 182 | 4 |

Mail to zone 4.

b.

| *Code* | *Quantity* |
|---|---|
| 112 | 1 |
| 117 | 4 |
| 131 | 2 |
| 160 | 3 |
| 182 | 5 |

Mail to zone 3.

c.

| *Code* | *Quantity* |
|---|---|
| 117 | 3 |
| 131 | 1 |
| 155 | 2 |
| 160 | 4 |
| 182 | 1 |
| 199 | 3 |

Mail to zone 2.

Chris works at B & W Garage as an auto mechanic and has just completed an engine overhaul for a customer. To determine the cost of the repair job, Chris keeps a list of times worked and parts used. A parts list and a list of the times worked are shown below. Use this table for Exercises 136–139.

*Parts Used*

| *Item* | *Quantity* |
|---|---|
| Gasket set | 1 |
| Ring set | 1 |
| Valves | 8 |
| Wrist pins | 8 |
| Valve springs | 16 |
| Rod bearings | 8 |
| Main bearings | 5 |
| Valve seals | 16 |
| Timing chain | 1 |

*Time Spent*

| *Day* | *Hours* |
|---|---|
| Monday | 7.0 |
| Tuesday | 7.5 |
| Wednesday | 6.5 |
| Thursday | 8.5 |
| Friday | 9.0 |

*Price List*

| *Item Number* | *Description* | *Unit Price* |
|---|---|---|
| 27345 | Valve spring | $1.85 |
| 41257 | Main bearing | 3.40 |
| 54678 | Valve | 4.79 |
| 29753 | Ring set | 33.98 |
| 45837 | Gasket set | 48.99 |
| 23751 | Timing chain | 42.95 |
| 23765 | Fuel pump | 77.59 |
| 28632 | Wrist pin | 2.71 |
| 34922 | Rod bearing | 2.67 |
| 2871 | Valve seal | 0.42 |

**136.** Organize a table of data showing the parts used, the unit price for each, and the price of the quantity used. Hint: Use the following headings for the table.

*Quantity* *Item Number* *Description* *Unit Price* *Total*

**137.** Add up the numbers in the "Total" column to find the total cost of the parts.

**138.** If the charge for labor is $26.75 per hour, compute the cost of labor.

**139.** What is the total cost for parts and labor?

**140.** [W] Explain how the decimal point is placed when a number is multiplied by 10, 100, 1000, 10,000, etc.

**141.** [W] Explain how the decimal point is placed in the product of two decimals.

**142.** An emissions test for cars requires that of the total engine exhaust, less than 1 part per thousand $\left(\frac{1}{1000} = 0.001\right)$ be hydrocarbon emissions. Using this figure, determine which of the cars in the table below would fail the emissions test.

| *Car* | *Total Engine Exhaust* | *Hydrocarbon Emission* |
|---|---|---|
| 1 | 367,921 | 36 |
| 2 | 401,346 | 42 |
| 3 | 298,773 | 21 |
| 4 | 330,045 | 32 |
| 5 | 432,989 | 45 |

# 3.5 Division of Decimals

## *Objective A To divide decimals*

To divide decimals, move the decimal point in the divisor to the right to make the divisor a whole number. Move the decimal in the dividend the same number of places to the right. Place the decimal point in the quotient directly over the decimal point in the dividend, and then divide as in whole numbers.

➡ Divide: $3.25\overline{)15.275}$

$3.25.\overline{)15.27.5}$

Move the decimal point 2 places to the right in the divisor and then in the dividend. Place the decimal point in the quotient.

$$\begin{array}{r} 4.7 \\ 325.\overline{)\ 1527.5} \\ -1300\phantom{.5} \\ \hline 227\ 5 \\ -227\ 5 \\ \hline 0 \end{array}$$

Moving the decimal point the same number of decimal places in the divisor and dividend does not change the value of the quotient, because this process is the same as multiplying the numerator and denominator of a fraction by the same number. In the example above,

$$3.25\overline{)15.275} = \frac{15.275}{3.25} = \frac{15.275 \times 100}{3.25 \times 100} = \frac{1527.5}{325} = 325\overline{)1527.5}$$

When dividing decimals, we usually round the quotient off to a specified place value, rather than writing the quotient with a remainder.

➡ Divide: $0.3\overline{)0.56}$
Round to the nearest hundredth.

$$\begin{array}{r} 1.866 \approx 1.87 \\ 0.3.\overline{)\ 0.5.600} \\ -\ 3\phantom{.600} \\ \hline 2\ 6\phantom{00} \\ -2\ 4\phantom{00} \\ \hline 20\phantom{0} \\ -18\phantom{0} \\ \hline 20 \\ -18 \\ \hline \end{array}$$

The division must be carried to the thousandths' place to round the quotient to the nearest hundredth. Therefore, zeros must be inserted in the dividend so that the quotient has a digit in the thousandths' place.

➡ Divide: $57.93 \div 3.24$ Round to the nearest thousandth.

```
          17.8796 ≈ 17.880
3.24.) 57.93.0000
      -32 4
       25 53
      -22 68
        2 85 0
       -2 59 2
          25 80
         -22 68
           3 120
          -2 916
             2040
            -1944
```

Zeros must be inserted in the dividend so that the quotient has a digit in the ten-thousandths' place.

To divide a decimal by a power of 10 (10, 100, 1000,...), move the decimal point to the left the same number of places as there are zeros in the power of 10.

$34.65 \div 10 = 3.465$
1 zero — 1 decimal place

$34.65 \div 100 = 0.3465$
2 zeros — 2 decimal places

$34.65 \div 1000 = 0.03465$
3 zeros — 3 decimal places

Note that a zero must be inserted between the 3 and the decimal point.

$34.65 \div 10{,}000 = 0.003465$
4 zeros — 4 decimal places

Note that two zeros must be inserted between the 3 and the decimal point.

If the power of 10 is written in exponential notation, the exponent indicates how many places to move the decimal point.

$34.65 \div 10^1 = 3.465$ — 1 decimal place

$34.65 \div 10^2 = 0.3465$ — 2 decimal places

$34.65 \div 10^3 = 0.03465$ — 3 decimal places

$34.65 \div 10^4 = 0.003465$ — 4 decimal places

**Example 1** Divide: $0.1344 \div 0.032$

**Solution**

```
                4.2
0.032.)0.134.4
        -128
           6 4
          -6 4
             0
```

**You Try It 1** Divide: $0.1404 \div 0.052$

**Your solution**

*Solution on p. A21*

**Example 2** Divide: $58.092 \div 82$
Round to the nearest thousandth.

**Solution**

$$
\begin{array}{r}
0.7084 \approx 0.708 \\
82\overline{)\;58.0920} \\
-57\;4\phantom{000} \\
\hline
69\phantom{00} \\
-\;\;0\phantom{00} \\
\hline
692\phantom{0} \\
-656\phantom{0} \\
\hline
360 \\
-328 \\
\hline
\end{array}
$$

**You Try It 2** Divide: $37.042 \div 76$
Round to the nearest thousandth.

**Your solution**

**Example 3** Divide: $420.9 \div 7.06$
Round to the nearest tenth.

**Solution**

$$
\begin{array}{r}
59.61 \approx 59.6 \\
7.06.\overline{)\;420.90.00} \\
-353\;0\phantom{0.00} \\
\hline
67\;90\phantom{.00} \\
-63\;54\phantom{.00} \\
\hline
4\;36\;0\phantom{0} \\
-4\;23\;6\phantom{0} \\
\hline
12\;40 \\
-\;\;7\;06 \\
\hline
\end{array}
$$

**You Try It 3** Divide: $370.2 \div 5.09$
Round to the nearest tenth.

**Your solution**

**Example 4** Divide: $402.75 \div 1000$

**Solution** $402.75 \div 1000 = 0.40275$

**You Try It 4** Divide: $309.21 \div 10{,}000$

**Your solution**

**Example 5** What is 0.625 divided by $10^2$?

**Solution** $0.625 \div 10^2 = 0.00625$

**You Try It 5** What is 42.93 divided by $10^4$?

**Your solution**

*Solutions on p. A21*

## ESTIMATION

### Estimating the Quotient of Two Decimals

Estimate and then use your calculator to find $282.18 \div 0.485$.

To estimate a quotient, round each number so that there is one non-zero digit. Then divide.

$282.18 \div 0.485 \approx$
$300 \div 0.5 = 600$

The estimated answer is 600.

Now use your calculator to find the exact answer.

282.18 [÷] 0.485 [=] 581.814433

The exact answer is 581.814433.

## *Objective B To solve application problems*

The table at the right shows the total amount of gasoline tax paid per gallon of gas in selected cities in 1994. Use this table for Example 6 and You Try It 6.

*Gasoline Taxes in Dollars per Gallon*

| *City* | *Taxes* |
|---|---|
| Chicago | \$.5929 |
| Tampa | \$.4380 |
| Boston | \$.4013 |
| Detroit | \$.3924 |

**Example 6**

Sandra Lopez's car gets 24 miles per gallon. In a year in which she drove 10,500 miles in Chicago, how much did she pay, to the nearest dollar, in gasoline taxes?

**Strategy**

To find the amount she paid in gasoline taxes:

- Find the total number of gallons of gas she used by dividing her total miles (10,500) by the number of miles traveled per gallon of gasoline (24).
- Multiply the total tax (\$.5929) by the total number of gallons of gasoline used.

**Solution**

$10{,}500 \div 24 = 437.5$

$0.5929 \times 437.5 = 259.39375$

Sandra paid approximately \$259 for gasoline taxes.

**You Try It 6**

In 1994, Jefferson Beckwith drove his car 13,400 miles in Boston. If his car gets 26 miles per gallon, how much did he pay, to the nearest dollar, in gasoline taxes?

**Your strategy**

**Your solution**

**Example 7**

In 1994, manufacturers offered consumers approximately \$183.4 billion worth of discount coupons. Of those, only \$8.3 billion were redeemed. Find the quotient of the unredeemed amount.

**Strategy**

To find the quotient:

- Subtract 8.3 from 183.4
- Divide the difference by 183.4

**Solution**

$183.4 - 8.3 = 175.1$

$175.1 \div 183.4 \approx 0.9547$

The quotient of the unredeemed amount is 0.9547.

**You Try It 7**

A Nielsen survey of the number of people (in millions) who watch television during the week is given in the table below.

| *Mon.* | *Tues.* | *Wed.* | *Thu.* | *Fri.* | *Sat.* | *Sun.* |
|---|---|---|---|---|---|---|
| 91.9 | 89.8 | 90.6 | 93.9 | 78.0 | 77.1 | 87.7 |

Find the average number of people watching television per day.

**Your strategy**

**Your solution**

*Solutions on p. A22*

## 3.5 Exercises

### *Objective A*

Divide.

1. $3\overline{)2.46}$
2. $7\overline{)3.71}$
3. $0.8\overline{)3.84}$
4. $0.9\overline{)6.93}$
5. $0.7\overline{)62.3}$
6. $0.4\overline{)52.8}$
7. $0.4\overline{)24}$
8. $0.5\overline{)65}$
9. $0.7\overline{)59.01}$
10. $0.9\overline{)8.721}$
11. $0.5\overline{)16.15}$
12. $0.8\overline{)77.6}$
13. $0.7\overline{)3.542}$
14. $0.6\overline{)2.436}$
15. $6.3\overline{)8.19}$
16. $3.2\overline{)7.04}$
17. $3.6\overline{)0.396}$
18. $2.7\overline{)0.648}$
19. $6.9\overline{)26.22}$
20. $1.7\overline{)84.66}$

Divide. Round to the nearest tenth.

21. $55.62 \div 8.8$
22. $25.43 \div 5.4$
23. $5.427 \div 9.5$
24. $1.837 \div 1.4$
25. $18.4 \div 7.3$
26. $52.9 \div 8.1$
27. $0.183 \div 0.17$
28. $0.381 \div 0.47$
29. $6.924 \div 0.053$

Divide. Round to the nearest hundredth.

30. $4.817 \div 16$
31. $6.467 \div 8$
32. $0.0418 \div 0.53$
33. $19.08 \div 0.45$
34. $21.792 \div 0.96$
35. $38.665 \div 0.95$
36. $13.97 \div 25.4$
37. $27.738 \div 60.3$
38. $3.171 \div 45.3$

Divide. Round to the nearest thousandth.

**39.** 1.028 ÷ 54

**40.** 6.729 ÷ 27

**41.** 0.0437 ÷ 0.5

**42.** 75.469 ÷ 77.8

**43.** 34.31 ÷ 95.3

**44.** 0.2695 ÷ 2.67

**45.** 0.4871 ÷ 4.72

**46.** 0.1142 ÷ 17.2

**47.** 0.2307 ÷ 26.7

Divide. Round to the nearest whole number.

**48.** 16.5 ÷ 4

**49.** 89.76 ÷ 90

**50.** 1.94 ÷ 0.3

**51.** 1.0478 ÷ 0.413

**52.** 2.148 ÷ 0.519

**53.** 0.79 ÷ 0.778

**54.** 3.092 ÷ 0.075

**55.** 392 ÷ 6.9

**56.** 8.729 ÷ 0.075

Divide.

**57.** 4.07 ÷ 10

**58.** 0.039 ÷ 10

**59.** 42.67 ÷ 10

**60.** 389.7 ÷ 100

**61.** 1.037 ÷ 100

**62.** 237.835 ÷ 100

**63.** 8.295 ÷ 1000

**64.** 82,547 ÷ 1000

**65.** 825.37 ÷ 1000

**66.** 8.35 ÷ 10

**67.** 0.32 ÷ 10

**68.** 87.65 ÷ 10

**69.** $23.627 \div 10^2$

**70.** $2.954 \div 10^2$

**71.** $0.0053 \div 10^2$

**72.** $289.32 \div 10^3$

**73.** $1.8932 \div 10^3$

**74.** $0.139 \div 10^3$

75. Divide 44.208 by 2.4.

76. Divide 0.04664 by 0.44.

77. Find the quotient of 723.15 and 45.

78. Find the quotient of 3.3463 and 3.07.

79. Divide 13.5 by $10^3$.

80. Divide 0.045 by $10^5$.

81. Find the quotient of 23.678 and 1000.

82. Find the quotient of 7.005 and 10,000.

83. What is 0.0056 divided by 0.05?

84. What is 123.8 divided by 0.02?

Estimate and then use your calculator to divide. Round your calculated answer to the nearest ten-thousandth.

85. $42.42 \div 3.8$

86. $69.8 \div 7.2$

87. $389 \div 0.44$

88. $642 \div 0.83$

89. $6.394 \div 3.5$

90. $8.429 \div 4.2$

91. $1.235 \div 0.021$

92. $7.456 \div 0.072$

93. $95.443 \div 1.32$

94. $423.0925 \div 4.0927$

95. $1.000523 \div 429.07$

96. $0.03629 \div 0.00054$

## *Objective B* Application Problems

97. Ramon, a high school football player, gained 162 yards on 26 carries in a high school football game. Find the number of yards gained per carry. (Round to two decimal places.)

98. Ross Lapointe earns $39,440.64 for 12 months' work as a park ranger. How much does he earn in 1 month?

99. You pay $947.60 a year in car insurance. The insurance is paid in four equal payments. Find the amount of each payment.

100. A case of diet cola costs $6.79. If there are 24 cans in a case, find the cost per can. Round to the nearest cent.

101. Anne is building bookcases that are 3.4 feet long. How many complete shelves can be cut from a 12-foot board?

102. You travel 295 miles on 12.5 gallons of gasoline. How many miles can you travel on 1 gallon of gasoline?

103. An oil company had issued 3,541,221,500 shares of stock. The company paid $6,090,990,120 in dividends. Find the dividend for each share of stock. Round to two decimal places.

**104.** The total budget for the United States in 1993 was $1.47 trillion. If each person in the United States were to pay the same amount of taxes, how much would each person pay to raise that amount of money? Assume that there are 250 million people in the United States. (Round to the nearest dollar.)

**105.** You buy a home entertainment center for $1242.58. The down payment is $400, and the balance is to be paid in 18 equal monthly payments.
a. Find the amount to be paid in monthly payments.
b. Find the amount of each monthly payment.

### *APPLYING THE CONCEPTS*

**106.** [W] Explain how the decimal point is moved when dividing a number by 10, 100, 1,000, 10,000 etc.

**107.** A ball point pen priced at 50¢ was not selling. When the price was reduced to a different whole number of cents, the entire stock sold for $31.95. How many cents were charged per pen when the price was reduced? (*Hint*: There is more than one possible answer.)

**108.** [W] Explain how baseball batting averages are determined. Then find Tony Gwynn's batting average with 175 hits out of 489 at bats. (Round to three decimal places).

**109.** [W] Explain how the decimal point is placed in the quotient when dividing by a decimal.

The graph at the right shows the increase in the consumption of yogurt in the U.S. for a 20-year period. Use this graph for Exercises 110 and 111.

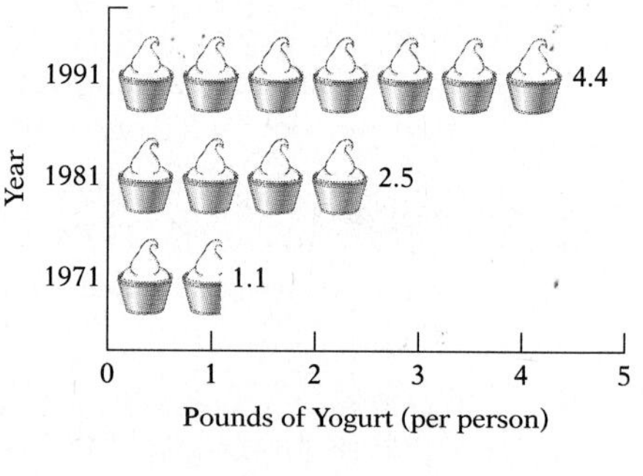


**110.** What was the average annual increase in yogurt consumption between 1971 and 1981?

**111.** If the average annual increase in yogurt consumption between 1981 and 1991 continued at the same rate beyond 1991, what would be the average yogurt consumption in 1995?

For each of the problems below, insert a +, −, ×, or ÷ into the space left for it so that the statement is true.

**112.** 3.45  0.5 = 6.9 **113.** 3.46  0.24 = 0.8304 **114.** 6.009  4.68 = 1.329

**115.** 0.064  1.6 = 0.1024 **116.** 9.876  23.12 = 32.996 **117.** 3.0381  1.23 = 2.47

Fill in the space to make a true statement.

**118.** 6.47 −  = 1.253 **119.** 6.47 +  = 9 **120.** 0.009 ÷  = 0.36

# 3.6 Comparing and Converting Fractions and Decimals

## Objective A To convert fractions to decimals

Every fraction can be written as a decimal. To write a fraction as a decimal, divide the numerator of the fraction by the denominator. The quotient can be rounded to the desired place value.

➡ Convert $\frac{3}{7}$ to a decimal.

$7\overline{)3.00000}$ = 0.42857

$\frac{3}{7}$ rounded to the nearest hundredth is 0.43.

$\frac{3}{7}$ rounded to the nearest thousandth is 0.429.

$\frac{3}{7}$ rounded to the nearest ten-thousandth is 0.4286.

➡ Convert $3\frac{2}{9}$ to a decimal. Round to the nearest thousandth.

$3\frac{2}{9} = \frac{29}{9}$ $\quad 9\overline{)29.0000}$ = 3.2222 $\quad 3\frac{2}{9}$ rounded to the nearest thousandth is 3.222.

**Example 1** Convert $\frac{3}{8}$ to a decimal. Round to the nearest hundredth.

**Solution** $8\overline{)3.000}$ = 0.375 $\approx$ 0.38

**You Try It 1** Convert $\frac{9}{16}$ to a decimal. Round to the nearest tenth.

**Your solution**

**Example 2** Convert $2\frac{3}{4}$ to a decimal. Round to the nearest tenth.

**Solution** $2\frac{3}{4} = \frac{11}{4}$ $\quad 4\overline{)11.00}$ = 2.75 $\approx$ 2.8

**You Try It 2** Convert $4\frac{1}{6}$ to a decimal. Round to the nearest hundredth.

**Your solution**

*Solutions on p. A22*

## Objective B To convert decimals to fractions

To convert a decimal to a fraction, remove the decimal point and place the decimal part over a denominator equal to the place value of the last digit in the decimal.

hundredths: $0.47 = \frac{47}{100}$

hundredths: $7.45 = 7\frac{45}{100} = 7\frac{9}{20}$

thousandths: $0.275 = \frac{275}{1000} = \frac{11}{40}$

hundredths: $0.16\frac{2}{3} = \frac{16\frac{2}{3}}{100} = 16\frac{2}{3} \div 100 = \frac{50}{3} \times \frac{1}{100} = \frac{1}{6}$

**Example 3** Convert 0.82 and 4.75 to fractions.

**Solution** $0.82 = \frac{82}{100} = \frac{41}{50}$

$4.75 = 4\frac{75}{100} = 4\frac{3}{4}$

**You Try It 3** Convert 0.56 and 5.35 to fractions.

**Your solution**

**Example 4** Convert $0.15\frac{2}{3}$ to a fraction.

**Solution** $0.15\frac{2}{3} = \frac{15\frac{2}{3}}{100} = 15\frac{2}{3} \div 100$

$= \frac{47}{3} \times \frac{1}{100} = \frac{47}{300}$

**You Try It 4** Convert $0.12\frac{7}{8}$ to a fraction.

**Your solution**

*Solutions on p. A22*

***Objective C To identify the order relation between two decimals or between a decimal and a fraction***

Decimals, like whole numbers and fractions, can be graphed as points on the number line. The number line can be used to show the order of decimals. A decimal that appears to the right of a given number is greater than the given number. A decimal that appears to the left of a given number is less than the given number.

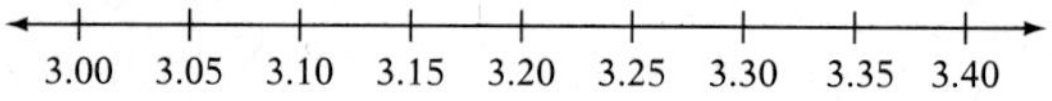


Note that 3, 3.0, and 3.00 represent the same number.

➡ Find the order relation between $\frac{3}{8}$ and 0.38.

$\frac{3}{8} = 0.375 \qquad 0.38 = 0.380$

$0.375 < 0.38$

$\frac{3}{8} < 0.38$

**Example 5** Place the correct symbol, < or >, between the numbers.

$\frac{5}{16} \quad 0.32$

**Solution** $\frac{5}{16} \approx 0.313$

$0.313 < 0.32$

$\frac{5}{16} < 0.32$

**You Try It 5** Place the correct symbol, < or >, between the numbers.

$0.63 \quad \frac{5}{8}$

**Your solution**

*Solution on p. A22*

## 3.6 Exercises

### *Objective A*

Convert the fraction to a decimal. Round to the nearest thousandth.

**1.** $\frac{5}{8}$ **2.** $\frac{7}{12}$ **3.** $\frac{2}{3}$ **4.** $\frac{5}{6}$ **5.** $\frac{1}{6}$ **6.** $\frac{7}{8}$

**7.** $\frac{5}{12}$ **8.** $\frac{9}{16}$ **9.** $\frac{7}{4}$ **10.** $\frac{5}{3}$ **11.** $1\frac{1}{2}$ **12.** $2\frac{1}{3}$

**13.** $\frac{16}{4}$ **14.** $\frac{36}{9}$ **15.** $\frac{3}{1000}$ **16.** $\frac{5}{10}$ **17.** $7\frac{2}{25}$ **18.** $16\frac{7}{9}$

**19.** $37\frac{1}{2}$ **20.** $87\frac{1}{2}$ **21.** $\frac{3}{8}$ **22.** $\frac{11}{16}$ **23.** $\frac{5}{24}$ **24.** $\frac{4}{25}$

**25.** $3\frac{1}{3}$ **26.** $8\frac{2}{5}$ **27.** $5\frac{4}{9}$ **28.** $3\frac{1}{12}$ **29.** $\frac{5}{16}$ **30.** $\frac{11}{12}$

### *Objective B*

Convert the decimal to a fraction.

**31.** 0.8 **32.** 0.4 **33.** 0.32 **34.** 0.48 **35.** 0.125

**36.** 0.485 **37.** 1.25 **38.** 3.75 **39.** 16.9 **40.** 17.5

**41.** 8.4 **42.** 10.7 **43.** 8.437 **44.** 9.279 **45.** 2.25

**46.** 7.75 **47.** $0.15\frac{1}{3}$ **48.** $0.17\frac{2}{3}$ **49.** $0.87\frac{7}{8}$ **50.** $0.12\frac{5}{9}$

Convert the decimal to a fraction.

**51.** 1.68 **52.** 7.38 **53.** 0.045 **54.** 0.085 **55.** 16.72

**56.** 82.32 **57.** 0.33 **58.** 0.57 **59.** $0.33\frac{1}{3}$ **60.** $0.66\frac{2}{3}$

### Objective C

Place the correct symbol, $<$ or $>$, between the numbers.

**61.** 0.15 0.5 **62.** 0.6 0.45 **63.** 6.65 6.56 **64.** 3.89 3.98

**65.** 2.504 2.054 **66.** 0.025 0.105 **67.** $\frac{3}{8}$ 0.365 **68.** $\frac{4}{5}$ 0.802

**69.** $\frac{2}{3}$ 0.65 **70.** 0.85 $\frac{7}{8}$ **71.** $\frac{5}{9}$ 0.55 **72.** $\frac{7}{12}$ 0.58

**73.** 0.62 $\frac{7}{15}$ **74.** $\frac{11}{12}$ 0.92 **75.** 0.161 $\frac{1}{7}$ **76.** 0.623 0.6023

**77.** 0.86 0.855 **78.** 0.87 0.087 **79.** 1.005 0.5 **80.** 0.033 0.3

### APPLYING THE CONCEPTS

**81.** Which of the following is true?
a. $\frac{137}{300} = 0.456666667$ b. $\frac{137}{300} < 0.45666666$ c. $\frac{137}{300} > 0.45666666$

**82.** Is $\frac{19}{23}$ in decimal form a non-repeating decimal? Why or why not?

**83.** If a number is rounded to the nearest thousandth, is it always greater than if it was rounded to the nearest hundredth? Give examples to support your answer.

**84.** [W] Convert $\frac{1}{9}$, $\frac{2}{9}$, $\frac{3}{9}$, and $\frac{4}{9}$ to decimals. Describe the pattern. Use the pattern to convert $\frac{5}{9}$, $\frac{7}{9}$, and $\frac{8}{9}$ to decimals.

**85.** [W] Explain the difference between terminating, repeating, and non-repeating decimals. Give an example of each kind of decimal.

# Project in Mathematics

## Topographical Maps

Carpenters use fractions to measure the wood that is used to frame a house. For instance, a door entry may measure $42\frac{1}{2}$ inches. However, the grading contractor (the person who levels the lot on which the house is built) measures the height of the lot by using decimals. For instance, a certain place may be at a height of 554.2 feet. (The height is measured above sea level; so 554.2 means 554.2 feet above sea level.)

A surveyor provides the grading contractor with a grading plan that shows the elevation of each point of the lot. These plans are drawn so that the house can be sited in such a way that water will drain away from it. The diagram below is a **Topographical Map** for a lot. Along each closed curve, called a **contour curve,** the lot is at the same height above sea level. For instance, the curve in red means that every point on that curve is 556.5 feet above sea level.

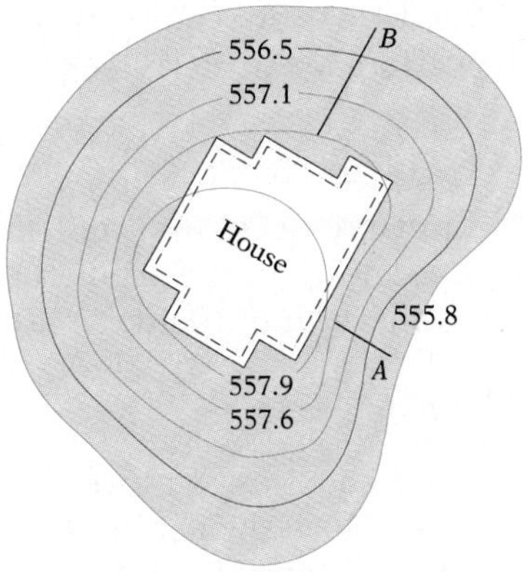


***Exercises:***

Using the map above, answer the following questions.

**1.** What is the elevation of the highest point on the lot?

**2.** What is the elevation of the lowest point on the lot?

**3.** What is the difference in elevation between the highest and lowest points on the lot?

**4.** Assuming this map is drawn to scale, describe the significance of how steep the slope is along line *A* compared to that along line *B*.

## Chapter Summary

**Key Words** A number written in *decimal notation* has three parts: a whole-number part, a decimal point, and a decimal part.

The position of a digit in a *decimal* determines the digit's *place value*.

**Essential Rules**

*To Write a Decimal in Words*

To write a decimal in words, write the decimal part as if it were a whole number. Then name the place value of the last digit.

*To Add Decimals*

To add decimals, write the numbers so that the decimal points are on a vertical line. Add as in whole numbers, and place the decimal point in the sum directly below the decimal point in the addends.

*To Subtract Decimals*

To subtract decimals, place the numbers so that the decimal points are on a vertical line. Subtract as for whole numbers, and write the decimal point in the difference directly below the decimal point in the subtrahend.

*To Multiply Decimals*

To multiply decimals, multiply the numbers as in whole numbers. Place the decimal point in the product so that the number of decimal places in the product is the sum of the decimal places in the factors.

*To Divide Decimals*

To divide decimals, move the decimal point in the divisor to make it a whole number. Move the decimal point in the dividend the same number of places to the right. Place the decimal point in the quotient directly over the decimal point in the dividend. Then divide as in whole numbers.

*To Write a Fraction as a Decimal*

To write a fraction as a decimal, divide the numerator of the fraction by the denominator. Round the quotient to the desired number of places.

*To Convert a Decimal to a Fraction*

To convert a decimal to a fraction, remove the decimal point and place the decimal part over a denominator equal to the place value of the last digit in the decimal.

# Chapter Review

SECTION 3.1

1. Write 22.0092 in words.

2. Write 342.37 in words.

3. Write thirty-four and twenty-five thousandths in standard form.

4. Write three and six thousand seven hundred fifty-three hundred-thousandths in standard form.

5. Round 7.93704 to the nearest hundredth.

6. Round 0.05678235 to the nearest hundred-thousandths.

SECTION 3.2

7. Add: $3.42 + 0.794 + 32.5$

8. Find the sum of 369.41, 88.3, 9.774, and 366.474.

9. You have $237.44 in your checking account. You make deposits of $56.88, $127.40, and $56.30. Find the amount in your checking account after you make the deposits.

SECTION 3.3

10. Subtract: $27.31 - 4.4465$

11. What is 7.796 decreased by 2.9175?

12. You had a balance of $895.68 in your checking account. You then wrote checks of $145.72 and $88.45. Find the new balance in your checking account.

## SECTION 3.4

**13.** Find the product of 3.08 and 2.9.

**14.** Multiply: $\begin{array}{r} 34.79 \\ \times\ 0.74 \\ \hline \end{array}$

**15.** The state income tax on the business you own is \$560 plus 0.08 times your profit. You made a profit of \$63,000 last year. Find the amount of income tax you paid last year.

## SECTION 3.5

**16.** Find the quotient of 3.6515 and 0.067.

**17.** Divide: $0.053\overline{)0.349482}$

**18.** A car costing \$5944.20 is bought with a down payment of \$1500 and 36 equal monthly payments. Find the amount of each monthly payment.

## SECTION 3.6

**19.** Convert $\frac{7}{9}$ to a decimal. Round to the nearest thousandth.

**20.** Convert $2\frac{1}{3}$ to a decimal. Round to the nearest hundredth.

**21.** Convert 0.375 to a fraction.

**22.** Convert $0.66\frac{2}{3}$ to a fraction.

**23.** Place the correct symbol, < or >, between the two numbers.
0.055 0.1

**24.** Place the correct symbol, < or >, between the two numbers.
$\frac{5}{8}$ 0.62

# Chapter Test

1. Write 45.0302 in words.

2. Write two hundred nine and seven thousand eighty-six hundred-thousandths in standard form.

3. Round 7.0954625 to the nearest thousandth.

4. Round 0.07395 to the nearest ten-thousandth.

5. What is the total of 62.3, 4.007, and 189.65?

6. Add:

$$\begin{array}{r} 270.93 \\ 97. \\ 1.976 \\ +\ 88.675 \\ \hline \end{array}$$

7. You received a salary of \$363.75, a commission of \$954.82, and a bonus of \$225. Find your total income.

8. Subtract:

$$\begin{array}{r} 13.027 \\ -\ 8.94 \\ \hline \end{array}$$

9. Find 9.23674 less than 37.003.

10. Find the missing dimension.

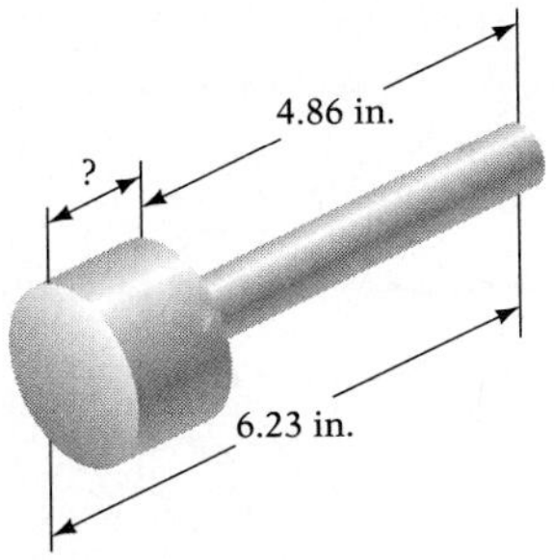


11. Multiply:

$$\begin{array}{r} 1.37 \\ 0.004 \\ \hline \end{array}$$

The table at the right shows the taxes on one gallon of gasoline in selected cities in 1994. Use this table for Questions 12 and 13.

| *City* | *Federal Tax* | *State Tax* | *Other Taxes* |
|---|---|---|---|
| Boston | \$.1863 | \$.12 | \$.0005 |
| Detroit | \$.1863 | \$.15 | \$.0561 |
| Chicago | \$.1863 | \$.19 | \$.2166 |
| Los Angeles | \$.1863 | \$.17 | \$.4747 |
| Dallas | \$.1863 | \$.20 | \$.0059 |
| New Haven | \$.1863 | \$.29 | \$.0528 |

**12.** Which city had the highest total tax for a gallon of gasoline?

**13.** Which city had the lowest total tax for a gallon of gasoline?

**14.** A long-distance telephone call costs \$.85 for the first 3 minutes and \$.42 for each additional minute. Find the cost of a 12-minute long-distance telephone call.

**15.** Divide: $0.006\overline{)1.392}$

**16.** Find 0.0569 divided by 0.037. Round to the nearest thousandth.

**17.** A car was bought for \$6392.60, with a down payment of \$1250. The balance was paid in 36 monthly payments. Find the amount of each monthly payment.

**18.** Convert $\frac{9}{13}$ to a decimal. Round to the nearest thousandth.

**19.** Convert 0.825 to a fraction.

**20.** Place the correct symbol, < or >, between the two numbers.
0.66   0.666

# Cumulative Review

1. Divide: $89\overline{)20{,}932}$

2. Simplify: $2^3 \cdot 4^2$

3. Simplify: $2^2 - (7 - 3) \div 2 + 1$

4. Find the LCM of 9, 12, and 24.

5. Write $\frac{22}{5}$ as a mixed number.

6. Write $4\frac{5}{8}$ as an improper fraction.

7. Build an equivalent fraction with the given denominator.
$\frac{5}{12} = \frac{\quad}{60}$

8. Add: $\frac{3}{8} + \frac{5}{12} + \frac{9}{16}$.

9. What is $5\frac{7}{12}$ increased by $3\frac{7}{18}$?

10. Subtract: $9\frac{5}{9} - 3\frac{11}{12}$

11. Multiply: $\frac{9}{16} \times \frac{4}{27}$

12. Find the product of $2\frac{1}{8}$ and $4\frac{5}{17}$.

13. Divide: $\frac{11}{12} \div \frac{3}{4}$

14. What is $2\frac{3}{8}$ divided by $2\frac{1}{2}$?

15. Simplify: $\left(\frac{2}{3}\right)^2 \cdot \left(\frac{3}{4}\right)^3$

16. Simplify: $\left(\frac{2}{3}\right)^2 - \left(\frac{2}{3} - \frac{1}{2}\right) + 2$

17. Write 65.0309 in words.

18. Add:
$$\begin{array}{r} 379.006 \\ 27.523 \\ 9.8707 \\ +\ 88.2994 \\ \hline \end{array}$$

19. What is 29.005 decreased by 7.9286?

20. Multiply: $\begin{array}{r} 9.074 \\ \times\ 6.09 \\ \hline \end{array}$

21. Divide: $8.09\overline{)17.42963}$. Round to the nearest thousandth.

22. Convert $\frac{11}{15}$ to a decimal. Round to the nearest thousandth.

23. Convert $0.16\frac{2}{3}$ to a fraction.

24. Place the correct symbol, < or >, between the two numbers.

$\frac{8}{9} \quad 0.98$

25. An airplane had 204 passengers aboard. During a stop, 97 passengers got off the plane and 127 passengers got on the plane. How many passengers were on the continuing flight?

26. An investor purchased stock at $\$32\frac{1}{8}$ per share. During the first 2 months of ownership, the stock lost $\$\frac{3}{4}$ and then gained $\$1\frac{1}{2}$. Find the value of each share of stock at the end of the 2 months.

27. You have a checking balance of $814.35. You then write checks for $42.98, $16.43, and $137.56. Find your checking account balance after you write the checks.

28. A machine lathe takes 0.017 inch from a brass bushing that is 1.412 inches thick. Find the resulting thickness of the bushing.

29. The state income tax on your business is $820 plus 0.08 times your profit. You made a profit of $64,860 last year. Find the amount of income tax you paid last year.

30. You bought a camera costing $210.96. The down payment is $20, and the balance is to be paid in 8 equal monthly payments. Find the monthly payment.

Chapter

# 4 Ratio and Proportion

## Objectives

### Section 4.1

To write the ratio of two quantities in simplest form

To solve application problems

### Section 4.2

To write rates

To write unit rates

To solve application problems

### Section 4.3

To determine whether a proportion is true

To solve proportions

To solve application problems

## Musical Scales

When a metal wire is stretched tight and then plucked, a sound is heard. Guitars, banjos, and violins are examples of instruments that use this principle to produce music. A piano is another example of this principle, but the sound is produced by the string being struck by a small hammerlike object.

After the string is plucked or struck, the string begins to vibrate. The number of times the string vibrates in 1 second is called the frequency of the vibration, or the pitch. Normally humans can hear vibrations as low as 16 cps (cycles per second) and as high as 20,000 cps. The longer the string, the lower the pitch of the sound; the shorter the string, the higher the pitch of the sound. In fact, a string half as long as another string vibrates twice as fast.

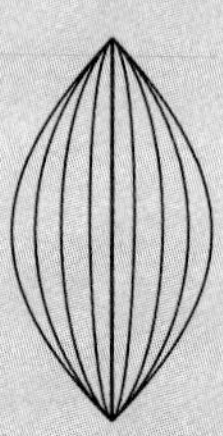

Vibrating String

Most music that is heard today is based on what is called the chromatic or twelve-tone scale. For this scale, a vibration of 261 cps is called middle C. A string half as long as the string for middle C vibrates twice as fast and produces a musical note one octave higher.

To produce the notes between the two C's, strings are placed that produce the desired pitch. Recall that as the string gets shorter, the pitch increases. A top view of a grand piano illustrates how the strings vary in length.

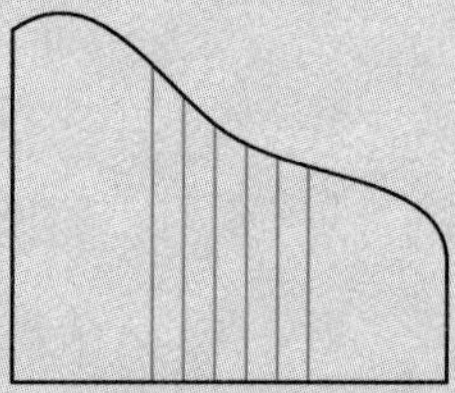

A well-tempered chromatic scale is one in which the string lengths are chosen so that the ratios of the frequencies of adjacent notes are the same.

$$\frac{C}{C\#} = \frac{C\#}{D} = \frac{D}{D\#} = \frac{D\#}{E} = \frac{E}{F} = \frac{F}{F\#} = \frac{F\#}{G} = \frac{G}{G\#} = \frac{G\#}{A} = \frac{A}{A\#} = \frac{A\#}{B} = \frac{B}{C}$$

The common ratio for the chromatic scale is approximately $\frac{1}{1.0595}$.

# 4.1 Ratio

## Objective A To write the ratio of two quantities in simplest form

Quantities such as 3 feet, 12 cents, and 9 cars are number quantities written with **units**.

3 feet
12 cents These are only some examples of units. Shirts, dollars, trees,
9 cars miles, and gallons are further examples.

units

A **ratio** is the comparison of two quantities that have the *same* units. This comparison can be written three different ways:

1. As a fraction
2. As two numbers separated by a colon (:)
3. As two numbers separated by the word *to*

The ratio of the lengths of two boards, one 8 feet long and the other 10 feet long, can be written as

1. $\frac{8 \text{ feet}}{10 \text{ feet}} = \frac{8}{10} = \frac{4}{5}$
2. 8 feet:10 feet = 8:10 = 4:5
3. 8 feet to 10 feet = 8 to 10 = 4 to 5

A ratio is in **simplest form** when the two numbers do not have a common factor. Note that in a ratio, the units are not written.

This ratio means that the smaller board is $\frac{4}{5}$ the length of the longer board.

**Example 1**
Write the comparison \$6 to \$8 as a ratio in simplest form using a fraction, a colon, and the word *to*.

**Solution** $\frac{\$6}{\$8} = \frac{6}{8} = \frac{3}{4}$
\$6:\$8 = 6:8 = 3:4
\$6 to \$8 = 6 to 8 = 3 to 4

**You Try It 1**
Write the comparison 20 pounds to 24 pounds as a ratio in simplest form using a fraction, a colon, and the word *to*.

**Your solution**

**Example 2**
Write the comparison 18 quarts to 6 quarts as a ratio in simplest form using a fraction, a colon, and the word *to*.

**Solution** $\frac{18 \text{ quarts}}{6 \text{ quarts}} = \frac{18}{6} = \frac{3}{1}$
18 quarts:6 quarts
18:6 = 3:1
18 quarts to 6 quarts
18 to 6 = 3 to 1

**You Try It 2**
Write the comparison 64 miles to 8 miles as a ratio in simplest form using a fraction, a colon, and the word *to*.

**Your solution**

*Solutions on p. A23*

## *Objective B To solve application problems*

Use the table below for Example 3 and You Try It 3.

| *Board Feet of Wood at a Lumber Store* | | | |
|---|---|---|---|
| *Pine* | *Ash* | *Oak* | *Cedar* |
| 20,000 | 18,000 | 10,000 | 12,000 |

**Example 3**

Find, as a fraction in simplest form, the ratio of the number of board feet of pine to board feet of oak.

**Strategy**

To find the ratio, write the ratio of board feet of pine (20,000) to board feet of oak (10,000) in simplest form.

**Solution**

$$\frac{20{,}000}{10{,}000} = \frac{2}{1}$$

The ratio is $\frac{2}{1}$.

**You Try It 3**

Find, as a fraction in simplest form, the ratio of the number of board feet of cedar to board feet of ash.

**Your strategy**

**Your solution**

**Example 4**

The cost of building a patio cover was \$250 for labor and \$350 for materials. What, as a fraction in simplest form, is the ratio of the cost of materials to the total cost for labor and materials?

**Strategy**

To find the ratio, write the ratio of the cost of materials (\$350) to the total cost (\$250 + \$350) in simplest form.

**Solution**

$$\frac{\$350}{\$250 + \$350} = \frac{350}{600} = \frac{7}{12}$$

The ratio is $\frac{7}{12}$.

**You Try It 4**

A company spends \$20,000 a month for television advertising and \$15,000 a month for radio advertising. What, as a fraction in simplest form, is the ratio of the cost of radio advertising to the total cost of radio and television advertising?

**Your strategy**

**Your solution**

*Solutions on p. A23*

## 4.1 Exercises

### *Objective A*

Write the comparison as a ratio in simplest form using a fraction, a colon (:), and the word *to*.

**1.** 3 pints to 15 pints

**2.** 6 pounds to 8 pounds

**3.** $40 to $20

**4.** 10 feet to 2 feet

**5.** 3 miles to 8 miles

**6.** 2 hours to 3 hours

**7.** 37 hours to 24 hours

**8.** 29 inches to 12 inches

**9.** 6 minutes to 6 minutes

**10.** 8 days to 12 days

**11.** 35 cents to 50 cents

**12.** 28 inches to 36 inches

**13.** 30 minutes to 60 minutes

**14.** 25 cents to 100 cents

**15.** 32 ounces to 16 ounces

**16.** 12 quarts to 4 quarts

**17.** 3 cups to 4 cups

**18.** 6 years to 7 years

**19.** $5 to $3

**20.** 30 yards to 12 yards

**21.** 12 quarts to 18 quarts

**22.** $20 to $28

**23.** 14 days to 7 days

**24.** 9 feet to 3 feet

### *Objective B* Application Problems

*Family Budget*

| *Housing* | *Food* | *Transportation* | *Taxes* | *Utilities* | *Miscellaneous* | *Total* |
|---|---|---|---|---|---|---|
| $800 | $400 | $300 | $350 | $150 | $400 | $2400 |

**25.** Use the table to find the ratio of housing cost to total income.

**26.** Use the table to find the ratio of food cost to total income.

**27.** Use the table to find the ratio of utilities cost to food cost.

**28.** Use the table to find the ratio of transportation cost to housing cost.

**29.** According to the National Collegiate Athletic Association (NCAA), for every 1200 high school senior football players, 40 will play football as seniors in college. Write the ratio of the number of college senior football players to high school senior football players.

**30.** The NCAA estimates that for every 2300 high school seniors that want to play basketball in college, only 40 will make a college team. Write the ratio of the number of students who make a college basketball team to the number of high school seniors who want to make a college basketball team.

**31.** A transformer has 40 turns in the primary coil and 480 turns in the secondary coil. State the ratio of the turns in the primary coil to the number of turns in the secondary coil.

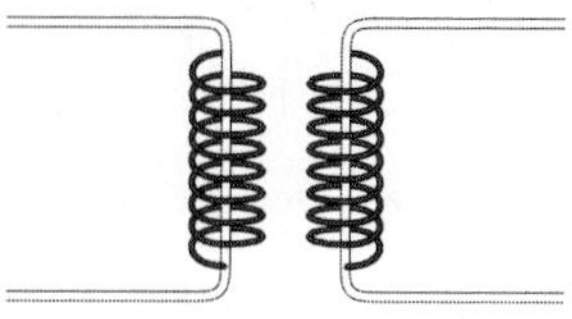

**32.** Rita Sterling bought a computer system for $2400. Five years later she sold the computer for $900. Find the ratio of the amount she received for the computer to the cost of the computer.

**33.** A house with an original value of $90,000 increased in value to $110,000 in 5 years.
a. Find the increase in the value of the house.
b. What is the ratio of the increase in value to the original value of the house?

**34.** A decorator bought one box of ceramic floor tile for $21 and a box of wood tile for $33.
a. What was the total cost of the box of ceramic tile and the box of wood tile?
b. What is the ratio of the cost of the box of wood tile to the total cost?

**35.** The price of gasoline jumped from $0.96 to $1.26 in 1 year. What is the ratio of the increase in price to the original price?

## APPLYING THE CONCEPTS

A bank uses the ratio of a borrower's total monthly debts to the borrower's total monthly income to determine the maximum monthly payment for a potential homeowner. This ratio is called the debt–income ratio. Use the homeowner's income–debt table at the right for Exercises 36 and 37.

| *Income* | *Debts* |
|---|---|
| $3500 | $900 |
| 250 | 170 |
| 140 | 160 |
| | 95 |

**36.** Compute the debt–income ratio for the potential homeowner.

**37.** If Central Trust Bank will make a loan to a customer whose debt–income ratio is less than $\frac{1}{3}$, will the potential homeowner qualify? Explain your answer.

**38.** To make a home loan, First National Bank requires a debt–income ratio that is less than $\frac{2}{5}$. Would the homeowner whose income–debt table is given at the right qualify for a loan using these standards?

| *Income* | | *Debts* | |
|---|---|---|---|
| Salary | 3400 | Mortgage | 1,800 |
| Interest | 83 | Property tax | 104 |
| Rent | 650 | Insurance | 35 |
| Dividends | 34 | Liabilities | 120 |
| | | Credit card | 234 |
| | | Car loan | 197 |

**39.** Is the value of a ratio always less than 1? Explain.
[W]

# 4.2 Rates

## Objective A To write rates

A **rate** is a comparison of two quantities that have *different* units. A rate is written as a fraction.

A distance runner ran 26 miles in 4 hours. The distance-to-time rate is written

$\frac{26 \text{ miles}}{4 \text{ hours}} = \frac{13 \text{ miles}}{2 \text{ hours}}$

A rate is in **simplest form** when the numbers that form the rate have no common factors. Note that the units are written as part of the rate.

**Example 1** Write "6 roof supports for every 9 feet" as a rate in simplest form.

**Solution** $\frac{6 \text{ supports}}{9 \text{ feet}} = \frac{2 \text{ supports}}{3 \text{ feet}}$

**You Try It 1** Write "15 pounds of fertilizer for 12 trees" as a rate in simplest form.

**Your solution**

*Solution on p. A23*

## Objective B To write unit rates

A **unit rate** is a rate in which the number in the denominator is 1.

$\frac{\$3.25}{1 \text{ pound}}$ or \$3.25/pound is read "\$3.25 per pound."

To find unit rates, divide the number in the numerator of the rate by the number in the denominator of the rate.

A car traveled 344 miles on 16 gallons of gasoline. To find the miles per gallon (unit rate), divide the numerator of the rate by the denominator of the rate.

$\frac{344 \text{ miles}}{16 \text{ gallons}}$ is the rate.

$16\overline{)344.0}$ = 21.5   21.5 miles/gallon is the unit rate.

**Example 2** Write "300 feet in 8 seconds" as a unit rate.

**Solution** $\frac{300 \text{ feet}}{8 \text{ seconds}}$   $8\overline{)300.0}$ = 37.5

37.5 feet/second

**You Try It 2** Write "260 miles in 8 hours" as a unit rate.

**Your solution**

*Solution on p. A23*

### *Objective C To solve application problems*

The table at the right shows typical air fare costs for long routes.

| *Long Routes* | *Miles* | *Fare* |
|---|---|---|
| New York–Los Angeles | 2,475 | \$683 |
| San Francisco–Dallas | 1,464 | \$536 |
| Denver–Pittsburgh | 1,302 | \$525 |
| Minneapolis–Hartford | 1,050 | \$483 |

Find the cost per mile of the four routes. Which route is the most expensive, and which is the least expensive, for each mile flown?

**Strategy**
To find the cost per mile, divide the miles flown by the fare for each route. Compare the costs per mile to determine the most expensive and least expensive routes per mile.

**Solution**

New York–Los Angeles $\frac{683}{2,475} \approx 0.28$

San Francisco–Dallas $\frac{536}{1,464} \approx 0.37$

Denver–Pittsburgh $\frac{525}{1,302} \approx 0.40$

Minneapolis–Hartford $\frac{483}{1,050} = 0.46$

The Minneapolis–Hartford route is the most expensive per mile, and the New York–Los Angeles route is the least expensive per mile.

**Example 3**
As an investor, Jung Ho purchased 100 shares of stock for \$1500. One year later, Jung sold the 100 shares for \$1800. What was his profit per share?

**Strategy**
To find Jung's profit per share:

- Find the total profit by subtracting the original cost (\$1500) from the selling price (\$1800).
- Find the profit per share (unit rate) by dividing the total profit by the number of shares of stock (100).

**Solution**

$$\begin{array}{r} \$1800 \\ -\ 1500 \\ \hline \$300 \end{array} \text{ total profit} \qquad 100\overline{)\$300} = \$3$$

Jung Ho's profit per share was \$3.

**You Try It 3**
Erik Peltier, a jeweler, purchased 5 ounces of gold for \$1625. Later, he sold the 5 ounces for \$1720. What was Erik's profit per ounce?

**Your strategy**

**Your solution**

*Solution on p. A23*

## 4.2 Exercises

*Objective A*

Write as a rate in simplest form.

1. 3 pounds of meat for 4 people
2. 30 ounces in 24 glasses
3. $80 for 12 boards
4. 84 cents for 6 bars of soap
5. 300 miles on 15 gallons
6. 88 feet in 8 seconds
7. 20 children in 8 families
8. 48 leaves on 9 plants
9. 16 gallons in 2 hours
10. 25 ounces in 5 minutes

*Objective B*

Write as a unit rate.

11. 10 feet in 4 seconds
12. 816 miles in 6 days
13. $1300 earned in 4 weeks
14. $27,000 earned in 12 months
15. 1100 trees planted on 10 acres
16. 3750 words on 15 pages
17. $32.97 earned in 7 hours
18. $315.70 earned in 22 hours
19. 628.8 miles in 12 hours
20. 388.8 miles in 8 hours
21. 344.4 miles on 12.3 gallons of gasoline
22. 409.4 miles on 11.5 gallons of gasoline
23. $349.80 for 212 pounds
24. $11.05 for 3.4 pounds

*Objective C* Application Problems

25. An automobile was driven 326.6 miles on 11.5 gallons of gas. Find the number of miles driven per gallon of gas.

26. You drive 246.6 miles in 4.5 hours. Find the number of miles driven per hour.

27. The Saturn-5 rocket uses 534,000 gallons of fuel in 2.5 minutes. How much fuel does the rocket use in 1 minute?

**28.** Shawna Monte, a ski instructor, worked 6 months at a ski resort and earned $15,900. What was her wage per month?

**29.** An investor owns 600 shares of Ford Motor Company and receives $960 in yearly dividends. Find the dividend per share.

**30.** An investor purchased 420 shares of Charcor Corporation for $7,980. What was the cost per share?

**31.** Assume that Apple Computer produced 5000 compact discs for $26,536.32. Of the discs made, 122 did not meet company standards.
a. How many compact discs did meet company standards?
b. What was the cost per disc for those discs that met company standards?

**32.** The Pierre family purchased a 250-pound side of beef for $365.75 and had it packaged. During the packaging, 75 pounds of beef were discarded as waste.
a. How many pounds of beef were packaged?
b. What was the cost per pound for the packaged beef?

**33.** The Bear Valley Fruit Stand purchased 250 boxes of strawberries for $162.50. All the strawberries were sold for $312.50. What was the profit per box of strawberries?

## APPLYING THE CONCEPTS

Television advertisers use the rate of $\frac{\text{cost of commercial}}{\text{one thousand viewers}}$. This rate is an indication of the cost-effectiveness of an advertisement. One graph below gives the cost of a 30-second commercial for various television shows. The second graph gives the number of people watching the show. Use these graphs for Exercises 34–37.

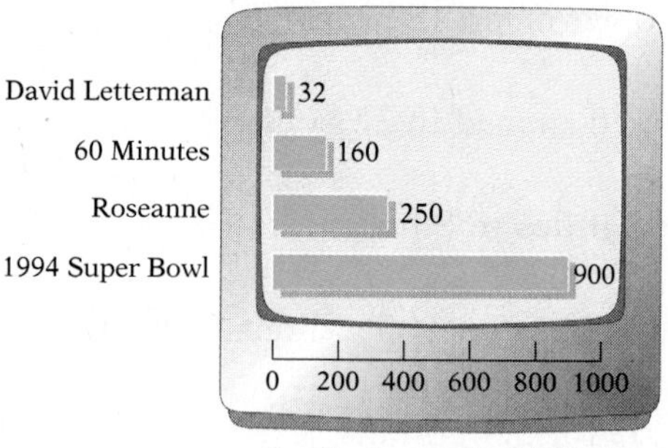

Cost (in thousands of dollars)

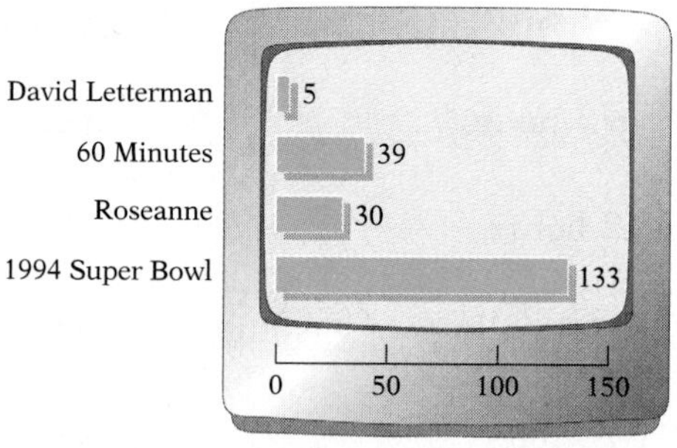

Number of Viewers (in millions)

**34.** What was the cost per thousand viewers for the 1994 Super Bowl?

**35.** What was the cost per thousand viewers for Roseanne?

**36.** [W] If you were an ad agency executive purchasing time for a commercial, would you try to have a low or a high cost per thousand viewers? Explain your answer.

**37.** Which of the four programs listed has the smallest cost per thousand viewers?

**38.** [W] The price–earnings ratio of a company's stock is one measure used by stock market analysts to assess the financial well-being of the company. Explain the meaning of the price–earnings ratio.

# 4.3 Proportions

### *Objective A To determine whether a proportion is true*

**POINT OF INTEREST**

Proportions were studied by the earliest mathematicians. Clay tablets uncovered by archeologists show evidence of proportions in Egyptian and Babylonian cultures dating from 1800 B.C.

A **proportion** is an expression of the equality of two ratios or rates.

$\frac{50 \text{ miles}}{4 \text{ gallons}} = \frac{25 \text{ miles}}{2 \text{ gallons}}$ Note that the units of the numerators are the same and the units of the denominators are the same.

$\frac{3}{6} = \frac{1}{2}$

A proportion is **true** if the fractions are equal when written in lowest terms.

In any true proportion, the "cross products" are equal.

➡ Is $\frac{2}{3} = \frac{8}{12}$ a true proportion?

$\frac{2}{3} \times \frac{8}{12}$ → $3 \times 8 = 24$, $2 \times 12 = 24$

Cross products *are* equal.

$\frac{2}{3} = \frac{8}{12}$ is a true proportion.

A proportion is **not true** if the fractions are not equal when reduced to lowest terms.

If the cross products are not equal, then the proportion is not true.

➡ Is $\frac{4}{5} = \frac{8}{9}$ a true proportion?

$\frac{4}{5} \times \frac{8}{9}$ → $5 \times 8 = 40$, $4 \times 9 = 36$

Cross products *are not* equal.

$\frac{4}{5} = \frac{8}{9}$ is not a true proportion.

**Example 1**

Use cross products to determine whether $\frac{5}{8} = \frac{10}{16}$ is a true proportion.

**Solution**

$\frac{5}{8} \times \frac{10}{16}$ → $8 \times 10 = 80$, $5 \times 16 = 80$

The proportion is true.

**You Try It 1**

Use cross products to determine whether $\frac{6}{10} = \frac{9}{15}$ is a true proportion.

**Your solution**

**Example 2**

Use cross products to determine whether $\frac{62 \text{ miles}}{4 \text{ gallons}} = \frac{33 \text{ miles}}{2 \text{ gallons}}$ is a true proportion.

**Solution**

$\frac{62}{4} \times \frac{33}{2}$ → $4 \times 33 = 132$, $62 \times 2 = 124$

The proportion is not true.

**You Try It 2**

Use cross products to determine whether $\frac{\$32}{6 \text{ hours}} = \frac{\$90}{8 \text{ hours}}$ is a true proportion.

**Your solution**

*Solutions on p. A24*

## *Objective B To solve proportions*

Sometimes one of the numbers in a proportion is unknown. In this case, it is necessary to *solve* the proportion.

To **solve** a proportion, find a number to replace the unknown so that the proportion is true.

➡ Solve: $\frac{9}{6} = \frac{3}{n}$

$\frac{9}{6} = \frac{3}{n}$

$9 \times n = 6 \times 3$ • Find the cross products.

$9 \times n = 18$

$n = 18 \div 9$ • Think of $9 \times n = 18$ as $9\overline{)18}$ with quotient $n$.

$n = 2$

Check:

$\frac{9}{6} \;\; \frac{3}{2}$ → $6 \times 3 = 18$ → $9 \times 2 = 18$

**Example 3**

Solve $\frac{n}{12} = \frac{25}{60}$ and check.

**Solution**

$n \times 60 = 12 \times 25$

$n \times 60 = 300$

$n = 300 \div 60$

$n = 5$

Check:

$\frac{5}{12} \;\; \frac{25}{60}$ → $12 \times 25 = 300$ → $5 \times 60 = 300$

**You Try It 3**

Solve $\frac{n}{14} = \frac{3}{7}$ and check.

**Your solution**

**Example 4**

Solve $\frac{4}{9} = \frac{n}{16}$. Write the answer to the nearest tenth.

**Solution**

$4 \times 16 = 9 \times n$

$64 = 9 \times n$

$64 \div 9 = n$

$7.1 \approx n$

*Note:* A rounded answer is an approximation. Therefore, the answer to a check will not be exact.

**You Try It 4**

Solve $\frac{5}{8} = \frac{n}{20}$. Write the answer to the nearest tenth.

**Your solution**

*Solutions on p. A24*

**Example 5**

Solve $\frac{28}{52} = \frac{7}{n}$ and check.

**Solution**

$28 \times n = 52 \times 7$
$28 \times n = 364$
$n = 364 \div 28$
$n = 13$

Check:

$\frac{28}{52} \quad \frac{7}{13}$ → $52 \times 7 = 364$
→ $28 \times 13 = 364$

**You Try It 5**

Solve $\frac{15}{20} = \frac{12}{n}$ and check.

**Your solution**

**Example 6**

Solve $\frac{15}{n} = \frac{8}{3}$. Write the answer to the nearest hundredth.

**Solution**

$15 \times 3 = n \times 8$
$45 = n \times 8$
$45 \div 8 = n$
$5.63 \approx n$

**You Try It 6**

Solve $\frac{12}{n} = \frac{7}{4}$. Write the answer to the nearest hundredth.

**Your solution**

**Example 7**

Solve $\frac{n}{9} = \frac{3}{1}$ and check.

**Solution**

$n \times 1 = 9 \times 3$
$n \times 1 = 27$
$n = 27 \div 1$
$n = 27$

Check:

$\frac{27}{9} \quad \frac{3}{1}$ → $9 \times 3 = 27$
→ $27 \times 1 = 27$

**You Try It 7**

Solve $\frac{n}{12} = \frac{4}{1}$ and check.

**Your solution**

**Example 8**

Solve $\frac{5}{9} = \frac{15}{n}$ and check.

**Solution**

$5 \times n = 9 \times 15$
$5 \times n = 135$
$n = 135 \div 5$
$n = 27$

Check:

$\frac{5}{9} \quad \frac{15}{27}$ → $9 \times 15 = 135$
→ $5 \times 27 = 135$

**You Try It 8**

Solve $\frac{3}{8} = \frac{12}{n}$ and check.

**Your solution**

*Solutions on p. A24*

## Objective C To solve application problems

**Example 9**

A mason determines that 9 cement blocks are required for a retaining wall 2 feet long. At this rate, how many cement blocks are required for a retaining wall that is 24 feet long?

**Strategy**

To find the number of cement blocks for a retaining wall 24 feet long, write and solve a proportion using $n$ to represent the number of blocks required.

**Solution**

$$\frac{9 \text{ cement blocks}}{2 \text{ feet}} = \frac{n \text{ cement blocks}}{24 \text{ feet}}$$

$$9 \times 24 = 2 \times n$$
$$216 = 2 \times n$$
$$216 \div 2 = n$$
$$108 = n$$

108 cement blocks are required for a 24-foot retaining wall.

**You Try It 9**

Twenty-four jars can be packed in 6 identical boxes. At this rate, how many jars can be packed in 15 boxes?

**Your strategy**

**Your solution**

**Example 10**

The dosage of a certain medication is 2 ounces for every 50 pounds of body weight. How many ounces of this medication are required for a person who weighs 175 pounds?

**Strategy**

To find the number of ounces of medication for a person weighing 175 pounds, write and solve a proportion using $n$ to represent the number of ounces of medication for a 175-pound person.

**Solution**

$$\frac{2 \text{ ounces}}{50 \text{ pounds}} = \frac{n \text{ ounces}}{175 \text{ pounds}}$$

$$2 \times 175 = 50 \times n$$
$$350 = 50 \times n$$
$$350 \div 50 = n$$
$$7 = n$$

7 ounces of medication are required for a 175-pound person.

**You Try It 10**

Three tablespoons of a liquid plant fertilizer are to be added to every 4 gallons of water. How many tablespoons of fertilizer are required for 10 gallons of water?

**Your strategy**

**Your solution**

*Solutions on p. A25*

## 4.3 Exercises

### *Objective A*

Determine whether the proportion is true or not true.

1. $\frac{4}{8} = \frac{10}{20}$

2. $\frac{39}{48} = \frac{13}{16}$

3. $\frac{7}{8} = \frac{11}{12}$

4. $\frac{15}{7} = \frac{17}{8}$

5. $\frac{27}{8} = \frac{9}{4}$

6. $\frac{3}{18} = \frac{4}{19}$

7. $\frac{45}{135} = \frac{3}{9}$

8. $\frac{3}{4} = \frac{54}{72}$

9. $\frac{16}{3} = \frac{48}{9}$

10. $\frac{15}{5} = \frac{3}{1}$

11. $\frac{7}{40} = \frac{7}{8}$

12. $\frac{9}{7} = \frac{6}{5}$

13. $\frac{50 \text{ miles}}{2 \text{ gallons}} = \frac{25 \text{ miles}}{1 \text{ gallon}}$

14. $\frac{16 \text{ feet}}{10 \text{ seconds}} = \frac{24 \text{ feet}}{15 \text{ seconds}}$

15. $\frac{6 \text{ minutes}}{5 \text{ cents}} = \frac{30 \text{ minutes}}{25 \text{ cents}}$

16. $\frac{16 \text{ pounds}}{12 \text{ days}} = \frac{20 \text{ pounds}}{14 \text{ days}}$

17. $\frac{\$15}{4 \text{ pounds}} = \frac{\$45}{12 \text{ pounds}}$

18. $\frac{270 \text{ trees}}{6 \text{ acres}} = \frac{90 \text{ trees}}{2 \text{ acres}}$

19. $\frac{300 \text{ feet}}{4 \text{ rolls}} = \frac{450 \text{ feet}}{7 \text{ rolls}}$

20. $\frac{1 \text{ gallon}}{4 \text{ quarts}} = \frac{7 \text{ gallons}}{28 \text{ quarts}}$

21. $\frac{\$65}{5 \text{ days}} = \frac{\$26}{2 \text{ days}}$

22. $\frac{80 \text{ miles}}{2 \text{ hours}} = \frac{110 \text{ miles}}{3 \text{ hours}}$

23. $\frac{7 \text{ tiles}}{4 \text{ feet}} = \frac{42 \text{ tiles}}{20 \text{ feet}}$

24. $\frac{15 \text{ feet}}{3 \text{ yards}} = \frac{90 \text{ feet}}{18 \text{ yards}}$

### Objective B

Solve. Round to the nearest hundredth.

25. $\frac{n}{4} = \frac{6}{8}$

26. $\frac{n}{7} = \frac{9}{21}$

27. $\frac{12}{18} = \frac{n}{9}$

28. $\frac{7}{21} = \frac{35}{n}$

29. $\frac{10}{12} = \frac{n}{24}$

30. $\frac{6}{n} = \frac{24}{36}$

31. $\frac{3}{n} = \frac{15}{10}$

32. $\frac{n}{45} = \frac{17}{135}$

33. $\frac{9}{4} = \frac{18}{n}$

34. $\frac{7}{15} = \frac{21}{n}$

35. $\frac{n}{6} = \frac{2}{3}$

36. $\frac{5}{12} = \frac{n}{144}$

37. $\frac{n}{5} = \frac{7}{8}$

38. $\frac{4}{n} = \frac{9}{5}$

39. $\frac{8}{5} = \frac{n}{6}$

40. $\frac{n}{11} = \frac{32}{4}$

41. $\frac{3}{4} = \frac{8}{n}$

42. $\frac{5}{12} = \frac{n}{8}$

43. $\frac{36}{20} = \frac{12}{n}$

44. $\frac{15}{n} = \frac{65}{100}$

45. $\frac{n}{15} = \frac{21}{12}$

46. $\frac{40}{n} = \frac{15}{8}$

47. $\frac{32}{n} = \frac{1}{3}$

48. $\frac{5}{8} = \frac{42}{n}$

49. $\frac{18}{11} = \frac{16}{n}$

50. $\frac{25}{4} = \frac{n}{12}$

51. $\frac{28}{8} = \frac{12}{n}$

52. $\frac{n}{30} = \frac{65}{120}$

53. $\frac{0.3}{5.6} = \frac{n}{25}$

54. $\frac{1.3}{16} = \frac{n}{30}$

55. $\frac{0.7}{9.8} = \frac{3.6}{n}$

56. $\frac{1.9}{7} = \frac{13}{n}$

### Objective C *Application Problems*

Solve. Round to the nearest hundredth.

57. A 6-ounce package of Puffed Wheat contains 600 calories. How many calories are in a 0.5-ounce serving of the cereal?

58. A 10-ounce container of egg noodles contains 700 milligrams of cholesterol. How many milligrams of cholesterol are in a 2-ounce serving of egg noodles?

59. Ron Stokes uses 2 pounds of fertilizer for every 100 square feet of lawn for landscape maintenance. At this rate, how many pounds of fertilizer did he use on a lawn that measures 2500 square feet?

Solve. Round to the nearest hundredth.

60. A nursery provides a liquid plant food by adding 1 gallon of water for each 2 ounces of plant food. At this rate, how many ounces of plant food are required for 25 gallons of water?

61. The property tax on a $90,000 home is $1800. At this rate, what is the property tax on a home worth $125,000?

62. The sales tax on a $90 purchase is $4.50. At this rate, what is the sales tax on a $300 purchase?

63. A brick wall 20 feet in length contains 1040 bricks. At the same rate, how many bricks would it take to build a wall 48 feet in length?

64. From past experience, a farmer knows that planting 50 acres of wheat will yield 2900 bushels of wheat. At the same rate, find the number of bushels of wheat the farmer can expect from another 240 acres of wheat.

65. The scale on a map is 1.25 inch equals 10 miles. Find the distance between Carlsbad and Del Mar, which are 2 inches apart on the map.

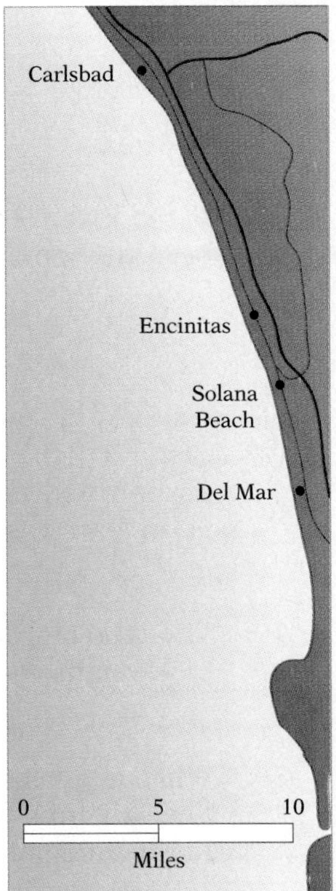


66. The scale on the plans for a new house is 1 inch equals 3 feet. Find the length and width of a room that measures 5 inches by 8 inches on the drawing.

67. The dosage for a medication is $\frac{1}{3}$ ounce for every 40 pounds of body weight. At this rate, how many ounces of medication should a physician prescribe for a patient who weighs 150 pounds?

68. Randall, who is on a reducing program, has lost 3 pounds in 5 weeks. At the same rate, how long will it take him to lose 27 pounds?

69. A pre-election survey showed that 2 out of every 3 eligible voters would cast ballots in the county election. At this rate, how many people in a county of 240,000 eligible voters would vote in the election?

70. A paint manufacturer suggests using 1 gallon of paint for every 400 square feet of a wall. At this rate, how many gallons of paint would be required for a room that has 1400 square feet of wall?

71. An automobile recall was based on tests that showed 24 braking defects in 1000 cars. At this rate, how many defects would be found in 40,000 cars?

72. Suppose a computer chip manufacturer knows from experience that in an average production run of 2000 circuit boards, 60 will be defective. What number of defective circuit boards can be expected from a run of 25,000 circuit boards?

73. You own 240 shares of a computer stock. The company declares a stock split of 5 shares for every 3 owned. How many shares of stock will you own after the stock split?

Solve. Round to the nearest hundredth.

**74.** Carlos Capasso owns 50 shares of Texas Utilities that pay dividends of $153. At this rate, what dividend would Carlos receive after buying 300 additional shares of Texas Utilities?

**75.** The director of data processing at a college estimates that the ratio of student time to administrative time on a certain computer is 3:2. During a month in which the computer was used 200 hours for administration, how many hours was it used by students?

### *APPLYING THE CONCEPTS*

The table at the right shows the gross domestic product, the federal budget, and the national debt for the years 1947 and 1992. Use the table for Exercises 76 and 77.

| *Year* | *Gross Domestic Product* | *Federal Budget* | *National Debt* |
|---|---|---|---|
| 1947 | 223 billion | 35 billion | 257 billion |
| 1992 | 5.87 trillion | 1.38 trillion | 4.00 trillion |

**76.** If the ratio of the national debt to the federal budget had been the same in 1992 as in 1947, what would the national debt have been in 1992? What is the difference between this value and the actual debt in 1992?

**77.** If the ratio of the national debt to the gross domestic product had been the same in 1992 as in 1947, what would the national debt have been in 1992? What is the difference between this value and the actual debt in 1992?

**78.** [W] A survey of voters in a city claimed that 2 people of every 5 who voted cast a ballot in favor of city amendment A and that 3 people of every 4 who voted cast a ballot against amendment A. Is this possible? Explain your answer.

**79.** The ratio of weight on the moon to weight on Earth is 1:6. If a bowling ball weighs 16 pounds on Earth, what would it weigh on the moon?

**80.** When engineers design a new car, they first build a model of the car. The ratio of a part on the model to the actual size is 2:5. If a door is 1.3 feet long on the model, what is the length of the door on the car?

**81.** [W] Write a word problem that requires solving a proportion to find the answer.

**82.** [W] Choose a local pizza restaurant and a particular type of pizza. Determine the size and cost of a medium pizza and of a large pizza. (Use regular prices and no special discounts.) Is $\frac{\text{cost of medium}}{\text{size of medium}}$ approximately equal to $\frac{\text{cost of large}}{\text{size of large}}$? Explain your answer.

# Project in Mathematics

## The Golden Ratio

There are certain designs that have been repeated over and over in both art and architecture. One of these involves the **golden rectangle.**

A golden rectangle is drawn at the right. Begin with a square that measures, say, 2 inches on a side. Now measure the distance from A to B. Place this length along the bottom of the square, starting at A. The resulting rectangle is a golden rectangle.

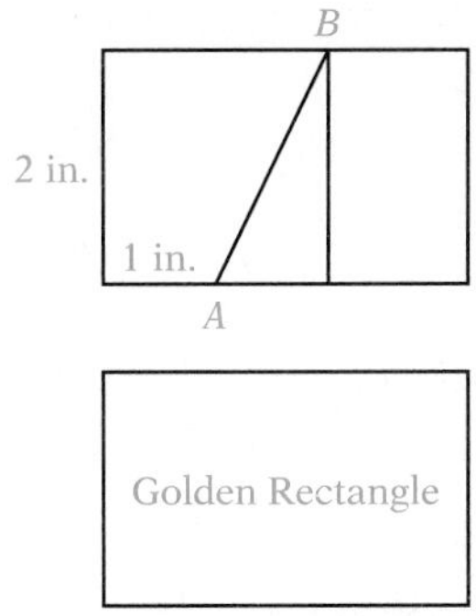


The **golden ratio** is the ratio of the length of the golden rectangle to its width. If you have drawn the rectangle following the procedure above, you will find that the golden ratio is approximately 1.6.

The golden ratio appears in many different situations. Some historians claim that some of the great pyramids of Egypt are based on the golden ratio. The drawing at the right is the Pyramid of Gizeh, which dates from approximately 2600 B.C. The ratio of the height to a side of the base is approximately 1.6.

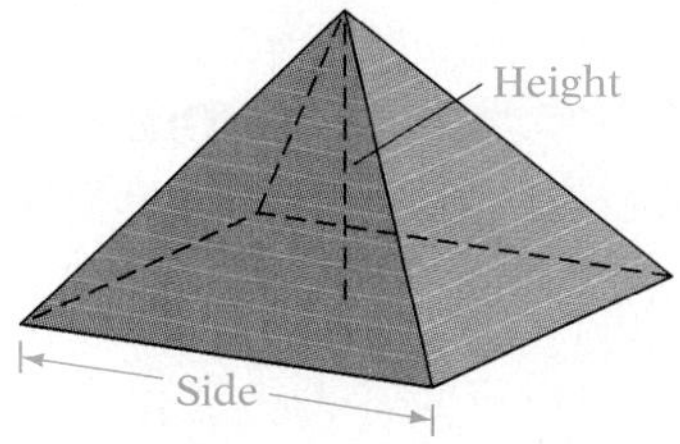


*Exercises:*

1. The canvas of the Mona Lisa painted by Leonardo da Vinci is a golden rectangle. However, there are other instances of the golden rectangle in the painting itself. Do some research on this painting and write a few paragraphs summarizing your findings.
2. What do $3 \times 5$ and $5 \times 8$ index cards have to do with the golden rectangle?
3. When was the United Nations building in New York built? What does the front of that building have to do with a golden rectangle?
4. When was the Parthenon in Athens, Greece, built? What does the front of that building have to do with a golden rectangle?

## Chapter Summary

**Key Words**

Quantities such as 8 feet and 60 miles are number quantities written with *units*.

A *ratio* is the comparison of two quantities that have the same units.

A ratio is in *simplest form* when the two numbers that form the ratio have no common factors.

A *rate* is the comparison of two quantities that have different units.

A rate is in *simplest form* when the numbers that form the rate have no common factors.

A *unit rate* is a rate in which the number in the denominator is 1.

A *proportion* is an expression of the equality of two ratios or rates.

**Essential Rules**

*To Find Unit Rates*

To find unit rates, divide the number in the numerator of the rate by the number in the denominator of the rate.

*To Solve a Proportion*

One of the numbers in a proportion may be unknown. To solve a proportion, find a number to replace the unknown so that the proportion is true.

*Ways to Express a Ratio*

A ratio can be written three different ways:

1. As a fraction
2. As two numbers separated by a colon (:)
3. As two numbers separated by the word *to*

# Chapter Review

## SECTION 4.1

1. Write the comparison 8 feet to 28 feet as a ratio in simplest form using a fraction, a colon (:), and the word *to*.

2. Write the comparison 6 inches to 15 inches as a ratio in simplest form using a fraction, a colon (:), and the word *to*.

3. Write the comparison 12 days to 12 days as a ratio in simplest form using a fraction, a colon (:), and the word *to*.

4. Write the comparison 32 dollars to 80 dollars as a ratio in simplest form using a fraction, a colon (:), and the word *to*.

5. The high temperature during a 24-hour period was 84 degrees and the low temperature was 42 degrees. Write the ratio of the high temperature to the low temperature for the 24-hour period.

6. A house had an original value of $80,000 but increased in value to $120,000 in 2 years. Find the ratio of the increase to the original value.

7. In 5 years the price of a calculator went from $40 to $24. What is the ratio of the decrease in price to the original price?

8. A retail computer store spends $30,000 a year on TV advertising and $12,000 on newspaper advertising. Find the ratio of TV advertising to newspaper advertising.

## SECTION 4.2

9. Write "100 miles in 3 hours" as a rate in simplest form.

10. Write "$15 in 4 hours" as a rate in simplest form.

11. Write "250 miles in 4 hours" as a unit rate.

12. Write "$300 earned in 40 hours" as a unit rate.

13. Write "326.4 miles on 12 gallons" as a unit rate.

14. Write "$8.75 for 5 pounds" as a unit rate.

15. Pascal Hollis purchased 80 shares of stock for $3580. What is the cost per share?

16. Mahesh drove 198.8 miles in 3.5 hours. Find the average number of miles he drove per hour.

17. A 15-pound turkey costs $10.20. What is the cost per pound?

18. The total cost of manufacturing 1000 radios was $36,600. Of the radios made, 24 did not pass inspection. Find the cost per radio of the radios that did pass inspection.

## SECTION 4.3

19. Determine whether the proportion is true or not true.
$\frac{5}{7} = \frac{25}{35}$

20. Determine whether the proportion is true or not true.
$\frac{2}{9} = \frac{10}{45}$

21. Determine whether the proportion is true or not true.
$\frac{8}{15} = \frac{32}{60}$

22. Determine whether the proportion is true or not true.
$\frac{3}{8} = \frac{10}{24}$

23. Solve the proportion.
$\frac{n}{8} = \frac{9}{2}$

24. Solve the proportion.
$\frac{16}{n} = \frac{4}{17}$

25. Solve the proportion. Round to hundredths.
$\frac{24}{11} = \frac{n}{30}$

26. Solve the proportion. Round to hundredths.
$\frac{18}{35} = \frac{10}{n}$

27. Monique used 1.5 pounds of fertilizer for every 200 square feet of lawn. How many pounds of fertilizer will she have to use on a lawn that measures 3000 square feet?

28. An insurance policy costs $3.87 for every $1000 of insurance. At this rate, what is the cost of $50,000 of insurance?

29. The property tax on a $45,000 home is $900. At the same rate, what is the property tax on a home valued at $120,000?

30. A brick wall 40 feet in length contains 448 concrete blocks. At the same rate, how many blocks would it take to build a wall that is 120 feet in length?

# Chapter Test

1. Write the comparison 12 days to 8 days as a ratio in simplest form using a fraction, a colon (:), and the word *to*.

2. Write the comparison 18 feet to 30 feet as a ratio in simplest form using a fraction, a colon (:), and the word *to*.

3. Write the comparison 27 dollars to 81 dollars as a ratio in simplest form using a fraction, a colon (:), and the word *to*.

4. Write the comparison 40 miles to 240 miles as a ratio in simplest form using a fraction, a colon (:), and the word *to*.

5. The average summer temperature in a California desert is 112 degrees. In a city 100 miles away, the average summer temperature is 86 degrees. Write the ratio of the city temperature to the desert temperature.

6. An automobile sales company spends \$25,000 each month for television advertising and \$40,000 each month for radio advertising. Find, as a fraction in simplest form, the ratio of the cost of radio advertising to the total cost of advertising.

7. Write "18 supports for every 8 feet" as a rate in simplest form.

8. Write "\$81 for 12 boards" as a rate in simplest form.

9. Write "\$22,036.80 earned in 12 months" as a unit rate.

10. Write "256.2 miles on 8.4 gallons of gas" as a unit rate.

**11.** 40 feet of lumber costs \$69.20. What is the per-foot cost of the lumber?

**12.** A plane travels 2421 miles in 4.5 hours. Find the plane's speed in miles per hour.

**13.** Determine whether the proportion is true or not true.

$$\frac{5}{14} = \frac{25}{70}$$

**14.** Determine whether the proportion is true or not true.

$$\frac{40}{125} = \frac{5}{25}$$

**15.** Solve the proportion.

$$\frac{n}{18} = \frac{9}{4}$$

**16.** Solve the proportion.

$$\frac{5}{12} = \frac{60}{n}$$

**17.** A research scientist estimates that the human body contains 88 pounds of water for every 100 pounds of body weight. At this rate, estimate the number of pounds of water in a college student who weighs 150 pounds.

**18.** The dosage of a medicine is $\frac{1}{4}$ ounce for every 50 pounds of body weight. How many ounces of this medication are required for a person who weighs 175 pounds?

**19.** The property tax on a house valued at \$75,000 is \$1500. At the same rate, find the property tax on a house valued at \$140,000.

**20.** Fifty shares of a utility stock pay a dividend of \$62.50. At the same rate, find the dividend paid on 500 shares of the utility stock.

# Cumulative Review

1. Subtract: $\begin{array}{r} 20{,}095 \\ -\ 10{,}937 \\ \hline \end{array}$

2. Write $2 \cdot 2 \cdot 2 \cdot 2 \cdot 3 \cdot 3 \cdot 3$ in exponential notation.

3. Simplify: $4 - (5 - 2)^2 \div 3 + 2$

4. Find the prime factorization of 160.

5. Find the LCM of 9, 12, and 18.

6. Find the GCF of 28 and 42.

7. Reduce $\frac{40}{64}$ to simplest form.

8. Find $4\frac{7}{15}$ more than $3\frac{5}{6}$.

9. What is $4\frac{5}{9}$ less than $10\frac{1}{6}$?

10. Multiply: $\frac{11}{12} \times 3\frac{1}{11}$

11. Find the quotient of $3\frac{1}{3}$ and $\frac{5}{7}$.

12. Simplify: $\left(\frac{2}{5} + \frac{3}{4}\right) \div \frac{3}{2}$

13. Write 4.0709 in words.

14. Round 2.09762 to the nearest hundredth.

15. Divide: $8.09\overline{)16.0976}$
Round to the nearest thousandth.

16. Convert $0.06\frac{2}{3}$ to a fraction.

**17.** Write the comparison 25 miles to 200 miles as a ratio in simplest form.

**18.** Write "87 cents for 6 bars of soap" as a rate in simplest form.

**19.** Write "250.5 miles on 7.5 gallons of gas" as a unit rate.

**20.** Solve $\frac{40}{n} = \frac{160}{17}$. Round to the nearest hundredth.

**21.** A car traveled 457.6 miles in 8 hours. Find the car's speed in miles per hour.

**22.** Solve the proportion.
$\frac{12}{5} = \frac{n}{15}$

**23.** You had \$1024 in your checking account. You then wrote checks for \$192 and \$88. What is your new checking account balance?

**24.** Malek Khatri buys a tractor for \$22,760. A down payment of \$5000 is required. The balance remaining is paid in 48 equal monthly installments. What is the monthly payment?

**25.** Yuko is assigned to read a book containing 175 pages. She reads $\frac{2}{5}$ of the book during Thanksgiving vacation. How many pages of the assignment remain to be read?

**26.** A building contractor bought $2\frac{1}{3}$ acres of land for \$84,000. What was the cost of each acre?

**27.** Benjamin Eli bought a shirt for \$21.79 and a tie for \$8.59. He used a \$50 bill to pay for the purchases. Find the amount of change.

**28.** A college baseball player had 42 hits in 155 at-bats. Find the baseball player's batting average. (Round to three decimal places.)

**29.** A soil conservationist estimates that a river bank is eroding at the rate of 3 inches every 6 months. At this rate, how many inches will be eroded in 50 months?

**30.** The dosage of a medicine is $\frac{1}{2}$ ounce for every 50 pounds of body weight. How many ounces of this medication are required for a person who weighs 160 pounds?

Chapter

# 5

# Percents

## Objectives

### Section 5.1

To write a percent as a fraction or a decimal

To write a fraction or decimal as a percent

### Section 5.2

To find the amount when the percent and the base are given

To solve application problems

### Section 5.3

To find the percent when the base and amount are given

To solve application problems

### Section 5.4

To find the base when the percent and amount are given

To solve application problems

### Section 5.5

To solve percent problems using proportions

To solve application problems

# Percent Symbol

The idea of using percent dates back many hundreds of years. Percents are used in business for all types of consumer and business loans, in chemistry to measure the percent concentration of an acid, in economics to measure the increases or decreases of the consumer price index (CPI), and in many other areas that affect our daily lives.

The word *percent* comes from the Latin phrase *per centum,* which means "by the hundred." The symbol that is used today for percent is %, but this was not always the symbol.

The present symbol apparently is a result of abbreviations for the word "percent."

One abbreviation was p. cent; later, p. 100 and p. $\overset{o}{c}$ were used.

From p. $\overset{o}{c}$, the abbreviation changed to p$\frac{o}{o}$ around the 17th century. This probably was a result of continual writing of p. $\overset{o}{c}$ and the eventual closing of the "c" to make an "o." By the 19th century, the "p" in front of the symbol p$\frac{o}{o}$ was no longer written. The bar that separated the o's became a slash, and the modern symbol % became widely used.

# 5.1 Introduction to Percents

## *Objective A To write a percent as a fraction or a decimal*

**Percent** means "parts of 100." In the figure at the right, there are 100 parts. Because 13 of the 100 parts are shaded, 13% of the figure is shaded.

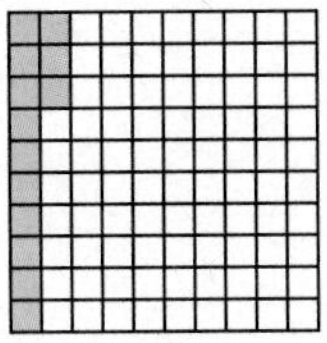

In most applied problems involving percents, it is necessary either to rewrite a percent as a fraction or a decimal or to rewrite a fraction or a decimal as a percent.

To write a percent as a fraction, remove the percent sign and multiply by $\frac{1}{100}$.

$$13\% = 13 \times \frac{1}{100} = \frac{13}{100}$$

To rewrite a percent as a decimal, remove the percent sign and multiply by 0.01.

$$13\% = 13 \times 0.01 = 0.13$$

Move the decimal point two places to the left. Then remove the percent sign.

---

**Example 1** Write 120% as a fraction and as a decimal.

**Solution**

$$120\% = 120 \times \frac{1}{100} = \frac{120}{100} = 1\frac{1}{5}$$

$$120\% = 120 \times 0.01 = 1.2$$

Note that percents larger than 100 are greater than 1.

**You Try It 1** Write 125% as a fraction and as a decimal.

**Your solution**

**Example 2** Write $16\frac{2}{3}\%$ as a fraction.

**Solution**

$$16\frac{2}{3}\% = 16\frac{2}{3} \times \frac{1}{100} = \frac{50}{3} \times \frac{1}{100} = \frac{50}{300} = \frac{1}{6}$$

**You Try It 2** Write $33\frac{1}{3}\%$ as a fraction.

**Your solution**

**Example 3** Write 0.5% as a decimal.

**Solution** $0.5\% = 0.5 \times 0.01 = 0.005$

**You Try It 3** Write 0.25% as a decimal.

**Your solution**

*Solutions on p. A25*

## *Objective B To write a fraction or a decimal as a percent*

A fraction or a decimal can be written as a percent by multiplying by 100%.

➡ Write $\frac{3}{8}$ as a percent.

$$\frac{3}{8} = \frac{3}{8} \times 100\% = \frac{3}{8} \times \frac{100}{1}\% = \frac{300}{8}\% = 37\frac{1}{2}\% \text{ or } 37.5\%$$

➡ Write 0.37 as a percent.

$$0.37 = 0.37 \times 100\% = 37\%$$

Move the decimal point two places to the right. Then write the percent sign.

**Example 4** Write 0.015 as a percent.

**Solution** $0.015 = 0.015 \times 100\%$
$= 1.5\%$

**You Try It 4** Write 0.048 as a percent.

**Your solution**

**Example 5** Write 2.15 as a percent.

**Solution** $2.15 = 2.15 \times 100\% = 215\%$

**You Try It 5** Write 3.67 as a percent.

**Your solution**

**Example 6** Write $0.33\frac{1}{3}$ as a percent.

**Solution** $0.33\frac{1}{3} = 0.33\frac{1}{3} \times 100\%$
$= 33\frac{1}{3}\%$

**You Try It 6** Write $0.62\frac{1}{2}$ as a percent.

**Your solution**

**Example 7** Write $\frac{2}{3}$ as a percent. Write the remainder in fractional form.

**Solution** $\frac{2}{3} = \frac{2}{3} \times 100\% = \frac{200}{3}\%$
$= 66\frac{2}{3}\%$

**You Try It 7** Write $\frac{5}{6}$ as a percent. Write the remainder in fractional form.

**Your solution**

**Example 8** Write $2\frac{2}{7}$ as a percent. Round to the nearest tenth.

**Solution** $2\frac{2}{7} = \frac{16}{7} = \frac{16}{7} \times 100\%$
$= \frac{1600}{7}\% \approx 228.6\%$

**You Try It 8** Write $1\frac{4}{9}$ as a percent. Round to the nearest tenth.

**Your solution**

*Solutions on p. A25*

## 5.1 Exercises

### *Objective A*

Write as a fraction and as a decimal.

1. 25%
2. 40%
3. 130%
4. 150%
5. 100%
6. 87%
7. 73%
8. 45%
9. 383%
10. 425%
11. 70%
12. 55%
13. 88%
14. 64%
15. 32%
16. 18%

Write as a fraction.

17. $66\frac{2}{3}\%$
18. $12\frac{1}{2}\%$
19. $83\frac{1}{3}\%$
20. $3\frac{1}{8}\%$
21. $11\frac{1}{9}\%$
22. $\frac{3}{8}\%$
23. $45\frac{5}{11}\%$
24. $15\frac{3}{8}\%$
25. $4\frac{2}{7}\%$
26. $5\frac{3}{4}\%$
27. $6\frac{2}{3}\%$
28. $8\frac{2}{3}\%$

Write as a decimal.

29. 6.5%
30. 12.3%
31. 0.55%
32. 2%
33. 8.25%
34. 5.05%
35. 6.75%
36. 3.08%
37. 0.45%
38. 6.4%
39. 80.4%
40. 16.7%

### *Objective B*

Write as a percent.

41. 0.16
42. 0.73
43. 0.05
44. 0.13
45. 0.01
46. 0.95
47. 0.70
48. 1.07
49. 1.24
50. 2.07
51. 0.004
52. 0.37
53. 0.006
54. 1.012
55. 3.106
56. 0.12

Write as a percent. Round to the nearest tenth.

57. $\frac{27}{50}$
58. $\frac{37}{100}$
59. $\frac{1}{3}$
60. $\frac{2}{5}$

**61.** $\frac{5}{8}$ **62.** $\frac{1}{8}$ **63.** $\frac{1}{6}$ **64.** $1\frac{1}{2}$

**65.** $\frac{7}{40}$ **66.** $1\frac{2}{3}$ **67.** $1\frac{7}{9}$ **68.** $\frac{7}{8}$

Write as a percent. Write the remainder in fractional form.

**69.** $\frac{15}{50}$ **70.** $\frac{12}{25}$ **71.** $\frac{7}{30}$ **72.** $\frac{1}{3}$

**73.** $2\frac{3}{8}$ **74.** $1\frac{2}{3}$ **75.** $2\frac{1}{6}$ **76.** $\frac{7}{8}$

## *APPLYING THE CONCEPTS*

**77.** Determine whether the statement is true or false. If the statement is false, give an example to show that the statement is false.
a. Multiplying a number by a percent always decreases the number.
b. Dividing by a percent always increases the number.
c. The word *percent* means "per hundred."
d. A percent is always less than one.

**78.** Write the part of the square that is shaded as a fraction, as a decimal, and as a percent. Write the part of the square that is not shaded as a fraction, as a decimal, and as a percent.

**79.** Explain in your own words how to change a percent to a decimal and
[W] a decimal to a percent.

**80.** A sale on computers advertised $\frac{1}{3}$ off the regular price. What percent of the regular price does this represent?

**81.** A suit was priced at 50% off the regular price. What fraction of the regular price does this represent?

**82.** If $\frac{2}{5}$ of the population voted in an election, what percent of the population did not vote?

**83.** Is $\frac{1}{2}$% the same as 0.5 If not, what is the difference between 0.5 and the decimal equivalent of $\frac{1}{2}$%?

**84.** Is 9.4% the same as $9\frac{2}{5}$%? If not, what is the difference between the decimal equivalent of 9.4% and $9\frac{2}{5}$%?

**85.** How can you recognize a fraction that represents a number that is less than 1%?

# 5.2 Percent Equations: Part I

## Objective A To find the amount when the percent and base are given

A real estate broker receives a payment that is 4% of an $85,000 sale. To find the amount the broker receives requires answering the question "4% of $85,000 is what?"

This sentence can be written using mathematical symbols and then solved for the unknown number.

| 4% | of | $85,000 | is | what? |
|---|---|---|---|---|
| ↓ | ↓ | ↓ | ↓ | ↓ |
| percent 4% | × | base $85,000 | = | amount $n$ |

*of* is written as × (times)
*is* is written as = (equals)
*what* is written as $n$ (the unknown number)

$0.04 \times \$85{,}000 = n$
$\$3400 = n$

Note that the percent is written as a decimal.

The broker receives a payment of $3400.

The solution was found by solving the basic percent equation for amount.

**The Basic Percent Equation**

| Percent | × | base | = | amount |
|---|---|---|---|---|

In most cases, the percent is written as a decimal before the basic percent equation is solved. However, some percents are more easily written as a fraction than as a decimal. For example,

$33\frac{1}{3}\% = \frac{1}{3}$ $\quad 66\frac{2}{3}\% = \frac{2}{3}$ $\quad 16\frac{2}{3}\% = \frac{1}{6}$ $\quad 83\frac{1}{3}\% = \frac{5}{6}$

**Example 1** Find 5.7% of 160.

**Solution** $n = 0.057 \times 160$
$n = 9.12$

Note that the words "what is" are missing from the problem but are implied by the word "Find."

**You Try It 1** Find 6.3% of 150.

**Your solution**

**Example 2** What is $33\frac{1}{3}\%$ of 90?

**Solution** $n = \frac{1}{3} \times 90$
$n = 30$

**You Try It 2** What is $16\frac{2}{3}\%$ of 66?

**Your solution**

*Solutions on p. A26*

## *Objective B To solve application problems*

Solving percent problems requires identifying the three elements of the basic percent equation. Recall that these three parts are percent, base, and amount. Usually the base follows the phrase "percent of."

The table at the right shows the numbers of visitors to the Rocky Mountains in 1985 from selected countries. By 1992, travel to the Rocky Mountains from Japan had increased by 53%, from Germany by 187%, from the U.K. by 148%, and from Australia by 119%.

| | *1985* |
|---|---|
| Japan | 160,094 |
| Germany | 134,920 |
| U.K. | 129,986 |
| Australia | 56,295 |

➡ Find the number of tourists who came to the Rocky Mountains from Germany in 1992.

**Strategy**
Write and solve the basic percent equation to find the increase in the number of tourists. The percent is 187%; the base is 134,920.

Add the increase to the number of tourists in 1985.

**Solution** Percent · base = amount

$$187\% \times 134{,}920 = n$$
$$1.87 \times 134{,}920 = n$$
$$252{,}300 \approx n$$
$$252{,}300 + 134{,}920 = 387{,}220$$

There were approximately 387,220 tourists from Germany who visited the Rocky Mountains in 1992.

**Example 3**
A quality control inspector found that 1.2% of 2500 telephones inspected were defective. How many telephones inspected were not defective?

**Strategy**
To find the number of nondefective phones:
- Find the number of defective phones. Write and solve a basic percent equation using $n$ to represent the number of defective phones (amount). The percent is 1.2% and the base is 2500.
- Subtract the number of defective phones from the number of phones inspected (2500).

**Solution**

$$1.2\% \times 2500 = n$$
$$0.012 \times 2500 = n$$
$$30 = n \text{ defective phones}$$
$$2500 - 30 = 2470$$

2470 telephones were not defective.

**You Try It 3**
An electrician's hourly wage was $13.50 before an 8% raise. What is the new hourly wage?

**Your strategy**

**Your solution**

*Solution on p. A26*

## 5.2 Exercises

### *Objective A*

Solve.

1. 8% of 100 is what?
2. 16% of 50 is what?
3. 27% of 40 is what?
4. 52% of 95 is what?
5. 0.05% of 150 is what?
6. 0.075% of 625 is what?
7. 125% of 64 is what?
8. 210% of 12 is what?
9. Find 10.7% of 485.
10. Find 12.8% of 625.
11. What is 0.25% of 3000?
12. What is 0.06% of 250?
13. 80% of 16.25 is what?
14. 26% of 19.5 is what?
15. What is $1\frac{1}{2}$% of 250?
16. What is $5\frac{3}{4}$% of 65?
17. $16\frac{2}{3}$% of 120 is what?
18. $83\frac{1}{3}$% of 246 is what?
19. What is $33\frac{1}{3}$% of 630?
20. What is $66\frac{2}{3}$% of 891?
21. Which is larger: 5% of 95, or 75% of 6?
22. Which is larger: 112% of 5, or 0.45% of 800?
23. Which is larger: 82% of 16, or 20% of 65?
24. Which is smaller: 15% of 80, or 95% of 15?
25. Which is larger: 22% of 120, or 84% of 32?
26. Which is smaller: 2% of 1500, or 72% of 40?
27. Find 31.294% of 82,460.
28. Find 123.94% of 275,976.

### *Objective B* Application Problems

29. Natasha Gomez receives a salary of $2240 per month. Of this amount 18% is deducted for income tax. Find the amount deducted for income tax.

30. In a city election for mayor, the successful candidate received 53% of the 385,000 votes cast. How many of the votes did the successful candidate receive?

31. Hidden Valley Antique Emporium expects to receive $16\frac{2}{3}\%$ of the shop's sales as profit. What is the profit in a month when the total sales are $24,000?
32. A student survey in a community college found that $33\frac{1}{3}\%$ of its 7500 beginning students did not complete their first year of college. How many of the beginning students did not complete the first year of college?
33. An RV sales center offers a 7% rebate on the Roadtrek model. What rebate would a buyer receive on a Roadtrek that costs $24,000?
34. A farmer is given an income tax credit of 10% of the cost of farm machinery. What tax credit would the farmer receive on farm equipment that cost $85,000?
35. A sales tax of 6% of the cost of a car was added to the purchase price of $9500.
    a. How much was the sales tax?
    b. What is the total cost of the car including sales tax?
36. During the packaging process for oranges, spoiled oranges are discarded by an inspector. In 1 day an inspector found that 4.8% of the 20,000 pounds of oranges inspected were spoiled.
    a. How many pounds of oranges were spoiled?
    b. How many pounds of oranges were not spoiled?
37. Funtimes Amusement Park has 550 employees and must hire an additional 22% for the vacation season. What is the total number of employees needed for the vacation season?

## *APPLYING THE CONCEPTS*

The graph at the right shows where permanent injuries to 154 football players occurred. Use this graph for Exercises 38 and 39.

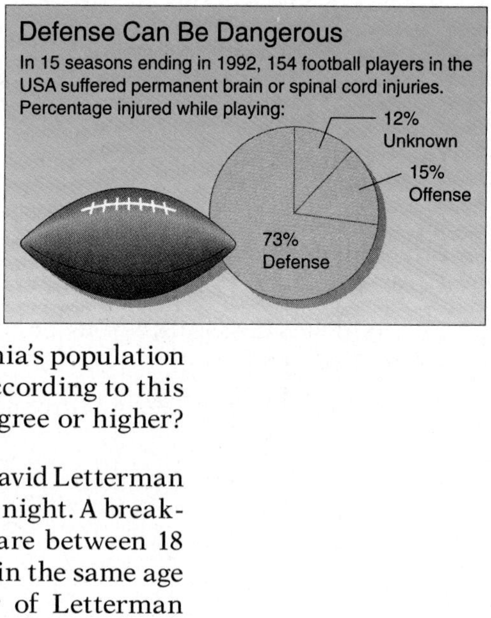


38. How many permanent injuries were obtained by playing defense?
39. How many permanent injuries were obtained by playing offense?
40. According to the U.S. Census Bureau, 23.4% of California's population of 29,760,000 people had a college degree or higher. According to this information, how many Californians had a college degree or higher?
41. According to statistics on late-night television shows, David Letterman and Ted Koppel each have 4.7 million viewers on a given night. A breakdown by age shows that 41% of Letterman's viewers are between 18 and 34 years old, whereas 25% of Koppel's viewers are in the same age group. What is the difference between the number of Letterman viewers in this age group and the number of Koppel viewers in this age group?
42. A chef purchased a new knife for $37 and a stock pot for $55. If the sales tax is 7.75% of the total, how much change did the chef receive from a $100 bill?
43. A car company increased the average price of its new models by 2.6%. If the average price of a car was $7856, what was the amount of the increase?

# 5.3 Percent Equations: Part II

## Objective A To find the percent when the base and amount are given

A recent promotional game at a grocery store listed the probability of winning a prize as "1 chance in 2." A percent can be used to describe the chance of winning. This requires answering the question "What percent of 2 is 1?"

The chance of winning can be found by solving the basic percent equation for *percent*.

What percent of 2 is 1?

$$\boxed{\begin{matrix}\text{percent}\\ n\end{matrix}} \times \boxed{\begin{matrix}\text{base}\\ 2\end{matrix}} = \boxed{\begin{matrix}\text{amount}\\ 1\end{matrix}}$$

$$\begin{aligned} n \times 2 &= 1 \\ n &= 1 \div 2 \\ n &= 0.5 \\ n &= 50\% \end{aligned}$$

• The solution must be written as a percent to answer the question.

There is a 50% chance of winning a prize.

**POINT OF INTEREST**

Inflation increased by approximately 275% between 1971 and 1994. The price of a first class stamp has increased by 300%, not much more than inflation.

**Example 1** What percent of 40 is 30?

**Solution**

$$\begin{aligned} n \times 40 &= 30 \\ n &= 30 \div 40 \\ n &= 0.75 \\ n &= 75\% \end{aligned}$$

**You Try It 1** What percent of 32 is 16?

**Your solution**

**Example 2** What percent of 12 is 27?

**Solution**

$$\begin{aligned} n \times 12 &= 27 \\ n &= 27 \div 12 \\ n &= 2.25 \\ n &= 225\% \end{aligned}$$

**You Try It 2** What percent of 15 is 48?

**Your solution**

**Example 3** 25 is what percent of 75?

**Solution**

$$\begin{aligned} 25 &= n \times 75 \\ 25 \div 75 &= n \\ 0.33\frac{1}{3} &= n \\ 33\frac{1}{3}\% &= n \end{aligned}$$

**You Try It 3** 30 is what percent of 45?

**Your solution**

*Solutions on p. A26*

## *Objective B To solve application problems*

To solve percent problems, remember that it is necessary to identify the percent, base, and amount. Usually the base follows the phrase "percent of."

**Example 4**
The Kaminski family has an income of \$1500 each month and makes a car payment of \$180 a month. What percent of the monthly income is the car payment?

**Strategy**
To find what percent of the income the car payment is, write and solve the basic percent equation using $n$ to represent the unknown percent. The base is \$1500 and the amount is \$180.

**Solution**

$$n \times \$1500 = \$180$$
$$n = \$180 \div \$1500$$
$$n = 0.12 = 12\%$$

The car payment is 12% of the monthly income.

**You Try It 4**
Tomo Nagata had an income of \$20,000 and paid \$3000 in income tax. What percent of the income is the income tax?

**Your strategy**

**Your solution**

**Example 5**
Claudia Sloan missed 25 questions out of 200 on a state nursing license exam. What percent of the questions did she answer correctly?

**Strategy**
To find what percent of the questions she answered correctly:

- Find the number of questions the nurse answered correctly (200 − 25).
- Write and solve a basic percent equation using $n$ to represent the unknown percent. The base is 200 and the amount is the number of questions answered correctly.

**Solution** 200 − 25 = 175 questions answered correctly.

$$n \times 200 = 175$$
$$n = 175 \div 200$$
$$n = 0.875 = 87.5\%$$

Claudia answered 87.5% of the questions correctly.

**You Try It 5**
A survey of 1000 people showed that 667 people favored Minna Oliva for governor of the state. What percent of the people surveyed did not favor her?

**Your strategy**

**Your solution**

*Solutions on p. A27*

## 5.3 Exercises

### *Objective A*

Solve.

1. What percent of 75 is 24?
2. What percent of 80 is 20?
3. 15 is what percent of 90?
4. 24 is what percent of 60?
5. What percent of 12 is 24?
6. What percent of 6 is 9?
7. What percent of 16 is 6?
8. What percent of 24 is 18?
9. 18 is what percent of 100?
10. 54 is what percent of 100?
11. 5 is what percent of 2000?
12. 8 is what percent of 2500?
13. What percent of 6 is 1.2?
14. What percent of 2.4 is 0.6?
15. 16.4 is what percent of 4.1?
16. 5.3 is what percent of 50?
17. 1 is what percent of 40?
18. 0.3 is what percent of 20?
19. What percent of 48 is 18?
20. What percent of 11 is 88?
21. What percent of 2800 is 7?
22. What percent of 400 is 12?
23. 4.2 is what percent of 175?
24. 41.79 is what percent of 99.5?
25. What percent of 86.5 is 8.304?
26. What percent of 1282.5 is 2.565?

### *Objective B* Application Problems

27. A company spends \$4500 of its \$90,000 budget for advertising. What percent of the budget is spent for advertising?

28. Last month a thrift shop had an income of \$2812.50. The thrift store pays \$900 a month for rent. What percent of last month's income was spent for rent?

29. Kodak stock is trading at $\$62\frac{1}{2}$ and pays a dividend of \$2.00. What percent of the stock price is the dividend?

30. A home was valued at $180,000 in 1990. The value of the home in 1993 was $162,000. What percent of the 1990 value is the 1993 value?

31. A television survey of 4000 families found that 2500 liked a new TV show. What percent of the families surveyed liked the new TV show?

32. A house in Phoenix was bought for $124,000. Three years later the house was sold for $155,000.
a. Find the increase in price.
b. What percent of the original price is the increase?

33. A 3.5-pound beef roast contains 0.7 pound of fat.
a. What percent of the beef roast is fat?
b. At the same rate, how many pounds of fat would there be in a side of beef weighing 550 pounds?

34. To receive a license to sell insurance, an insurance account executive must answer correctly 70% of the 250 questions on a test. Nicholas Mosley answered 177 questions correctly. Did he pass the test?

35. A test of the breaking strength of concrete slabs for freeway construction found that 3 of the 200 tested did not meet safety requirements. What percent of the slabs did meet safety requirements?

### *APPLYING THE CONCEPTS*

36. Make a table showing the decimal and percent equivalents of $\frac{1}{3}$, $\frac{5}{6}$, $\frac{3}{8}$, and $\frac{5}{8}$.

37. [W] Write a short essay on the use of percents in our everyday life.

38. Borland International offers its customers a $50 rebate on the purchase of a new computer program costing $189.95 before the rebate. If you purchase this program, what percent of your final cost is, to the nearest tenth of a percent, the rebate?

The graph at the right shows how the federal government spent $1.4 trillion in 1993. Use this graph for Exercises 39 and 40.

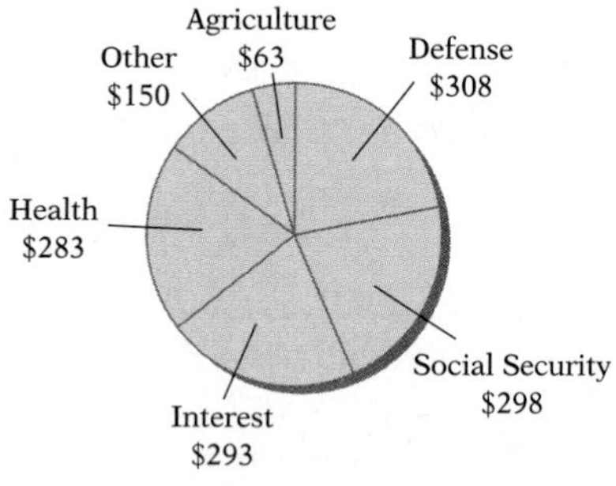

Federal Budget (in billions)

39. What percent of the total amount was spent for interest on the public debt? Round to the nearest tenth of a percent.

40. What percent of the total amount was spent for Social Security? Round to the nearest tenth of a percent.

41. Public utility companies will provide consumers with an analysis of their energy bills. For one customer, it was determined that $76 of a total bill of $134 was spent for home heating. What percent of the total bill was home heating?

42. The original cost of the Statue of Liberty was approximately $24,000. The cost to refurbish the statue today is approximately $780,000. What percent of the original cost is the cost to refurbish the Statue of Liberty?

# 5.4 Percent Equations: Part III

## *Objective A To find the base when the percent and amount are given*

**POINT OF INTEREST**

The sales of CD-ROM players is projected to increase to 22 million by 1997. That is a 900% increase over the 2.2 million units that were sold in 1992.

Each year an investor receives a payment that equals 12% of the value of an investment. This year that payment amounted to \$480. To find the value of the investment this year, the investor must answer the question "12% of what value is \$480?"

The value of the investment can be found by solving the basic percent equation for the base.

| 12% | of | what | is | \$480 |
|---|---|---|---|---|
| ↓ | ↓ | ↓ | ↓ | ↓ |
| percent 12% | × | base $n$ | = | amount \$480 |

$$0.12 \times n = \$480$$
$$n = \$480 \div 0.12$$
$$n = \$4000$$

This year the investment is worth \$4000.

**Example 1** 18% of what is 900?

**Solution** $0.18 \times n = 900$
$n = 900 \div 0.18$
$n = 5000$

**You Try It 1** 86% of what is 215?

**Your solution**

**Example 2** 30 is 1.5% of what?

**Solution** $30 = 0.015 \times n$
$30 \div 0.015 = n$
$2000 = n$

**You Try It 2** 15 is 2.5% of what?

**Your solution**

**Example 3** $33\frac{1}{3}$% of what is 7?

**Solution** $\frac{1}{3} \times n = 7$ • Note the percent is written as a fraction.
$n = 7 \div \frac{1}{3}$
$n = 21$

**You Try It 3** $16\frac{2}{3}$% of what is 5?

**Your solution**

*Solutions on p. A27*



## *Objective B To solve application problems*

To solve percent problems, it is necessary to identify the percent, base, and amount. Usually the base follows the phrase "percent of."

**Example 4**
A business office bought a used copy machine for \$450, which was 75% of the original cost. What was the original cost of the copier?

**Strategy**
To find the original cost of the copier, write and solve the basic percent equation using $n$ to represent the original cost (base). The percent is 75% and the amount is \$450.

**Solution**

$$75\% \times n = \$450$$
$$0.75 \times n = \$450$$
$$n = \$450 \div 0.75$$
$$n = \$600$$

The original cost of the copier was \$600.

**You Try It 4**
A used car has a value of \$3876, which is 51% of the car's original value. What was the car's original value?

**Your strategy**

**Your solution**

**Example 5**
A carpenter's wage this year is \$19.80 per hour, which is 110% of last year's wage. What was the increase in the hourly wage over last year?

**Strategy**
To find the increase in the hourly wage over last year:

- Find last year's wage. Write and solve the basic percent equation using $n$ to represent last year's wage (base). The percent is 110% and the amount is \$19.80.
- Subtract last year's wage from this year's wage (\$19.80).

**Solution**

$$110\% \times n = \$19.80$$
$$1.10 \times n = \$19.80$$
$$n = \$19.80 \div 1.10$$
$$n = \$18.00.$$ • last year's wage

$$\$19.80 - \$18.00 = \$1.80$$

The increase in the hourly wage was \$1.80.

**You Try It 5**
Chang's Sporting Goods has a tennis racket on sale for \$44.80, which is 80% of the original price. What is the difference between the original price and the sale price?

**Your strategy**

**Your solution**

*Solutions on p. A28*

## 5.4 Exercises

*Objective A*

Solve.

1. 12% of what is 9?

2. 38% of what is 171?

3. 8 is 16% of what?

4. 54 is 90% of what?

5. 10 is 10% of what?

6. 37 is 37% of what?

7. 30% of what is 25.5?

8. 25% of what is 21.5?

9. 2.5% of what is 30?

10. 10.4% of what is 52?

11. 125% of what is 24?

12. 180% of what is 21.6?

13. 18 is 240% of what?

14. 24 is 320% of what?

15. 4.8 is 15% of what?

16. 87.5 is 50% of what?

17. 25.6 is 12.8% of what?

18. 45.014 is 63.4% of what?

19. 0.7% of what is 0.56?

20. 0.25% of what is 1?

21. 30% of what is 2.7?

22. 78% of what is 3.9?

23. 84 is $16\frac{2}{3}$% of what?

24. 120 is $33\frac{1}{3}$% of what?

25. $66\frac{2}{3}$% of what is 72?

26. $83\frac{1}{3}$% of what is 13.5?

27. 6.59% of what is 469.35?

28. 182.3% of what is 46,253?

**Objective B** *Application Problems*

29. A mechanic estimates that the brakes of an RV still have 6000 miles of wear. This amounts to 12% of the estimated safe-life use of the brakes. What is the estimated life of the brakes?

30. A used mobile home was purchased for $18,000. This amount was 64% of the cost of a new mobile home. What was the cost of a new mobile home?

31. Redwood City has a population of 42,000. This is 75% of what the population was 5 years ago. What was the city's population 5 years ago?

32. Solitude Ski Resort had a snowfall of 198 inches during 1993. This is 110% of the snowfall in 1992. What was the snowfall in 1992?

33. Circuit City advertised a scientific calculator for $55.80. This amount was 120% of the cost at the May Company. What is the price of the calculator at the May Company?

34. A salesperson received a commission of $820 for selling a car. This was 5% of the selling price of the car. What was the selling price of the car?

35. During a quality control test, Micronics found that 24 computer boards were defective. This amount was 0.8% of the computer boards tested.
    a. How many computer boards were tested?
    b. How many computer boards tested were not defective?

36. Of the calls a directory assistance operator received, 441 were requests for telephone numbers listed in the current directory. This accounted for 98% of the calls for assistance that the operator received.
    a. How many calls did the operator receive?
    b. How many telephone numbers requested were not listed in the current directory?

## APPLYING THE CONCEPTS

The table at the right contains nutrition information about a breakfast cereal. Solve Exercises 37 and 38 with information taken from this table.

**NUTRITION INFORMATION**

SERVING SIZE: 1.4 OZ WHEAT FLAKES WITH 0.4 OZ. RAISINS: 39.4 g. ABOUT 1/2 CUP

SERVINGS PER PACKAGE: ...............14

| PERCENTAGE OF U.S. RECOMMENDED DAILY ALLOWANCES (U.S. RDA) | CEREAL & RAISINS | WITH 1/2 CUP VITAMINS A & D SKIM MILK |
|---|---|---|
| PROTEIN | 4 | 15 |
| VITAMIN A | 15 | 20 |
| VITAMIN C | ** | 2 |
| THIAMIN | 25 | 30 |
| RIBOFLAVIN | 25 | 35 |
| NIACIN | 25 | 35 |
| CALCIUM | ** | 15 |
| IRON | 100 | 100 |
| VITAMIN D | 10 | 25 |
| VITAMIN $B_6$ | 25 | 25 |
| FOLIC ACID | 25 | 25 |
| VITAMIN $B_{12}$ | 25 | 30 |
| PHOSPHOROUS | 10 | 15 |
| MAGNESIUM | 10 | 20 |
| ZINC | 25 | 30 |
| COPPER | 2 | 4 |

* 2% MILK SUPPLIES AN ADDITIONAL 20 CALORIES. 2 g FAT, AND 10 mg CHOLESTEROL.

** CONTAINS LESS THAN 2% OF THE U.S. RDA OF THIS NUTRIENT

37. The recommended daily amount of protein for an 80-pound child is 44 grams. Find the amount of protein in one serving of this cereal with skim milk.

38. The recommended daily amount of zinc for an adult male is 18 mg. Find the amount of zinc in one serving of cereal with skim milk.

39. [W] Increase a number by 10%. Now decrease the result by 10%. Is the result the original number? Explain.

40. When the Resolution Trust Corporation (RTC) takes over a failed bank, it may be necessary to sell some of the assets of the bank at less than the amount owed. As an example, a buyer may be able to purchase a building for "25 cents on the dollar." What percent of the owed price is the buyer paying?

# 5.5 Percent Problems: Proportion Method

## *Objective A To solve percent problems using proportions*

Problems that can be solved using the basic percent equation can also be solved using proportions.

The proportion method is based on writing two ratios. One ratio is the percent ratio, written as $\frac{\text{percent}}{100}$. The second ratio is the amount-to-base ratio, written as $\frac{\text{amount}}{\text{base}}$. These two ratios form the proportion

$$\frac{\textbf{percent}}{\textbf{100}} = \frac{\textbf{amount}}{\textbf{base}}$$

To use the proportion method, first identify the percent, the amount, and the base (the base usually follows the phrase "percent of").

What is 23% of 45?

$$\frac{23}{100} = \frac{n}{45}$$

$$\begin{aligned} 23 \times 45 &= 100 \times n \\ 1035 &= 100 \times n \\ 1035 \div 100 &= n \\ 10.35 &= n \end{aligned}$$

What percent of 25 is 4?

$$\frac{n}{100} = \frac{4}{25}$$

$$\begin{aligned} n \times 25 &= 100 \times 4 \\ n \times 25 &= 400 \\ n &= 400 \div 25 \\ n &= 16\% \end{aligned}$$

12 is 60% of what number?

$$\frac{60}{100} = \frac{12}{n}$$

$$\begin{aligned} 60 \times n &= 100 \times 12 \\ 60 \times n &= 1200 \\ n &= 1200 \div 60 \\ n &= 20 \end{aligned}$$

**Example 1** 15% of what is 7? Round to the nearest hundredth.

**Solution**

$$\frac{15}{100} = \frac{7}{n}$$

$$\begin{aligned} 15 \times n &= 100 \times 7 \\ 15 \times n &= 700 \\ n &= 700 \div 15 \\ n &\approx 46.67 \end{aligned}$$

**You Try It 1** 26% of what is 22? Round to the nearest hundredth.

**Your solution**

**Example 2** 30% of 63 is what?

**Solution**

$$\frac{30}{100} = \frac{n}{63}$$

$$\begin{aligned} 30 \times 63 &= n \times 100 \\ 1890 &= n \times 100 \\ 1890 \div 100 &= n \\ 18.90 &= n \end{aligned}$$

**You Try It 2** 16% of 132 is what?

**Your solution**

*Solutions on p. A28*

## *Objective B To solve application problems*

**Example 3**
An antiques dealer found that 86% of the 250 items that were sold during one month sold for under \$1000. How many items sold for under \$1000?

**Strategy**
To find the number of items that sold for under \$1000, write and solve a proportion using $n$ to represent the number of items sold (amount) for less than \$1000. The percent is 86% and the base is 250.

**Solution**

$$\frac{86}{100} = \frac{n}{250}$$
$$86 \times 250 = 100 \times n$$
$$21{,}500 = 100 \times n$$
$$21{,}500 \div 100 = n$$
$$215 = n$$

215 items sold for under \$1000.

**You Try It 3**
Last year it snowed 64% of the 150 days of the ski season at a resort. How many days did it snow?

**Your strategy**

**Your solution**

**Example 4**
In a test of the strength of nylon rope, 5 pieces of the 25 pieces tested did not meet the test standards. What percent of the nylon ropes tested did meet the standards?

**Strategy**
To find the percent of ropes tested that met the standards:

- Find the number of ropes that met the test standards (25 − 5).
- Write and solve a proportion using $n$ to represent the percent of ropes that met the test standards. The base is 25. The amount is the number of ropes that met the standards.

**Solution** $25 - 5 = 20$ ropes met test standards

$$\frac{n}{100} = \frac{20}{25}$$
$$n \times 25 = 100 \times 20$$
$$n \times 25 = 2000$$
$$n = 2000 \div 25$$
$$n = 80$$

80% of the ropes tested did meet the test standards.

**You Try It 4**
Five ballpoint pens in a box of 200 were found to be defective. What percent of the pens were not defective?

**Your strategy**

**Your solution**

*Solutions on p. A29*

## 5.5 Exercises

### *Objective A*

Solve.

1. 26% of 250 is what?
2. What is 18% of 150?
3. 37 is what percent of 148?
4. What percent of 150 is 33?
5. 68% of what is 51?
6. 126 is 84% of what?
7. What percent of 344 is 43?
8. 750 is what percent of 50?
9. 82 is 20.5% of what?
10. 2.4% of what is 21?
11. What is 6.5% of 300?
12. 96% of 75 is what?
13. 7.4 is what percent of 50?
14. What percent of 1500 is 693?
15. 50.5% of 124 is what?
16. What is 87.4% of 255?
17. 120% of what is 6?
18. 14 is 175% of what?
19. What is 250% of 18?
20. 325% of 4.4 is what?
21. 33 is 220% of what?
22. 160% of what is 40?

### *Objective B* Application Problems

23. A charity organization spent $2940 for administrative expenses. This amount is 12% of the money it collected. What is the total amount of money that the club collected?

24. A manufacturer of an anti-inflammatory drug claims that the drug will be effective for 6 hours. An independent testing service determined that the drug was effective only 80% of the length of time claimed by the manufacturer. Find the length of time the drug will be effective as determined by the testing service.

25. A calculator can be purchased for $28.50. This amount is 40% of the cost of the calculator 8 years ago. What was the cost of the calculator 8 years ago?

26. The Rincon Fire Department received 24 false alarms out of a total of 200 alarms received. What percent of the alarms received were false alarms?

**27.** Enrico made errors on 5 words on a typing test. The total length of the test was 250 words. What percent of the words did he type correctly?

**28.** It cost a mechanic $174 to repair a car. Of this cost, 55% was for labor and 45% was for parts.
a. What was the labor cost?
b. What was the total cost for parts?

**29.** In 1994, the National Football League increased the amount the players in the Super Bowl would receive. The winner's share was $38,000 per player, which was a $2000 increase from the previous year. What percent of the winner's share the previous year was the increase? Round to the nearest tenth of a percent.

**30.** In 1994, the National Football League increased the amount the players in the Super Bowl would receive. The loser's share was $23,500 per player, which was a $5500 increase from the previous year. What percent of the previous year's loser's share was the increase? Round to the nearest tenth of a percent.

## *APPLYING THE CONCEPTS*

The figure at the right shows the projections of the reduced corn harvest caused by the 1993 midwestern floods. Use the figure to answer Exercises 31 and 32.

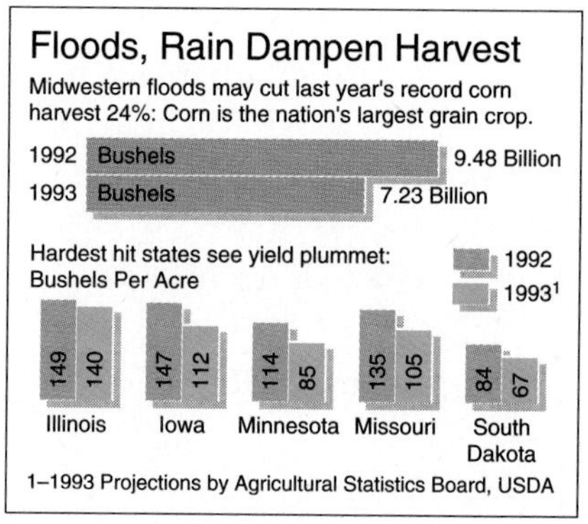


**31.** Find the percent decrease in the total harvest of the 5 states. Round to the nearest tenth of a percent.

**32.** Which of the states shown had the largest decrease? Which had the smallest decrease?

**33.** A $10,000, 8% bond has a quoted price of $10,500. Look up in a business math book the meaning of *current yield.* Then find the current yield on the $10,000 bond.

The table at the right shows the projected population increase for selected countries by 2010. Use this table for Exercises 34 and 35.

| *Country* | *Current Population* | *Projected Increase* |
|---|---|---|
| Netherlands | 15,200,000 | 1,400,000 |
| France | 56,900,000 | 1,700,000 |
| Kenya | 26,200,000 | 18,600,000 |
| United States | 260,000,000 | 35,500,000 |
| Mexico | 87,700,000 | 32,000,000 |

**34.** What percent of the Netherlands current population is the projected increase? Round to the nearest tenth of a percent.

**35.** What percent of the current population of the United States is the projected increase? Round to the nearest tenth of a percent.

**36.** The land area of North America is approximately 9,400,000 square miles. This represents about 16% of the total land area of the world. What is the approximate land area of the world?

**37.** A certain solution of iodine is used to study the function of the thyroid gland. Approximately every 8 days, the potency of the solution is reduced by 50% of the amount present at that time. By what percent has the potency decreased after 24 days?

**38.** A certain solution of iron is used to study the formation of blood cells. Approximately every 15 hours, the potency of the solution is reduced by 50% of the amount present at that time. By what percent has the potency decreased after 60 hours?

# Project in Mathematics

## Consumer Price Index

The Consumer Price Index (CPI) is a percent that is written without the percent sign. For instance, a CPI of 134.6 means 134.6%. This number means that an item that cost $100 between 1982 and 1984 (the base years) would cost $134.60 today. Determining the cost is an application of the basic percent equation.

$$\text{Base} \times \text{percent} = \text{amount}$$
$$\text{Cost in base years} \times \text{CPI} = \text{cost today}$$
$$100 \times 1.346 = 134.60$$

• 134.6% = 1.346

The table below gives the CPI for various products at the end of 1993.

| *Product* | *CPI* |
|---|---|
| Food | 144.3 |
| Fuel oil | 94.6 |
| Medical care | 195.7 |
| Rent | 116.5 |
| Tobacco | 223.1 |
| Clothes | 139.4 |

***Exercises:***

Use this table for the following exercises.

1. Of the items listed, are there any items that cost less in 1993 than they did during the base years? If so, which ones? Explain how you arrived at your answer.
2. Of the items listed, are there any items that in 1993 cost more than twice as much as they cost during the base years? If so, which ones? Explain how you arrived at your answer.
3. If it cost a family of four $122 per week for food during the base years, how much did that family spend in 1993?
4. If a raincoat cost $86 in 1993, what would a comparable raincoat have cost during the base years?

# Chapter Summary

**Key Words** *Percent* means "parts of 100."

**Essential Rules**

*To Write a Percent as a Fraction*

To write a percent as a fraction, remove the percent sign and multiply by $\frac{1}{100}$.

*To Write a Percent as a Decimal*

To write a percent as a decimal, remove the percent sign and multiply by 0.01.

*To Write a Decimal as a Percent*

To write a decimal as a percent, multiply by 100%.

*To Write a Fraction as a Percent*

To write a fraction as a percent, multiply by 100%.

*Basic Percent Equation* Percent × base = amount

*Proportion Method to Solve Percents* $\frac{\text{percent}}{100} = \frac{\text{amount}}{\text{base}}$

# Chapter Review

### SECTION 5.1

1. Write 12% as a fraction.

2. Write $16\frac{2}{3}\%$ as a fraction.

3. Write 42% as a decimal.

4. Write 7.6% as a decimal.

5. Write 0.38 as a percent.

6. Write $1\frac{1}{2}$ as a percent.

### SECTION 5.2

7. What is 30% of 200?

8. What is 7.5% of 72?

9. Find 22% of 88.

10. Find 125% of 62.

11. A company used 7.5% of its $60,000 expense budget for TV advertising. How much of the company's expense budget was spent for TV advertising?

12. Joshua purchased a video camera for $980 and paid a sales tax of 6.25% of the cost. What was the total cost of the video camera?

### SECTION 5.3

13. What percent of 20 is 30?

14. 16 is what percent of 80?

15. What percent of 30 is 2.2? Round to the nearest tenth of a percent.

16. What percent of 15 is 92? Round to the nearest tenth of a percent.

**17.** The stock of Advanced Board Games is trading at \$65 and paying a dividend of \$2.99. What percent of the stock price is the dividend?

**18.** A house in southern California had a value of \$125,000. Two years later the house was valued at \$205,000. Find the percent increase in value during the two years.

## SECTION 5.4

**19.** 20% of what is 15?

**20.** 78% of what is 8.5? Round to the nearest tenth.

**21.** $66\frac{2}{3}\%$ of what is 105?

**22.** $16\frac{2}{3}\%$ of what is 84?

**23.** A city's population this year was 157,500, which is 225% of what the city's population was 10 years ago. What was the city's population 10 years ago?

**24.** A baseball player has a batting average of 0.315. This average is an increase of 125% over the previous year's batting average. Find the previous year's batting average.

## SECTION 5.5

**25.** What is 62% of 320?

**26.** What percent of 25 is 40?

**27.** A computer system can be purchased for \$1800. This price is 60% of what the computer cost 4 years ago. What was the cost of the computer 4 years ago?

**28.** Trent missed 9 out of 60 questions on a history exam. What percent of the questions did he answer correctly?

# Chapter Test

1. Write 97.3% as a decimal.

2. Write $16\frac{2}{3}\%$ as a fraction.

3. Write 0.3 as a percent.

4. Write 1.63 as a percent.

5. Write $\frac{3}{2}$ as a percent.

6. Write $\frac{2}{3}$ as a percent.

7. What is 77% of 65?

8. 47.2% of 130 is what?

9. Which is larger:
7% of 120, or 76% of 13?

10. Which is smaller:
13% of 200, or 212% of 12?

11. A fast-food company uses 6% of its $75,000 budget for advertising. What amount of the budget is spent for advertising?

12. During the packaging process for vegetables, spoiled vegetables are discarded by an inspector. In one day an inspector found that 6.4% of the 1250 pounds of vegetables were spoiled. How many pounds of vegetables were not spoiled?

The table at the right contains nutrition information about a breakfast cereal. Solve Exercises 13 and 14 with information taken from this table.

**NUTRITION INFORMATION**

SERVING SIZE: 1.4 OZ WHEAT FLAKES WITH 0.4 OZ. RAISINS: 39.4 g. ABOUT 1/2 CUP
SERVINGS PER PACKAGE: ...............14

| | CEREAL & RAISINS | WITH 1/2 CUP VITAMINS A & D SKIM MILK |
|---|---|---|
| CALORIES ............... | 120 | 180 |
| PROTEIN, g ............. | 3 | 7 |
| CARBOHYDRATE, g .... | 28 | 34 |
| FAT, TOTAL, g .......... | 1 | 1* |
| UNSATURATED, g .... 1 | | |
| SATURATED, g ........ 0 | | |
| CHOLESTEROL, mg .... | 0 | 0* |
| SODIUM, mg ............ | 125 | 190 |
| POTASSIUM, mg ........ | 240 | 440 |

* 2% MILK SUPPLIES AN ADDITIONAL 20 CALORIES. 2 g FAT, AND 10 mg CHOLESTEROL.
** CONTAINS LESS THAN 2% OF THE U.S. RDA OF THIS NUTRIENT

13. The recommended amount of potassium for an adult per day is 3000 milligrams (mg). What percent, to the nearest tenth of a percent, of the daily recommended amount of potassium is provided by one serving of cereal with skim milk?

14. The daily recommended number of calories for a 190-pound man is 2200 calories. What percent, to the nearest tenth of a percent, of the daily recommended number of calories is provided by one serving of cereal with 2% milk?

**15.** The Urban Center Department Store has 125 permanent employees and must hire an additional 20 temporary employees for the holiday season. What percent of the permanent employees is the number hired as temporary employees for the holiday season?

**16.** Conchita missed 7 out of 80 questions on a math exam. What percent of the questions did she answer correctly? (Round to the nearest tenth of a percent.)

**17.** 12 is 15% of what?

**18.** 42.5 is 150% of what? Round to the nearest tenth.

**19.** A manufacturer of transistors found 384 defective transistors during a quality control study. This amount was 1.2% of the transistors tested. Find the number of transistors tested.

**20.** A new house was bought for \$95,000; 5 years later the house sold for \$152,000. The increase was what percent of the original price?

**21.** 123 is 86% of what number? Round to the nearest tenth.

**22.** What percent of 12 is 120?

**23.** A secretary receives a wage of \$9.52 per hour. This amount is 112% of last year's salary. What is the dollar increase in the hourly wage over last year?

**24.** A city has a population of 71,500; 10 years ago the population was 32,500. The population now is what percent of what the population was 10 years ago?

**25.** The annual license fee on a car is 1.4% of the value of the car. If the license fee during a year was \$91.00, what is the value of the car?

# Cumulative Review

1. Simplify $18 \div (7 - 4)^2 + 2$.

2. Find the LCM of 16, 24, and 30.

3. Find the sum of $2\frac{1}{3}$, $3\frac{1}{2}$, and $4\frac{5}{8}$.

4. Subtract: $27\frac{5}{12} - 14\frac{9}{16}$

5. Multiply: $7\frac{1}{3} \times 1\frac{5}{7}$

6. What is $\frac{14}{27}$ divided by $1\frac{7}{9}$?

7. Simplify: $\left(\frac{3}{4}\right)^3 \cdot \left(\frac{8}{9}\right)^2$

8. Simplify: $\left(\frac{2}{3}\right)^2 - \left(\frac{3}{8} - \frac{1}{3}\right) \div \frac{1}{2}$

9. Round 3.07973 to the nearest hundredth.

10. Subtract: $\begin{array}{r} 3.0902 \\ -\ 1.9706 \\ \hline \end{array}$

11. Divide: $0.032\overline{)1.097}$
Round to the nearest ten-thousandth.

12. Convert $3\frac{5}{8}$ to a decimal.

13. Convert 1.75 to a fraction.

14. Place the correct symbol, < or >, between the two numbers.
$\frac{3}{8}$ 0.87

15. Solve the proportion $\frac{3}{8} = \frac{20}{n}$. Round to the nearest tenth.

16. Write "$76.80 earned in 8 hours" as a unit rate.

17. Write $18\frac{1}{3}\%$ as a fraction.

18. Write $\frac{5}{6}$ as a percent.

19. 16.3% of 120 is what? Round to the nearest hundredth.

20. 24 is what percent of 18?

21. 12.4 is 125% of what?

22. What percent of 35 is 120? Round to the nearest tenth.

23. Sergio has an income of $740 per week. One-fifth of the income is deducted for income tax payments. Find his take-home pay.

24. Eunice bought a car for $4321, with a down payment of $1000. The balance was paid in 36 equal monthly payments. Find the monthly payment.

25. The gasoline tax is $0.19 a gallon. Find the number of gallons of gasoline used during a month in which $79.80 was paid in gasoline taxes.

26. The real estate tax on a $72,000 home is $1440. At the same rate, find the real estate tax on a home valued at $150,000.

27. Ken purchased a stereo set for $490 and pays $29.40 in sales tax. What percent of the purchase price was the sales tax?

28. A survey of 300 people showed that 165 people favored a certain candidate for mayor. What percent of the people surveyed did not favor this candidate?

29. The value of a home in the northern part of the United States was $62,000 in 1985. The same home in 1990 had a value of $155,000. What percent of the 1985 value is the 1990 value?

30. The Environmental Protection Agency found that 990 out of 5500 children tested had levels of lead in their blood exceeding federal guidelines. What percent of the children tested had levels of lead in the blood that exceeded federal standards?

Chapter

# 6

# Applications for Business and Consumers: A Calculator Approach

## Objectives

### Section 6.1

To find unit cost

To find the most economical purchase

To find total cost

### Section 6.2

To find percent increase

To apply percent increase to business—markup

To find percent decrease

To apply percent decrease to business—discount

### Section 6.3

To calculate simple interest

To calculate compound interest

### Section 6.4

To calculate the initial expenses of buying a home

To calculate ongoing expenses of owning a home

### Section 6.5

To calculate the initial expenses of buying a car

To calculate ongoing expenses of owning a car

### Section 6.6

To calculate commissions, total hourly wages, and salaries

### Section 6.7

To calculate checkbook balances

To balance a checkbook

# A Penny a Day

A fictitious job offer in a newspaper claimed that it would pay a salary of 1¢ on the first of the month and 2¢ on the second of the month. Next month, the salary would be 4¢ on the first and 8¢ on the second of the month. The next month and each succeeding month for 12 months, the procedure of doubling the previous payment would be continued the same way. Do you think you would want the job under these conditions?

This problem is an example of compounding interest, which is similar to the type of interest that is earned on bank or savings deposits. The big difference is that banks and savings and loans do not compound the interest as quickly as in the given example.

The table below shows the amount of salary you would earn for each month and then gives the total annual salary.

| | |
|---|---|
| January | \$0.01 + \$0.02 = \$0.03 |
| February | \$0.04 + \$0.08 = \$0.12 |
| March | \$0.16 + \$0.32 = \$0.48 |
| April | \$0.64 + \$1.28 = \$1.92 |
| May | \$2.56 + \$5.12 = \$7.68 |
| June | \$10.24 + \$20.48 = \$30.72 |
| July | \$40.96 + \$81.92 = \$122.88 |
| August | \$163.84 + \$327.68 = \$491.52 |
| September | \$655.36 + \$1310.72 = \$1966.08 |
| October | \$2621.44 + \$5242.88 = \$7864.32 |
| November | \$10,485.76 + \$20,971.52 = \$31,457.28 |
| December | \$41,943.04 + \$83,886.08 = \$125,829.12 |
| | Total Annual Salary = \$167,772.15 |

Not a bad annual salary!

# 6.1 Applications to Purchasing

## Objective A To find unit cost

Frequently stores advertise items for purchase as, say, 3 shirts for \$36 or 5 pounds of potatoes for 90¢.

The **unit cost** is the cost of 1 shirt or 1 pound of potatoes. To find the unit cost, divide the total cost by the number of units.

| | |
|---|---|
| 3 shirts for \$36 | 5 pounds of potatoes for 90¢ |
| 36 ÷ 3 = 12 | 90 ÷ 5 = 18 |
| \$12 is the cost of 1 shirt. | 18¢ is the cost of 1 pound of potatoes. |
| Unit cost: \$12 per shirt | Unit cost: 18¢ per pound |

**Example 1**
Find the unit cost. Round to the nearest tenth of a cent.
a) 10 gallons of gasoline for \$15.49
b) 3 yards of material for \$10.

**Strategy**
To find the unit cost, divide the total cost by the number of units.

**Solution**
a) 15.49 ÷ 10 = 1.549
\$1.549 per gallon
b) 10 ÷ 3 ≈ 3.3333
\$3.333 per yard

**You Try It 1**
Find the unit cost. Round to the nearest tenth of a cent.
a) 5 quarts of oil for \$6.25
b) 4 ears of corn for 85¢

**Your strategy**

**Your solution**

*Solution on p. A29*

## Objective B To find the most economical purchase

**POINT OF INTEREST**
The unit cost to send a payload into orbit on a Titan IV rocket is approximately \$15,000 per pound.

Comparison shoppers often find the most economical buy by comparing unit costs.

One store sells 6 cans of cola for \$2.04, and another store sells 24 cans of the same brand for \$7.92. To find the better buy, compare the unit costs.

| | |
|---|---|
| 2.04 ÷ 6 = 0.34 | 7.92 ÷ 24 = 0.33 |
| Unit cost: \$0.34 per can | Unit cost: \$0.33 per can |

Because \$0.33 < \$0.34, the better buy is 24 cans for \$7.92.

**Example 2**
Find the more economical purchase: 5 pounds of nails for $3.25, or 4 pounds of nails for $2.58.

**Strategy**
To find the more economical purchase, compare the unit costs.

**Solution**
3.25 ÷ 5 = 0.65
2.58 ÷ 4 = 0.645
$.645 < $.65

The more economical purchase is 4 pounds for $2.58.

**You Try It 2**
Find the more economical purchase: 6 cans of fruit for $2.52, or 4 cans of fruit for $1.66.

**Your strategy**

**Your solution**

*Solution on p. A29*

## *Objective C To find total cost*

An installer of floor tile found the unit cost of identical floor tiles at three stores.

| *Store 1* | *Store 2* | *Store 3* |
|---|---|---|
| $1.22 per tile | $1.18 per tile | $1.28 per tile |

By comparing the unit costs, the installer determined that store 2 would provide the most economical purchase.

The installer also uses the unit cost to find the total cost of purchasing 300 floor tiles at store 2. The **total cost** is found by multiplying the unit cost by the number of units purchased.

| Unit cost | × | number of units | = | total cost |
|---|---|---|---|---|
| 1.18 | × | 300 | = | 354 |

The total cost is $354.

**Example 3**
Clear redwood lumber costs $2.43 per foot. How much would 25 feet of clear redwood cost?

**Strategy**
To find the total cost, multiply the unit cost ($2.43) by the number of units (25).

**Solution**

| Unit cost | × | number of units | = | total cost |
|---|---|---|---|---|
| 2.43 | × | 25 | = | 60.75 |

The total cost is $60.75.

**You Try It 3**
Pine saplings costs $4.96 each. How much would 7 pine saplings cost?

**Your strategy**

**Your solution**

*Solution on p. A29*

## 6.1 Exercises

### *Objective A* Application Problems

Find the unit cost. Round to the nearest tenth of a cent.

**1.** Syrup, 16 ounces for \$0.89

**2.** Wood, 8 feet for \$11.36

**3.** Lunch meat, 12 ounces for \$1.79

**4.** Light bulbs, 6 bulbs for \$2

**5.** Wood screws, 15 screws for \$0.50

**6.** Bay leaves, $\frac{1}{2}$ ounce for \$0.96

**7.** Antenna wire, 10 feet for \$1.95

**8.** Sponges, 4 for \$5.00

**9.** Catsup, 24 ounces for \$0.89

**10.** Strawberry jam, 18 ounces for \$1.23

**11.** Metal clamps, 2 for \$5.99

**12.** Coffee, 3 pounds for \$7.65

### *Objective B* Application Problems

Find the more economical purchase.

**13.** Cheese, 12 ounces for \$1.79, or 16 ounces for \$2.40

**14.** Powder, 4 ounces for \$0.53, or 9 ounces for \$0.95

**15.** Shampoo, 8 ounces for \$1.89, or 12 ounces for \$2.64

**16.** Mayonnaise, 32 ounces for \$2.37, or 16 ounces for \$1.30

**17.** Fruit drink, 46 ounces for \$0.69, or 50 ounces for \$0.78

**18.** Detergent, 42 ounces for \$2.57, or 20 ounces for \$1.29

Find the more economical purchase.

**19.** Oil: 2 quarts for $4.79, or 5 quarts for $12.15

**20.** Potato chips: 8 ounces for $0.99, or 10 ounces for $1.23

**21.** Vitamins: 200 tablets for $9.98, or 150 tablets for $7.65

**22.** Wax: 16 ounces for $2.68, or 27 ounces for $4.39

**23.** Cereal: 18 ounces for $1.51, or 16 ounces for $1.32

**24.** Tires: 2 for $129.98, or 4 for $258.98

## *Objective C* *Application Problems*

Solve.

**25.** Steak costs $2.79 per pound. Find the total cost of 3 pounds.

**26.** Red brick costs $0.56 per brick. Find the total cost of 50 bricks.

**27.** Flowering plants cost $0.49 each. Find the total cost of 6 plants.

**28.** Chicken costs $0.89 per pound. Find the total cost of 3.6 pounds. Round to the nearest cent.

**29.** Tea costs $0.54 per ounce. Find the total cost of 6.5 ounces.

**30.** Cheese costs $2.89 per pound. Find the total cost of 0.65 pounds. Round to the nearest cent.

**31.** Tomatoes cost $0.22 per pound. Find the total cost of 1.8 pounds. Round to the nearest cent.

**32.** Ham costs $1.49 per pound. Find the total cost of 3.4 pounds. Round to the nearest cent.

**33.** Candy costs $2.68 per pound. Find the total cost of $\frac{3}{4}$ pound.

**34.** Photocopying costs $0.05 per page. Find the total cost for photocopying 250 pages.

## *APPLYING THE CONCEPTS*

**35.** Explain in your own words the meaning of unit pricing.
[W]

**36.** What is the UPC (Universal Product Code) and how is it used?
[W]

# 6.2 Percent Increase and Percent Decrease

## *Objective A* *To find percent increase*

**Percent increase** is used to show how much a quantity has increased over its original value. The statements "car prices will show a 3.5% increase over last year's prices" and "employees were given an 11% pay increase" are illustrations of the use of percent increase.

➡ A city's population increased from 50,000 to 51,500 in 1 year. Find the percent increase of the city's population.

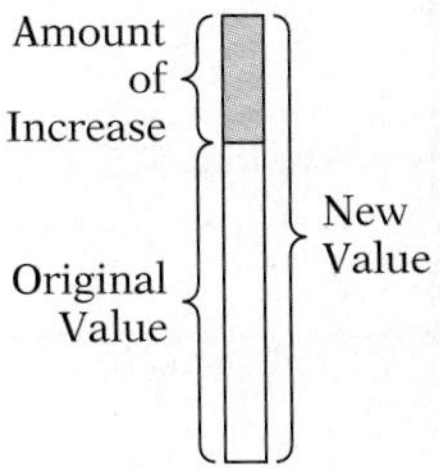


| New value | − | original value | = | amount of increase |
|---|---|---|---|---|
| 51,500 | − | 50,000 | = | 1500 |

Now solve the basic percent equation for percent.

Percent × base = amount

| Percent increase | × | original value | = | amount of increase |
|---|---|---|---|---|
| $n$ | × | 50,000 | = | 1500 |

$n = 1500 \div 50{,}000$

$n = 0.03 = 3\%$

The population increased by 3% over the previous year.

**POINT OF INTEREST**

According to the U.S. Census Bureau, the number of persons aged 65 and over in the United States will increase by 67% from 1992 to 2020.

**Example 1**

The average price of gasoline rose from \$0.92 to \$1.15 in 4 months. What was the percent increase in the price of gasoline?

**Strategy**

To find the percent increase:

- Find the amount of increase.
- Solve the basic percent equation for *percent*.

**Solution**

| New value | − | original value | = | amount of increase |
|---|---|---|---|---|
| 1.15 | − | 0.92 | = | 0.23 |

Percent × base = amount

$n \times 0.92 = 0.23$

$n = 0.23 \div 0.92$

$n = 0.25 = 25\%$

The percent increase was 25%.

**You Try It 1**

A moving and storage company increased its number of employees from 2000 to 2250. What was the percent increase in the number of employees?

**Your strategy**

**Your solution**

*Solution on p. A30*

**Example 2**

A seamstress was making a wage of \$4.80 an hour before a 10% increase in pay. What is the new hourly wage?

**Strategy**

To find the new hourly wage:

- Solve the basic percent equation for *amount*.
- Add the amount of increase to the original wage.

**Solution**

Percent × base = amount

0.10 × 4.80 = *n*

0.48 = *n*

4.80 + 0.48 = 5.28

The new hourly wage is \$5.28.

**You Try It 2**

A baker was making a wage of \$6.50 an hour before a 14% increase in pay. What is the new hourly wage?

**Your strategy**

**Your solution**

*Solution on p. A30*

## *Objective B To apply percent increase to business—markup*

Many expenses are involved in operating a business. Employee salaries, rent of office space, and utilities are examples of operating expenses. To pay these expenses and earn a profit, a business must sell a product at a higher price than it paid for the product.

**Cost** is the price that a business pays for a product, and **selling price** is the price at which a business sells a product to a customer. The difference between selling price and cost is called **markup**.

| Selling price | − | cost | = | markup |

or

| Cost | + | markup | = | selling price |

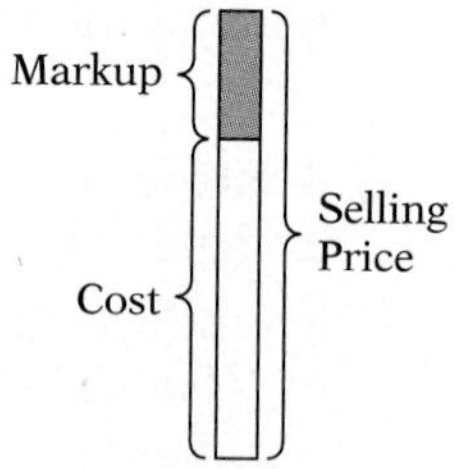


Markup is frequently expressed as a percent of a product's cost. This percent is called the **markup rate.**

| Markup rate | × | cost | = | markup |

➡ A bicycle store owner purchases a bicycle for \$105 and sells it for \$147. What markup rate does the owner use?

| Selling price | − | cost | = | markup |

147 − 105 = 42

- **First find the markup.**

| Percent | × | base | = | amount |
|---|---|---|---|---|
| Markup rate | × | cost | = | markup |
| $n$ | × | $105 | = | $42 |

$n = 42 \div 105 = 0.4$

- Then solve the basic percent equation for *percent*.

The markup rate is 40%.

**Example 3**
The manager of a sporting goods store determines that a markup rate of 36% is necessary to make a profit. What is the markup on a pair of skis that costs the store $125?

**Strategy**
To find the markup, solve the basic percent equation for *amount*.

**Solution**

| Percent | × | base | = | amount |
|---|---|---|---|---|
| Markup rate | × | cost | = | markup |
| 0.36 | × | 125 | = | $n$ |

$45 = n$

The markup is $45.

**You Try It 3**
A bookstore manager determines that a markup rate of 20% is necessary to make a profit. What is the markup on a book that costs the bookstore $8?

**Your strategy**

**Your solution**

**Example 4**
A plant nursery bought a citrus tree for $4.50 and used a markup rate of 46%. What is the selling price?

**Strategy**
To find the selling price:

- Find the markup by solving the basic percent equation for *amount*.
- Add the markup to the cost.

**Solution**

| Percent | × | base | = | amount |
|---|---|---|---|---|
| Markup rate | × | cost | = | markup |
| 0.46 | × | 4.50 | = | $n$ |

$2.07 = n$

| Cost | + | markup | = | selling price |
|---|---|---|---|---|
| 4.50 | + | 2.07 | = | 6.57 |

The selling price is $6.57.

**You Try It 4**
A clothing store bought a suit for $72 and used a markup rate of 55%. What is the selling price?

**Your strategy**

**Your solution**

*Solutions on p. A30*

## *Objective C To find percent decrease*

**Percent decrease** is frequently used to show how much a quantity has decreased from its original value. The statements "the unemployment rate decreased by 0.4% over last month" and "there has been a 12% decrease in the number of industrial accidents" are illustrations of the use of the percent decrease.

➡ A family's electric bill decreased from \$60 per month to \$52.80 per month. Find the percent decrease in the family's electric bill.

| Original value | − | new value | = | amount of decrease |
|---|---|---|---|---|
| 60 | − | 52.80 | = | 7.20 |

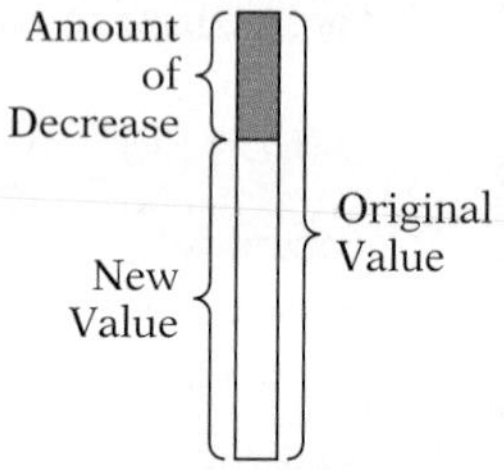


Now solve the basic percent equation for percent.

| Percent | × | base | = | amount |
|---|---|---|---|---|
| Percent decrease | × | original value | = | amount of decrease |
| $n$ | × | \$60 | = | \$7.20 |

$n = 7.20 \div 60$

$n = 0.12 = 12\%$

The electric bill decreased by 12% per month.

**POINT OF INTEREST**

A *bear market* is one in which the total value of all stocks decreases. During 1972 and 1973, the stock market decreased by approximately 55%. This means that an investment of \$100,000 would have decreased to \$45,000 during that period.

**Example 5**

Because of unusually high temperatures, the number of people visiting a desert resort dropped from 650 to 525. Find the percent decrease in attendance. Round to the nearest tenth of a percent.

**Strategy**

To find the percent decrease:

- Find the amount of decrease.
- Solve the basic percent equation for *percent*.

**Solution**

| Original value | − | new value | = | amount of decrease |
|---|---|---|---|---|
| 650 | − | 525 | = | 125 |

Percent × base = amount

$n \times 650 = 125$

$n = 125 \div 650$

$n \approx 0.192 = 19.2\%$

The percent decrease was 19.2%.

**You Try It 5**

Car sales at an automobile dealership dropped from 150 in June to 120 in July. Find the percent decrease in car sales.

**Your strategy**

**Your solution**

*Solution on p. A30*

**Example 6**
The total sales for December for a stationery store were \$16,000. For January, total sales showed an 8% decrease from December's sales. What were the total sales for January?

**Strategy**
To find the total sales for January:

- Find the amount of decrease by solving the basic percent equation for *amount*.
- Subtract the amount of decrease from the December sales.

**Solution**
Percent × base = amount

$0.08 \times 16{,}000 = n$

$1280 = n$

The decrease in sales was \$1280.

$16{,}000 - 1280 = 14{,}720$

The total sales for January were \$14,720.

**You Try It 6**
Fog decreased the normal 5-mile visibility at an airport by 40%. What was the visibility in the fog?

**Your strategy**

**Your solution**

*Solution on p. A30*

### *Objective D To apply percent decrease to business—discount*

To promote sales, a store may reduce the regular price of some of its products temporarily. The reduced price is called the **sale price.** The difference between the regular price and the sale price is called the **discount**.

Regular price − sale price = discount

or

Regular price − discount = sale price

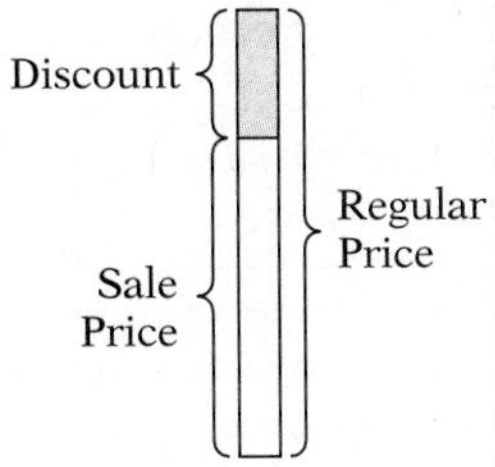


Discount is frequently stated as a percent of a product's regular price. This percent is called the **discount rate.**

Discount rate × regular price = discount

**Example 7**
An appliance store has a washing machine that regularly sells for $350 on sale for $297.50. What is the discount rate?

**Strategy**
To find the discount rate:
- Find the discount.
- Solve the basic percent equation for *percent*.

**Solution**

| Regular price | − | sale price | = | discount |
|---|---|---|---|---|
| 350 | − | 297.50 | = | 52.50 |

Percent × base = amount

| Discount rate | × | regular price | = | discount |
|---|---|---|---|---|
| $n$ | × | \$350 | = | \$52.50 |

$n = 52.50 \div 350$

$n = 0.15 = 15\%$

The discount rate is 15%.

**You Try It 7**
A lawn mower that regularly sells for $125 is on sale for $25 off the regular price. What is the discount rate?

**Your strategy**

**Your solution**

**Example 8**
A plumbing store is selling at 15% off the regular price sinks that regularly sell for $125. What is the sale price?

**Strategy**
To find the sale price:
- Find the discount by solving the basic percent equation for *amount*.
- Subtract to find the sale price.

**Solution**
Percent × base = amount

| Discount rate | × | regular price | = | discount |
|---|---|---|---|---|
| 0.15 | × | 125 | = | $n$ |

$18.75 = n$

| Regular price | − | discount | = | sale price |
|---|---|---|---|---|
| 125 | − | 18.75 | = | 106.25 |

The sale price is $106.25.

**You Try It 8**
A florist is selling at 20% off the regular price potted plants that regularly sell for $10.25. What is the sale price?

**Your strategy**

**Your solution**

*Solutions on p. A31*

## 6.2 Exercises

**Objective A** Application Problems

1. Elissa had a 0.275 batting average in her junior year in high school. The next year she had a batting average of 0.297. What percent increase has occurred in her batting average?

2. Rodney had a 3.80 rushing average for the Baker University football team. The next year he averaged 4.75 yards per carry. What percent increase does this amount represent?

3. Josif increased his typing speed of 60 words per minute by 20 words per minute. By what percent did he increase his typing speed?

4. You have increased your reading speed by 70 words per minute from 280 words per minute. What is your percent increase in reading speed?

5. A teacher's union negotiated a new contract calling for a 9% increase in pay over this year's pay.
   a. Find the increase in pay for a teacher who had a salary of $32,500.
   b. Find the new salary for a teacher who had a salary of $32,500.

6. A new labor contract for Westside Lumber called for an 8.5% increase in pay for all employees.
   a. What is the increase in pay for an employee who makes $324 per week?
   b. What is the weekly wage of this employee after the wage increase?

7. AFLAG, whose stock is selling for $30\frac{3}{8}$ per share, plans to raise its $1.28 dividend by 25%. Find the dividend per share of AFLAG stock after the increase.

8. An amusement park increased its summer staff from 500 to 625 employees. What percent increase does this represent?

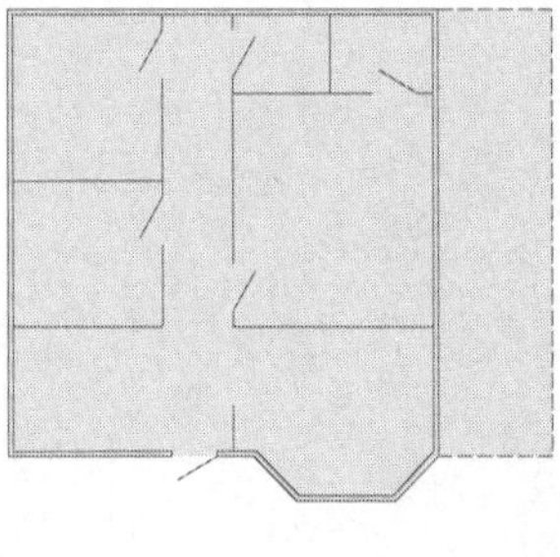

9. A family increased its 1440-square-foot home $16\frac{2}{3}\%$ by adding a recreation room. How much larger, in square feet, is the home now?

10. A hunter's lodge increased its weekly hours of operation from 60 hours to 70 hours. What percent increase does this amount represent?

**Objective B** Application Problems

11. An air conditioner cost the plumbing and air conditioning contractor $285. Find the markup on this air conditioner if the markup rate is 25%.

12. The owner of Kerr electronics uses a markup rate of 42%. What is the markup on a cassette player that costs Kerr $85?

13. The ski and tennis store uses a markup rate of 30%. What is the markup on a tennis racket that costs $42?

14. The manager of a department store determines that a markup of 38% is necessary for a profit to be made. What is the markup on a novelty item that costs $45?

**15.** Computer Inc. uses a markup of $975 on a computer system that costs $3250. What markup rate does this amount represent?

**16.** Saizon Pen & Office Supply uses a markup of $12 on a calculator that costs $20. What markup rate does this amount represent?

**17.** Giant Photo Service uses a markup rate of 48% on its Model ZA cameras, which cost the shop $162.
a. What is the markup?
b. What is the selling price?

**18.** The Circle R golf pro shop uses a markup rate of 45% on a set of golf clubs that costs the shop $210.
a. What is the markup?
b. What is the selling price?

**19.** Hauser Furniture uses a markup rate of 35% on a piece of lawn furniture that costs the store $62.
a. What is the markup?
b. What is the selling price?

**20.** Harvest Time Produce Inc. uses a 55% markup rate and pays $0.60 for a box of strawberries.
a. What is the markup?
b. What is the selling price of a box of strawberries?

**21.** Resner Builders' Hardware uses a markup rate of 42% for a table saw that costs $160. What is the selling price of the table saw?

**22.** Brad Burt's Magic Shop uses a markup rate of 48%. What is the selling price of an item that costs $50?

## *Objective C* *Application Problems*

**23.** A new bridge reduced the normal 45-minute travel time between two cities by 18 minutes. What percent decrease does this represent?

**24.** In an election, 25,400 voters went to the polls to elect a mayor. Four years later, 1778 fewer voters participated in the election for mayor. What was the percent decrease in the number of voters who went to the polls?

**25.** By installing energy-saving equipment, the Pala Rey Youth Camp reduced its normal $800-per-month utility bill by $320. What percent decrease does this amount represent?

**26.** During the last 40 years, the consumption of eggs in the United States has dropped from 400 eggs per person per year to 260 eggs per person per year. What percent decrease does this amount represent?

**27.** It is estimated that the value of a new car is reduced 30% after 1 year of ownership. Using this estimate, find how much value a $11,200 new car loses after 1 year.

**28.** Assume that a Kmart store employs 1200 people during the holiday. At the end of the holiday season, Kmart reduces the number of employees by 45%. What is the decrease in the number of employees?

**29.** Because of a decrease in demand for super-8 video cameras, Kit's Cameras reduced the orders for these models from 20 per month to 8 per month.
a. What is the amount of decrease?
b. What percent decrease does this amount represent?

**30.** A new computer system reduced the time for printing the payroll from 52 minutes to 39 minutes.
a. What is the amount of decrease?
b. What percent decrease does this amount represent?

**31.** Juanita's average expense for gasoline was $76. After joining a car pool, she was able to reduce the expense by 20%.
a. What was the amount of decrease?
b. What is the average monthly gasoline bill now?

**32.** An oil company paid a dividend of $1.60 per share. After reorganization, the company reduced the dividend by 37.5%.
a. What was the amount of decrease?
b. What is the new dividend?

**33.** Because of an improved traffic pattern at a sports stadium, the average amount of time a fan waits to park decreased from 3.5 minutes to 2.8 minutes. What percent decrease does this amount represent?

### *Objective D* *Application Problems*

**34.** The Austin College Bookstore is giving a discount of $8 on calculators that normally sell for $24. What is the discount rate?

**35.** A discount clothing store is selling a $72 sport jacket for $24 off the regular price. What is the discount rate?

**36.** A disk player that regularly sells for $340 is selling for 20% off the regular price. What is the discount?

**37.** Dacor Appliances is selling its $450 washing machine for 15% off the regular price. What is the discount?

**38.** An electric grill that regularly sells for $140 is selling for $42 off the regular price. What is the discount rate?

**39.** You bought a computer system for $350 off the regular price of $1400. What is the discount rate?

**40.** Quick Service Gas Station has its regularly priced $45 tune-up on sale for 16% off the regular price.
a. What is the discount?
b. What is the sale price?

**41.** Turkey that regularly sells for $.85 per pound is on sale for 20% off the regular price.
a. What is the discount?
b. What is the sale price?

**42.** An outdoor supply store has regularly priced \$160 sleeping bags on sale for \$120.
a. What is the discount?
b. What is the discount rate?

**43.** Standard Brands paint that regularly sells for \$16 per gallon is on sale for \$12 per gallon.
a. What is the discount?
b. What is the discount rate?

## APPLYING THE CONCEPTS

The figure at the right shows the closing price and the change for the day for the most active stocks on the NASDAQ stock exchange for October 13, 1993. Use this figure for Exercises 44–47. Round to the nearest tenth of a percent.

| NASDAQ | Vol. | Last | Chg |
|---|---|---|---|
| Intel s ...... | 13,277,800 | 65 | -1/2 |
| Clsco s ....... | 4,916,400 | 50 7/8 | +4 7/8 |
| Wellfit s ..... | 3,910,800 | 53 1/4 | +6 1/8 |
| WstwOn .... | 3,058,000 | 5 3/8 | +7/8 |
| SynOpt s .... | 2,981,500 | 24 1/2 | +7/8 |
| Centocor .... | 2,943,700 | 14 1/4 | +3 |
| GtARc ........ | 2,773,200 | 1 3/8 | +1/4 |
| Medgrp ...... | 2,736,800 | 1 17/32 | +15/32 |
| AppleC ....... | 2,735,200 | 24 | +1/4 |
| TelCmA ...... | 2,671,000 | 28 3/8 | -1/4 |

**44.** Find the one-day percent increase for Cisco.

**45.** Find the one-day percent increase for Centocor.

**46.** Find the one-day percent decrease for Intel.

**47.** One share of Wellfit increased by $\$6\frac{1}{8}$ and one share of WstwOn increased by $\frac{7}{8}$. Which of these two companies had the greater percent increase in price?

**48.** A welder earning \$12 per hour is given a 10% raise. To find the new wage, we can multiply \$12 by 0.10 and add the product to \$12. Can the new wage be found by multiplying \$12 by 1.10? Try both methods and compare your answers.

**49.** Grocers, florists, bakers, and other businesses must consider spoilage when deciding the markup of a product. For instance, a loaf of bread will become stale after a few days, and the baker will not be able to sell it. A florist purchased 200 roses at a cost of \$.86 per rose. The florist wants a markup rate of 50% of the total cost of all the roses and expects 7% of the roses to wilt and therefore not be saleable. Find the selling price per rose by answering each of the following questions.
a. What is the florist's total cost for the 200 roses?
b. Find the total selling price without spoilage.
c. Find the number of roses the florist expects to sell. *Hint:* The number of roses the florist expects to sell is

% of saleable roses × number of roses purchased

d. To find the selling price per rose, divide the total selling price per rose by the number of roses the florist expects to sell. Round to the nearest cent.

**50.** [W] A promotional sale at a department store offers 25% off the sale price. The sale price is 25% off the regular price. Is this the same as a sale that offers 50% off the regular price? If not, which sale gives the better price? Explain your answer.

**51.** [W] In your own words, explain how to find a percent increase or a percent decrease.

# 6.3 Interest

## Objective A To calculate simple interest

When money is deposited in a bank account, the bank pays the depositor for the privilege of using that money. The amount paid to the depositor is called **interest.** When money is borrowed from a bank, the borrower pays for the privilege of using that money. The amount paid to the bank is also called interest.

The original amount deposited or borrowed is called the **principal.** The amount of interest is a percent of the principal. The percent used to determine the amount of interest is the **interest rate.** Interest rates are given for specific periods of time, usually months or years.

Interest computed on the original principal is called **simple interest.** To calculate simple interest, multiply the principal by the interest rate per period by the number of time periods.

➡ Calculate the simple interest on \$1500 deposited for 2 years at an annual interest rate of 7.5%.

Principal × annual interest rate × time (in years) = interest

1500 × 0.075 × 2 = 225

The interest earned is \$225.

**Example 1**

Kamal borrowed \$500 from a savings and loan association for 6 months at an annual interest rate of 7%. What is the simple interest due on the loan?

**Strategy**

To find the simple interest, multiply:

Principal × annual interest rate × time (in years)

**Solution** 500 × 0.07 × 0.5 = 17.5

The interest due is \$17.50.

**You Try It 1**

A company borrowed \$15,000 from a bank for 18 months at an annual interest rate of 8%. What is the simple interest due on the loan?

**Your strategy**

**Your solution**

**Example 2**

A credit card company charges a customer 1.5% per month on the unpaid balance of charges on the credit card. What is the interest due in a month when the customer has an unpaid balance of \$54?

**Strategy**

To find the interest due, multiply the principal by the monthly interest rate by the time (in months).

**Solution** 54 × 0.015 × 1 = 0.81

The interest charge is \$0.81.

**You Try It 2**

A bank offers short-term loans at a simple interest rate of 1.2% per month. What is the interest due on a short-term loan of \$400 for 2 months?

**Your strategy**

**Your solution**

*Solutions on p. A31*

### *Objective B To calculate compound interest*

Usually the interest paid on money deposited or borrowed is compound interest. **Compound interest** is computed not only on the original principal but also on interest already earned. Compound interest is usually compounded annually (once a year), semiannually (twice a year), quarterly (four times a year), or daily.

\$100 is invested for 3 years at an annual interest rate of 9% compounded annually. The interest earned over the 3 years is calculated by first finding the interest earned each year.

| | | |
|---|---|---|
| 1st year | Interest earned: | $0.09 \times \$100.00 = \$9.00$ |
| | New principal: | $\$100.00 + \$9.00 = \$109.00$ |
| 2nd year | Interest earned: | $0.09 \times \$109.00 = \$9.81$ |
| | New principal: | $\$109.00 + \$9.81 = \$118.81$ |
| 3rd year | Interest earned: | $0.09 \times \$118.81 \approx \$10.69$ |
| | New principal: | $\$118.81 + \$10.69 = \$129.50$ |

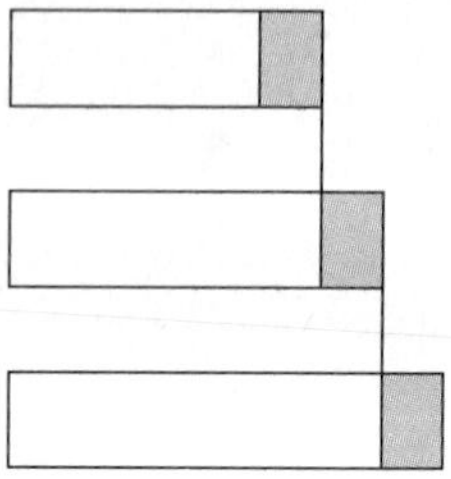

To find the interest earned, subtract the original principal from the new principal.

| New principal | − | original principal | = | interest earned |
|---|---|---|---|---|
| 129.50 | − | 100 | = | 29.5 |

The interest earned is \$29.50.

Note that the compound interest earned is \$29.50. The simple interest earned on the investment would have been only $\$100 \times 0.09 \times 3 = \$27$.

Calculating compound interest can be very tedious, so there are tables that can be used to simplify these calculations. A portion of a compound interest table is given on pages A4 and A5.

**Example 3**
An investment of \$650 pays 8% annual interest compounded semiannually. What is the interest earned in 5 years?

**Strategy** To find the interest earned:

- Find the new principal by multiplying the original principal by the factor (1.48024) found in the compound interest table.
- Subtract the original principal from the new principal.

**Solution**
$650 \times 1.48024 \approx 962.16$

The new principal is \$962.16.

$962.16 - 650 = 312.16$

The interest earned is \$312.16.

**You Try It 3**
An investment of \$1000 pays 6% annual interest compounded quarterly. What is the interest earned in 20 years?

**Your strategy**

**Your solution**

*Solution on p. A31*

## 6.3 Exercises

### *Objective A* Application Problems

1. To finance the purchase of 15 new cars, the Tropical Car Rental Agency borrowed \$100,000 for 9 months at an annual interest rate of 9%. What is the simple interest due on the loan?
2. A home builder obtained a pre-construction loan of \$50,000 for 8 months at an annual interest rate of 9.5%. What is the simple interest due on the loan? (Round to the nearest cent.)
3. The Mission Valley Credit Union charges its customers an interest rate of 2% per month on money that is transferred into an account that is overdrawn. Find the interest owed to the credit union for 1 month when \$800 is transferred into an overdrawn account.
4. Assume that Visa charges Francesca 1.6% per month on her unpaid balance. Find the interest owed to Visa when her unpaid balance for the month is \$1250.
5. Action Machining Company purchased a robot-controlled lathe for \$225,000 and financed the full amount at 8% simple annual interest for 4 years.
   a. Find the interest on the loan.
   b. Find the monthly payment.
   $\left(\text{Monthly payment} = \dfrac{\text{loan amount} + \text{interest}}{\text{number of months}}\right)$
6. For the purchase of an entertainment center, an \$1800 loan is obtained for 2 years at a simple interest rate of 9.4%.
   a. Find the interest due on the loan.
   b. Find the monthly payment.
   $\left(\text{Monthly payment} = \dfrac{\text{loan amount} + \text{interest}}{\text{number of months}}\right)$
7. To attract new customers, Heller Ford is offering car loans at a simple interest rate of 4.5%.
   a. Find the interest charged to a customer who finances a car loan of \$12,000 for 2 years.
   b. Find the monthly payment.
8. Cimarron Homes Inc. purchased a small plane for \$57,000 and financed the full amount for 5 years at a simple annual interest rate of 9%.
   a. Find the interest due on the loan.
   b. Find the monthly payment.
9. Dennis Pappas decided to build onto an existing structure instead of buying a new home. He borrowed \$42,000 for $3\frac{1}{2}$ years at a simple interest rate of 9.5%. Find the monthly payment.

### *Objective B* Application Problems

Solve. Use the table on pages A4 and A5. Round to the nearest cent.

10. North Island Federal Credit Union pays 4% annual interest, compounded daily, on time savings deposits. Find the value of \$750 deposited in this account after 1 year.

Solve. Use the table on pages A4 and A5. Round to the nearest cent.

**11.** What is the value after 5 years of $1000 invested at 7% annual interest compounded quarterly?

**12.** An investment club invested $50,000 in a certificate of deposit that pays 5% annual interest compounded quarterly. Find the value of this investment after 10 years.

**13.** Tanya invested $2500 in a tax-sheltered annuity that pays 8% annual interest compounded daily. Find the value of her investment after 20 years.

**14.** Sal Trovato invested $3000 in a corporate retirement account that pays 6% annual interest compounded semiannually. Find the value of his investment after 15 years.

**15.** To replace equipment, a farmer invested $20,000 in an account that pays 7% annual interest compounded semiannually. What is the value of the investment after 5 years?

**16.** Green River Lodge invests $75,000 in a trust account that pays 8% interest compounded quarterly.
a. What will the value of the investment be in 5 years?
b. How much interest will be earned in the 5 years?

**17.** To save for retirement, a couple deposited $3000 in an account that pays 7% annual interest compounded daily.
a. What will the value of the investment be in 10 years?
b. How much interest will be earned in the 10 years?

**18.** To save for a child's college education, the Petersens deposited $2500 into an account that pays 6% annual interest compounded daily. Find the amount of interest earned in this account over a 20-year period.

## APPLYING THE CONCEPTS

**19.** [W] Explain the fundamental difference between simple interest and compound interest.

**20.** [W] Visit a brokerage business and obtain information on annuities. Explain the advantages of an annuity as a retirement option.

**21.** Suppose you have a savings account that earns interest at the rate of 6% per year compounded monthly. On January 1, you open this account with a deposit of $100.
a. On February 1, you deposit an additional $100 into the account. What is the value of the account after the deposit?
b. On March 1, you deposit an additional $100 into the account. What is the value of the account after the deposit. *Note:* This type of savings plan, wherein equal amounts ($100) are saved at equal time intervals (every month), is called an annuity.

# 6.4 Real Estate Expenses

## *Objective A To calculate the initial expenses of buying a home*

One of the largest investments most people ever make is the purchase of a home. The major initial expense in the purchase is the down payment. The amount of the down payment is normally a percent of the purchase price. This percent varies among banks, but it usually ranges from 5% to 25%.

The **mortgage** is the amount that is borrowed to buy real estate. The mortgage amount is the difference between the purchase price and the down payment.

➡ A home is purchased for $85,000, and a down payment of $12,750 is made. Find the mortgage.

| Purchase price | − | down payment | = | mortgage |
|---|---|---|---|---|
| 85,000 | − | 12,750 | = | 72,250 |

The mortgage is $72,250.

Another large initial expense in buying a home is the loan origination fee, which is a fee the bank charges for processing the mortgage papers. The loan origination fee is usually a percent of the mortgage and is expressed in **points,** which is the term banks use to mean percent. For example, "5 points" means "5 percent."

| Points | × | mortgage | = | loan origination fee |
|---|---|---|---|---|

**Example 1**

A house is purchased for $87,000, and a down payment, which is 20% of the purchase price, is made. Find the mortgage.

**Strategy**

To find the mortgage:

- Find the down payment by solving the basic percent equation for *amount*.
- Subtract the down payment from the purchase price.

**Solution**

| Percent | × | base | = | amount |
|---|---|---|---|---|
| Percent | × | purchase price | = | down payment |
| 0.20 | × | 87,000 | = | $n$ |
| | | 17,400 | = | $n$ |

| Purchase price | − | down payment | = | mortgage |
|---|---|---|---|---|
| 87,000 | − | 17,400 | = | 69,600 |

The mortgage is $69,600.

**You Try It 1**

An office building is purchased for $216,000, and a down payment, which is 25% of the purchase price, is made. Find the mortgage.

**Your strategy**

**Your solution**

*Solution on p. A32*

**Example 2**
A home is purchased with a mortgage of \$65,000. The buyer pays a loan origination fee of $3\frac{1}{2}$ points. How much is the loan origination fee?

**Strategy**
To find the loan origination fee, solve the basic percent equation for *amount*.

**Solution**

Percent × base = amount

| Points | × | mortgage | = | fee |
|---|---|---|---|---|

0.035 × 65,000 = $n$

2275 = $n$

The loan origination fee is \$2275.

**You Try It 2**
The mortgage on a real estate investment is \$80,000. The buyer paid a loan origination fee of $4\frac{1}{2}$ points. How much was the loan origination fee?

**Your strategy**

**Your solution**

*Solution on p. A32*

## *Objective B To calculate ongoing expenses of owning a home*

Besides the initial expenses of buying a home, there are continuing monthly expenses involved in owning a home. The monthly mortgage payment, utilities, insurance, and taxes are some of these ongoing expenses. Of these expenses, the largest one is normally the monthly mortgage payment.

For a fixed-rate mortgage, the monthly mortgage payment remains the same throughout the life of the loan. The calculation of the monthly mortgage payment is based on the amount of the loan, the interest rate on the loan, and the number of years required to pay back the loan. Calculating the monthly mortgage payment is fairly difficult, so tables such as the one on page A6 are used to simplify these calculations.

➡ Find the monthly mortgage payment on a 30-year \$60,000 mortgage at an interest rate of 9%. Use the monthly payment table on page A6.

60,000 × 0.0080462 = 482.77
↓
from the table

The monthly mortgage payment is \$482.77.

The monthly mortgage payment includes the payment of both principal and interest on the mortgage. The interest charged during any one month is charged on the unpaid balance of the loan. Therefore, during the early years of the mortgage, when the unpaid balance is high, most of the monthly mortgage payment is interest charged on the loan. During the last few years of a mortgage, when the unpaid balance is low, most of the monthly mortgage payment goes toward paying off the loan.

**POINT OF INTEREST**

A person receiving a loan of $125,000 for 30 years at a fixed mortgage interest rate of 7% has a monthly payment of $649.18. If the loan is kept for 30 years, the total amount of interest paid is $108,704.80.

➡ Find the interest paid on a mortgage during a month when the monthly mortgage payment is $186.26 and $58.08 of that amount goes toward paying off the principal.

| Monthly mortgage payment | − | principal | = | interest |
|---|---|---|---|---|
| 186.26 | − | 58.08 | = | 128.18 |

The interest paid on the mortgage is $128.18.

Property tax is another ongoing expense of owning a house. Property tax is normally an annual expense that may be paid on a monthly basis. The monthly property tax, which is determined by dividing the annual property tax by 12, is usually added to the monthly mortgage payment.

➡ A home owner must pay $534 in property tax annually. Find the property tax that must be added each month to the home owner's monthly mortgage payment.

534 ÷ 12 = 44.5

Each month, $44.50 must be added to the monthly mortgage payment for property tax.

**Example 3**

Serge purchased some land for $120,000 and made a down payment of $25,000. The savings and loan association charges an annual interest rate of 8% on Serge's 25-year mortgage. Find the monthly mortgage payment.

**Strategy**

To find the monthly mortgage payment:

- Subtract the down payment from the purchase price to find the mortgage.
- Multiply the mortgage by the factor found in the monthly payment table on page A6.

**Solution**

| Purchase price | − | down payment | = | mortgage |
|---|---|---|---|---|
| 120,000 | − | 25,000 | = | 95,000 |

The mortgage is $95,000.

95,000 × 0.0077182 = 733.23

from the table

The monthly mortgage payment is $733.23.

**You Try It 3**

A new condominium project is selling townhouses for $75,000. A down payment of $15,000 is required, and a 20-year mortgage at an annual interest rate of 9% is available. Find the monthly mortgage payment.

**Your strategy**

**Your solution**

*Solution on p. A32*

**Example 4**

A home has a mortgage of $134,000 for 25 years at an annual interest rate of 7%. During a month when $375.88 of the monthly mortgage payment is principal, how much of the payment is interest?

**Strategy**

To find the interest:

- Multiply the mortgage by the factor found in the monthly payment table on page A6 to find the monthly mortgage payment.
- Subtract the principal from the monthly mortgage payment.

**Solution**

134,000 × 0.0070678 ≈ 947.09

from the table — monthly mortgage payment

| Monthly mortgage payment | − | principal | = | interest |
|---|---|---|---|---|
| 947.09 | − | 375.88 | = | 571.21 |

$571.21 is interest on the mortgage.

**You Try It 4**

An office building has a mortgage of $125,000 for 25 years at an annual interest rate of 9%. During a month when $492.65 of the monthly mortgage payment is principal, how much of the payment is interest?

**Your strategy**

**Your solution**

**Example 5**

The monthly mortgage payment for a home is $598.75. The annual property tax is $900. Find the total monthly payment for the mortgage and property tax.

**Strategy**

To find the monthly payment:

- Divide the annual property tax by 12 to find the monthly property tax.
- Add the monthly property tax to the monthly mortgage payment.

**Solution**

900 ÷ 12 = 75 monthly property tax

598.75 + 75 = 673.75

The total monthly payment is $673.75.

**You Try It 5**

The monthly mortgage payment for a home is $415.20. The annual property tax is $744. Find the total monthly payment for the mortgage and property tax.

**Your strategy**

**Your solution**

*Solutions on p. A32*

## 6.4 Exercises

***Objective A*** *Application Problems*

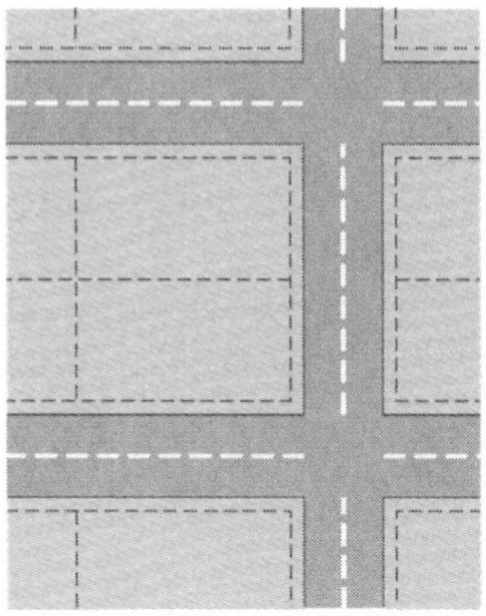

**1.** A condominium at Mt. Baldy Ski Resort was purchased for $97,000, and a down payment of $14,550 was made. Find the mortgage.

**2.** An insurance business was purchased for $173,000 and a down payment of $34,600 was made. Find the mortgage.

**3.** A building lot was purchased for $25,000. The lender requires a down payment of 30% of the purchase price. Find the down payment.

**4.** Ian Goldman purchased a new home for $88,500. The lender requires a down payment of 20% of the purchase price. Find the down payment.

**5.** Brian Stedman made a down payment of 25% of the $850,000 purchase price of an apartment building. How much was the down payment?

**6.** A clothing store was purchased for $125,000, and a down payment that was 25% of the purchase price was made. How much was the down payment?

**7.** A loan of $150,000 is obtained to purchase a home. The loan origination fee is $2\frac{1}{2}$ points. Find the amount of the loan origination fee.

**8.** Security Savings & Loan requires a borrower to pay $3\frac{1}{2}$ points for a loan. Find the amount of the loan origination fee for a loan of $90,000.

**9.** Baja Construction Inc. is selling homes for $95,000. A down payment of 5% is required.
a. Find the down payment.
b. Find the mortgage.

**10.** A cattle rancher purchased some land for $240,000. The bank requires a down payment of 15% of the purchase price.
a. Find the down payment.
b. Find the mortgage.

**11.** Vivian Tom purchased a home for $96,000. Find the mortgage if the down payment is 10% of the purchase price.

**12.** A mortgage lender requires a down payment of 5% of the $56,000 purchase price of a condominium. How much is the mortgage?

***Objective B*** *Application Problems*

Solve. Use the monthly payment table on page A6. Round to the nearest cent.

**13.** An investor obtained a loan of $150,000 to buy a car wash business. The monthly mortgage payment was based on 25 years at 8%. Find the monthly mortgage payment.

**14.** A beautician obtained a 20-year mortgage of $90,000 to expand the business. The credit union charges an annual interest rate of 9%. Find the monthly mortgage payment.

Solve. Use the monthly payment table of page A6. Round to the nearest cent.

15. A couple interested in buying a home determines that they can afford a monthly mortgage payment of $800. Can they afford to buy a home with a 30-year $110,000 mortgage at 8% interest?

16. A lawyer is considering purchasing a new office building with a 20-year $400,000 mortgage at 9% interest. The lawyer can afford a monthly mortgage payment of $4000. Can the lawyer afford the monthly mortgage payment on the new office building?

17. The county tax assessor has determined that the annual property tax on a $125,000 house is $1348.20. Find the monthly property tax.

18. The annual property tax on a $155,000 home is $1992. Find the monthly property tax.

19. Abacus Imports Inc. has a warehouse with a 25-year mortgage of $200,000 at an annual interest rate of 9%.
    a. Find the monthly mortgage payment.
    b. During a month when $941.72 of the monthly mortgage payment is principal, how much of the payment is interest?

20. A vacation home has a mortgage of $135,000 for 30 years at an annual interest rate of 7%.
    a. Find the monthly mortgage payment.
    b. During a month when $392.47 of the monthly mortgage payment is principal, how much of the payment is interest?

21. The annual mortgage payment on a duplex is $10,844.40. The owner must pay an annual property tax of $948. Find the total monthly payment for the mortgage and property tax.

22. The monthly mortgage payment on a home is $716.40, and the homeowner pays an annual property tax of $792. Find the total monthly payment for the mortgage and property tax.

23. Maria Hernandez purchased a home for $210,000 and made a down payment of $15,000. The balance was financed for 30 years at an annual interest rate of 8%. Find the monthly mortgage payment.

24. A customer of a savings and loan purchased a $185,000 home and made a down payment of $20,000. The savings and loan charges its customers an annual interest rate of 7% for 30 years for a home mortgage. Find the monthly mortgage payment.

## *APPLYING THE CONCEPTS*

25. A couple buying a home for $100,000 have a choice of loans. One loan is an 8% loan for 20 years and the other loan is at 9% for 30 years. Find the amount of interest that can be saved by choosing the 20-year loan.

26. [W] Find out what an adjustable-rate mortgage is. What is the difference between this type of loan and a fixed-rate mortgage. List some of the advantages and disadvantages of each.

# 6.5 Car Expenses

### *Objective A To calculate the initial expenses of buying a car*

The initial expenses in the purchase of a car usually include the down payment, the license fees, and the sales tax. The down payment may be very small or as much as 25% or 30% of the purchase price of the car, depending on the lending institution. License fees and sales tax are regulated by each state, so these expenses vary from state to state.

**Example 1**
A car is purchased for $8500, and the lender requires a down payment of 15% of the purchase price. Find the amount financed.

**Strategy**
To find the amount financed:

- Find the down payment by solving the basic percent equation for *amount*.
- Subtract the down payment from the purchase price.

**Solution**

Percent × base = amount

$$\boxed{\text{Percent}} \times \boxed{\text{purchase price}} = \boxed{\text{down payment}}$$

$$0.15 \times 8500 = n$$

$$1275 = n$$

$$8500 - 1275 = 7225$$

The amount financed is $7225.

**You Try It 1**
A down payment of 20% of the $9200 purchase price of a new car is made. Find the amount financed.

**Your strategy**

**Your solution**

**Example 2**
A sales clerk purchases a car for $6500 and pays a sales tax that is 5% of the purchase price. How much is the sales tax?

**Strategy**
To find the sales tax, solve the basic percent equation for *amount*.

**Solution**

Percent × base = amount

$$\boxed{\text{Percent}} \times \boxed{\text{purchase price}} = \boxed{\text{sales tax}}$$

$$0.05 \times 6500 = n$$

$$325 = n$$

The sales tax is $325.

**You Try It 2**
A car is purchased for $7350. The car license fee is 1.5% of the purchase price. How much is the license fee?

**Your strategy**

**Your solution**

*Solutions on p. A33*

## Objective B To calculate ongoing expenses of owning a car

Besides the initial expenses of buying a car, there are continuing expenses involved in owning a car. These ongoing expenses include car insurance, gas and oil, general maintenance, and the monthly car payment. The monthly car payment is calculated in the same manner as monthly mortgage payments on a home loan. A monthly payment table, such as the one on page A6, is used to simplify calculation of monthly car payments.

**Example 3**
At a cost of \$0.27 per mile, how much does it cost to operate a car during a year the car is driven 15,000 miles?

**Strategy**
To find the cost, multiply the cost per mile by the number of miles driven.

**Solution**
15,000 × 0.27 = 4050 The cost is \$4050.

**You Try It 3**
At a cost of \$0.22 per mile, how much does it cost to operate a car during a year the car is driven 23,000 miles?

**Your strategy**

**Your solution**

**Example 4**
During 1 month the total gasoline bill was \$84 and the car was driven 1200 miles. What was the cost per mile for gasoline?

**Strategy**
To find the cost per mile for gasoline, divide the cost for gasoline by the number of miles driven.

**Solution** 84 ÷ 1200 = 0.07
The cost per mile was \$0.07.

**You Try It 4**
In a year when the total car insurance bill was \$360 and the car was driven 15,000 miles, what was the cost per mile for car insurance?

**Your strategy**

**Your solution**

**Example 5**
A car is purchased for \$8500 with a down payment of \$1700. The balance is financed for 3 years at an annual interest rate of 9%. Find the monthly car payment.

**Strategy**
To find the monthly payment:

- Subtract the down payment from the purchase price to find the amount financed.
- Multiply the amount financed by the factor found in the monthly payment table on page A6.

**Solution**
8500 − 1700 = 6800

The amount financed is \$6800.

6800 × 0.0317997 ≈ 216.24

The monthly payment is \$216.24.

**You Try It 5**
A truck is purchased for \$15,900 with a down payment of \$3975. The balance is financed for 4 years at an annual interest rate of 8%. Find the monthly payment.

**Your strategy**

**Your solution**

*Solutions on p. A33*

## 6.5 Exercises

### *Objective A* Application Problems

1. Amanda has saved $780 to make a down payment on a car. The car dealer requires a down payment of 12% of the purchase price. Does she have enough money to make the down payment on a car that costs $7100?

2. A jeep was purchased for $9288. A down payment of 15% of the purchase price was required. How much was the down payment?

3. A drapery installer bought a minivan to carry drapery samples. The purchase price of the van was $16,500, and a 4.5% sales tax was paid. How much was the sales tax?

4. A delivery truck for the Dixieline Lumber Company was purchased for $18,500. A sales tax of 4% of the purchase price was paid. Find the sales tax.

5. A license fee of 2% of the purchase price of a truck is to be paid on a pickup truck costing $12,500. How much is the license fee for the truck?

6. Your state charges a license fee of 1.5% on the purchase price of a car. How much is the license fee for a car that costs $6998?

7. An electrician bought a $12,000 flatbed truck. A state license fee of $175 and a sales tax of 3.5% of the purchase price are required.
   a. Find the sales tax.
   b. Find the total cost of the sales tax and the license fee.

8. A physical therapist bought a car for $9375 and made a down payment of $1875. The sales tax is 5% of the purchase price.
   a. Find the sales tax.
   b. Find the total cost of the sales tax and the down payment.

9. Martin bought a motorcycle for $2200 and made a down payment that is 25% of the purchase price.
   a. Find the down payment.
   b. Find the amount financed.

10. A carpenter bought a utility van for $14,900 and made a down payment that is 15% of the purchase price.
    a. Find the down payment.
    b. Find the amount financed.

11. An author bought a sports car for $35,000 and made a down payment of 20% of the purchase price. Find the amount financed.

12. Tania purchased a new car for $13,500 and made a down payment of 25% of the cost. Find the amount financed.

### *Objective B* Application Problems

Solve. Use the monthly payment table on page A6. Round to the nearest cent.

13. A rancher financed $14,000 for the purchase of a truck through a credit union at 9% interest for 4 years. Find the monthly truck payment.

**14.** A car loan of $8000 is financed for 3 years at an annual interest rate of 10%. Find the monthly car payment.

**15.** An estimate of the cost of owning a compact car is $0.32 per mile. Using this estimate, how much does it cost to operate a car during a year in which the car is driven 16,000 miles.

**16.** An estimate of the cost of care and maintenance of automobile tires is $0.015 per mile. Using this estimate, how much would it cost for care and maintenance during a year in which the car is driven 14,000 miles?

**17.** A family spent $1600 on gas, oil, and car insurance during a period in which the car was driven 14,000 miles. Find the cost per mile for gas, oil, and car insurance.

**18.** Last year you spent $1050 for gasoline for your car. The car was driven 15,000 miles. What was your cost per mile for gasoline?

**19.** Elena's monthly car payment is $143.50. During a month in which $68.75 of the monthly payment is principal, how much of the payment is interest?

**20.** The cost for a pizza delivery truck for the year included $1870 in truck payments, $1200 for gasoline, and $675 for insurance. Find the total cost for truck payments, gasoline, and insurance for the year.

**21.** The city of Colton purchased a fire truck for $82,000 and made a down payment of $5400. The balance is financed for 5 years at an annual rate of 9%.
a. Find the amount financed.
b. Find the monthly truck payment.

**22.** A used car is purchased for $4995, and a down payment of $995 is made. The balance is financed for 3 years at an interest rate of 8%.
a. Find the amount financed.
b. Find the monthly car payment.

**23.** An artist purchased a new car costing $27,500 and made a down payment of $5500. The balance is financed for 3 years at an annual interest rate of 10%. Find the monthly car payment.

**24.** A half-ton truck with a camper is purchased for $19,500, and a down payment of $2500 is made. The balance is financed for 4 years at an annual interest rate of 9%. Find the monthly payment.

### *APPLYING THE CONCEPTS*

**25.** One bank offers a 4-year car loan at an annual interest rate of 10% plus a loan application fee of $45. A second bank offers 4-year car loans at an annual interest rate of 11% but charges no loan application fee. If you need to borrow $5800 to purchase a car, which of the two bank loans has the lesser loan costs? Assume you keep the car for 4 years.

**26.** How much interest is repaid on a 5-year car loan of $9000 if the interest rate is 9%?

# 6.6 Wages

## Objective A To calculate commissions, total hourly wages, and salaries

Commissions, hourly wage, and salary are three ways to receive payment for doing work.

**Commissions** are usually paid to salespersons and are calculated as a percent of total sales.

➡ As a real estate broker, Emma Smith receives a commission of 4.5% of the selling price of a house. Find the commission she earned for selling a home for $75,000.

To find the commission Emma earned, solve the basic percent equation for *amount*.

| Percent | × | base | = | amount |
|---|---|---|---|---|
| Commission rate | × | total sales | = | commission |
| 0.045 | × | 75,000 | = | 3375 |

The commission is $3375.

An employee who receives an **hourly wage** is paid a certain amount for each hour worked.

➡ A plumber receives an hourly wage of $13.25. Find the plumber's total wages for working 37 hours.

To find the plumber's total wages, multiply the hourly wage by the number of hours worked.

| Hourly wage | × | number of hours worked | = | total wages |
|---|---|---|---|---|
| 13.25 | × | 37 | = | 490.25 |

The plumber's total wages for working 37 hours are $490.25.

An employee who is paid a **salary** receives payment based on a weekly, biweekly (every other week), monthly, or annual time schedule. Unlike the employee who receives an hourly wage, the salaried worker does not receive additional pay for working more than the regularly scheduled work day.

➡ Ravi Basar is a computer operator who receives a weekly salary of $395. Find his salary for 1 month (4 weeks).

To find Ravi's salary for 1 month, multiply the salary per pay period by the number of pay periods.

| Salary per pay period | × | number of pay periods | = | total salary |
|---|---|---|---|---|
| 395 | × | 4 | = | 1580 |

Ravi's total salary for 1 month is $1580.

**Example 1**

A pharmacist's hourly wage is $28. On Saturday, the pharmacist earns time and a half ($1\frac{1}{2}$ times the regular hourly wage). How much does the pharmacist earn for working 6 hours on Saturday?

**Strategy**

To find the pharmacist's earnings:

- Find the hourly wage for working on Saturday by multiplying the hourly wage by $1\frac{1}{2}$.
- Multiply the product by the number of hours worked.

**Solution**

28 × 1.5 = 42     42 × 6 = 252

The pharmacist earns $252.

**You Try It 1**

A construction worker's hourly wage is $8.50. The worker earns double time (2 times the regular hourly wage) for working overtime. How much does the worker earn for working 8 hours of overtime?

**Your strategy**

**Your solution**

**Example 2**

An efficiency expert received a contract for $3000. The consultant spent 75 hours on the project. Find the consultant's hourly wage.

**Strategy**

To find the hourly wage, divide the total earnings by the number of hours worked.

**Solution**

3000 ÷ 75 = 40 The hourly wage was $40.

**You Try It 2**

A contractor for a bridge project receives an annual salary of $28,224. What is the contractor's salary per month?

**Your strategy**

**Your solution**

**Example 3**

Dani Greene receives $18,500 per year plus a $5\frac{1}{2}$% commission on sales over $100,000. During one year, Dani sold $150,000 worth of computers. Find Dani's total earnings for the year.

**Strategy**

To find the total earnings:

- Find the commission earned by multiplying the commission rate by sales over $100,000.
- Add the commission to the annual pay.

**Solution**

150,000 − 100,000 = 50,000

50,000 × 0.055 = 2750 commission

18,500 + 2750 = 21,250

Dani earned $21,250.

**You Try It 3**

An insurance agent receives $12,000 per year plus a $9\frac{1}{2}$% commission on sales over $50,000. During one year, the agent's sales totaled $175,000. Find the agent's total earnings for the year.

**Your strategy**

**Your solution**

*Solutions on p. A34*

## 6.6 Exercises

***Objective A*** *Application Problems*

1. Lewis works in a clothing store and earns $5.50 per hour. How much does he earn in a 40-hour week?

2. Sasha pays a gardener an hourly wage of $5.25. How much does she pay the gardener for working 25 hours?

3. A real estate agent receives a 3% commission for selling a house. Find the commission that the agent earned for selling a house for $98,500.

4. Ron Caruso works as an insurance agent and receives a commission of 40% of the first year's premium. Find Ron's commission for selling a life insurance policy with a first-year premium of $1050.

5. A stock broker receives a commission of 1.5% of the price of stock that is bought or sold. Find the commission on 100 shares of stock that were bought for $5600.

6. The owner of the Carousel Art Gallery receives a commission of 20% of paintings that are sold on consignment. Find the commission on a painting that sold for $22,500.

7. As an Italian Language teacher, Keisha Brown receives an annual salary of $38,928. How much does Keisha receive each month?

8. An apprentice plumber receives an annual salary of $19,680. How much does the plumber receive per month?

9. An electrician's hourly wage is $15.80. For working overtime, the electrician earns double time. What is the electrician's hourly wage for working overtime?

10. Carlos receives a commission of 12% of his weekly sales as a sales representative for a medical supply company. Find the commission he earned during a week in which sales were $4500.

11. A golf pro receives a commission of 25% for selling a golf set. Find what commission the pro earned for selling a golf set costing $450.

12. Steven receives $1.75 per square yard to install carpet. How much does he receive for installing 160 square yards of carpet?

13. A typist charged $1.75 per page for typing technical material. How much does the typist earn for typing a 225-page book?

**14.** A nuclear chemist received $15,000 for consulting fees while working on a nuclear power plant. The chemist worked 120 hours on the project. Find the consultant's hourly wage.

**15.** Maxine received $3400 for working on a project as a computer consultant for 40 hours. Find her hourly wage.

**16.** Gil Stratton's hourly wage is $10.78. For working overtime, he receives double time.
a. What is Gil's hourly wage for working overtime?
b. How much does he earn for working 16 hours of overtime?

**17.** Mark is a lathe operator and receives an hourly wage of $8.50. When working on Saturday, he receives time and a half.
a. What is Mark's hourly wage on Saturday?
b. How much does he earn for working 8 hours on Saturday?

**18.** A stock clerk at a supermarket earns $6.20 an hour. For working the night shift, the clerk's wage increases by 15%.
a. What is the increase in hourly pay for working the night shift?
b. What is the clerk's hourly wage for working the night shift?

**19.** A nurse earns $8.50 an hour. For working the night shift, the nurse receives a 10% increase in pay.
a. What is the increase in hourly pay for working the night shift?
b. What is the hourly pay for working the night shift?

**20.** Tony's hourly wage as a service station attendant is $6.40. For working the night shift, his wage is increased 25%. What is Tony's hourly wage for working the night shift?

**21.** Nicole Tobin, a door-to-door salesperson, receives a salary of $150 per week plus a commission of 15% on all sales over $1500. Find her earnings during a week in which sales totaled $3000.

### APPLYING THE CONCEPTS

The table at the right shows the *Money* magazine ranking of the ten most desirable careers of 1994. Complete Exercises 22–25 with information obtained from this table.

| *1994 Rank* | *1992 Rank* | *Occupation* | *Median Annual Earnings* | *11-Year Job Growth* |
|---|---|---|---|---|
| 1 | 31 | Computer systems analyst | $42,700 | 110% |
| 2 | 3 | Physician | 148,000 | 35 |
| 3 | 50 | Physical therapist | 37,200 | 88 |
| 4 | 13 | Electrical engineer | 59,100 | 24 |
| 5 | 9 | Civil engineer | 55,800 | 24 |
| 6 | 7 | Pharmacist | 47,500 | 29 |
| 7 | 29 | Psychologist | 53,000 | 48 |
| 8 | 2 | Geologist | 50,800 | 22 |
| 9 | 15 | High school teacher | 32,500 | 37 |
| 10 | 5 | School principal | 57,300 | 23 |

Source: *Money*, March 1994.

**22.** Find the monthly earnings for a civil engineer.

**23.** Find the difference in monthly salary between a physician and an electrical engineer.

**24.** What percent of a high school teacher's salary does a school principal receive?

**25.** In 1994, there were 455,000 computer systems analysts in the United States. Find the projected increase in the number of computer analyst jobs between now and 2005 (11 years from now).

# 6.7 Bank Statements

## Objective A To calculate checkbook balances

**POINT OF INTEREST**
There are a number of computer programs that serve as "electronic" checkbooks. With these programs, you can pay your bills by using a computer to write the check and then transmit the check over telephone lines using a modem. Paying your federal income tax is one of the few tasks that cannot be accomplished through these electronic checkbooks.

A checking account can be opened at most banks or savings and loan associations by depositing an amount of money in the bank. A checkbook contains checks and deposit slips and a checkbook register in which to record checks written and amounts deposited in the checking account. Each time a check is written, the amount of the check is subtracted from the amount in the account. When a deposit is made, the amount deposited is added to the amount in the account.

A portion of a checkbook register is shown below. The account holder had a balance of $587.93 before writing two checks, one for $286.87 and the other for $102.38, and making one deposit of $345.00.

RECORD ALL CHARGES OR CREDITS THAT AFFECT YOUR ACCOUNT

| NUMBER | DATE | DESCRIPTION OF TRANSACTION | PAYMENT/DEBIT (-) | | √ T | FEE (IF ANY) (-) | DEPOSIT/CREDIT (+) | | BALANCE | |
|---|---|---|---|---|---|---|---|---|---|---|
| | | | | | | | | | $ 587 | 93 |
| 108 | 8/4 | Plumber | $ 286 | 87 | | $ | $ | | 301 | 06 |
| 109 | 8/10 | Car Payment | 102 | 38 | | | | | 198 | 68 |
| | 8/14 | Deposit | | | | | 345 | 00 | 543 | 68 |
| | | | | | | | | | | |
| | | | | | | | | | | |
| | | | | | | | | | | |

To find the current checking account balance, subtract the amount of each check from the previous balance. Then add the amount of the deposit.

The current checking account balance is $543.68.

**Example 1** A mail carrier had a checking account balance of $485.93 before writing two checks, one for $18.98 and another for $35.72, and making a deposit of $250. Find the current checking account balance.

**Strategy** To find the current balance:

- Subtract the amount of each check from the old balance.
- Add the amount of the deposit.

**Solution**

| | |
|---|---|
| 485.93 | |
| − 18.98 | first check |
| 466.95 | |
| − 35.72 | second check |
| 431.23 | |
| + 250.00 | deposit |
| 681.23 | |

The current checking account balance is $681.23.

**You Try It 1** A cement mason had a checking account balance of $302.46 before writing a check for $20.59 and making two deposits, one in the amount of $176.86 and another in the amount of $94.73. Find the current checking account balance.

**Your strategy**

**Your solution**

*Solution on p. A35*

## *Objective B To balance a checkbook*

Each month a bank statement is sent to the account holder. The bank statement shows the checks that the bank has paid, the deposits received, and the current bank balance.

A bank statement and checkbook register are shown on the next page.

**Balancing the checkbook,** or determining if the checking account balance is accurate, requires a number of steps.

1. In the checkbook register, put a check (√) by each check paid by the bank and each deposit recorded by the bank.

RECORD ALL CHARGES OR CREDITS THAT AFFECT YOUR ACCOUNT

| NUMBER | DATE | DESCRIPTION OF TRANSACTION | PAYMENT/DEBIT (-) | √ T | FEE (IF ANY) (-) | DEPOSIT/CREDIT (+) | BALANCE |
|---|---|---|---|---|---|---|---|
| | | | | | | | $ 840 27 |
| 263 | 5/20 | Dentist | $ 25 00 | √ | $ | $ | 815 27 |
| 264 | 5/22 | Meat Market | 33 61 | √ | | | 781 66 |
| 265 | 5/22 | Gas Company | 67 14 | | | | 714 52 |
| | 5/29 | Deposit | | √ | | 192 00 | 906 52 |
| 266 | 5/29 | Pharmacy | 18 95 | √ | | | 887 57 |
| 267 | 5/30 | Telephone | 43 85 | | | | 843 72 |
| 268 | 6/2 | Groceries | 43 19 | √ | | | 800 53 |
| | 6/3 | Deposit | | √ | | 215 00 | 1015 53 |
| 269 | 6/7 | Insurance | 103 00 | √ | | | 912 53 |
| | 6/10 | Deposit | | | | 225 00 | 1137 53 |
| 270 | 6/15 | Clothing Store | 16 63 | √ | | | 1120 90 |
| 271 | 6/18 | Newspaper | 7 00 | | | | 1113 90 |

CHECKING ACCOUNT Monthly Statement — Account Number: 924-297-8

| Date | Transaction | Amount | Balance |
|---|---|---|---|
| 5/20 | OPENING BALANCE | | 840.27 |
| 5/21 | CHECK | 25.00 | 815.27 |
| 5/23 | CHECK | 33.61 | 781.66 |
| 5/29 | DEPOSIT | 192.00 | 973.66 |
| 6/1 | CHECK | 18.95 | 954.71 |
| 6/1 | INTEREST | 4.47 | 959.18 |
| 6/3 | CHECK | 43.19 | 915.99 |
| 6/3 | DEPOSIT | 215.00 | 1130.99 |
| 6/9 | CHECK | 103.00 | 1027.99 |
| 6/16 | CHECK | 16.63 | 1011.36 |
| 6/20 | SERVICE CHARGE | 3.00 | 1008.36 |
| 6/20 | CLOSING BALANCE | | 1008.36 |

**2.** Add to the current checkbook balance all checks that have been written but have not yet been paid by the bank and any interest paid on the account.

**3.** Subtract any service charges and any deposits not yet recorded by the bank. This is the checkbook balance.

| | |
|---|---|
| Current checkbook balance: | 1113.90 |
| Checks: 265 | 67.14 |
| 267 | 43.85 |
| 271 | 7.00 |
| Interest: | + 4.47 |
| | 1236.36 |
| Service charge: | − 3.00 |
| | 1233.36 |
| Deposit: | − 225.00 |
| Checkbook balance: | 1008.36 |

**4.** Compare the balance with the bank balance listed on the bank statement. If the two numbers are equal, the bank statement and checkbook balance.

| Checking bank balance from bank statement | | Checkbook balance |
|---|---|---|
| $1008.36 | = | $1008.36 |

The bank statement and checkbook balance.

RECORD ALL CHARGES OR CREDITS THAT AFFECT YOUR ACCOUNT

| NUMBER | DATE | DESCRIPTION OF TRANSACTION | PAYMENT/DEBIT (-) | | √ T | FEE (IF ANY) (-) | DEPOSIT/CREDIT (+) | | BALANCE | |
|---|---|---|---|---|---|---|---|---|---|---|
| | | | | | | | | | $ 1620 | 42 |
| 413 | 3/2 | Car Payment | $ 132 | 15 | √ | $ | $ | | 1488 | 27 |
| 414 | 3/2 | Utility | 67 | 14 | √ | | | | 1421 | 13 |
| 415 | 3/5 | Restaurant - Dinner for 4 | 78 | 14 | | | | | 1342 | 99 |
| | 3/8 | Deposit | | | √ | | 1842 | 66 | 3185 | 65 |
| 416 | 3/10 | House Payment | 672 | 14 | √ | | | | 2513 | 51 |
| 417 | 3/14 | Insurance | 177 | 10 | | | | | 2336 | 41 |

CHECKING ACCOUNT Monthly Statement — Account Number: 924-297-8

| Date | Transaction | Amount | Balance |
|---|---|---|---|
| 3/1 | OPENING BALANCE | | 1620.42 |
| 3/4 | CHECK | 132.15 | 1488.27 |
| 3/5 | CHECK | 67.14 | 1421.13 |
| 3/8 | DEPOSIT | 1842.66 | 3263.79 |
| 3/10 | INTEREST | 6.77 | 3270.56 |
| 3/12 | CHECK | 672.14 | 2598.42 |
| 3/25 | SERVICE CHARGE | 2.00 | 2596.42 |
| 3/30 | CLOSING BALANCE | | 2596.42 |

Balance the bank statement shown above.

1. In the checkbook register, put a check (√) by each check paid by the bank and each deposit recorded by the bank.

2. Add to the current checkbook balance all checks that have been written but have not yet been paid by the bank and any interest paid on the account.

| | |
|---|---|
| Current checkbook balance: | 2336.41 |
| Checks: 415 | 78.14 |
| 417 | 177.10 |
| Interest: | + 6.77 |
| | 2598.42 |

3. Subtract any service charges and any deposits not yet recorded by the bank. This is the checkbook balance.

| | |
|---|---|
| Service charge: | − 2.00 |
| | 2596.42 |
| Checkbook balance: | 2596.42 |

4. Compare the balance with the bank balance listed on the bank statement. If the two numbers are equal, the bank statement and checkbook balance.

| Closing bank balance from bank statement | | Checkbook balance: |
|---|---|---|
| $2596.42 | = | $2596.42 |

The bank statement and checkbook balance.

RECORD ALL CHARGES OR CREDITS THAT AFFECT YOUR ACCOUNT

| NUMBER | DATE | DESCRIPTION OF TRANSACTION | PAYMENT/DEBIT (-) | | √ T | FEE (IF ANY) (-) | DEPOSIT/CREDIT (+) | | BALANCE | |
|---|---|---|---|---|---|---|---|---|---|---|
| | | | | | | | | | $ 412 | 64 |
| 345 | 1/14 | Phone Bill | $ 34 | 75 | √ | $ | $ | | 377 | 89 |
| 346 | 1/19 | Magazine | 8 | 98 | √ | | | | 368 | 91 |
| 347 | 1/23 | Theatre Tickets | 45 | 00 | | | | | 323 | 91 |
| | 1/31 | Deposit | | | | | 947 | 00 | 1270 | 91 |
| 348 | 2/5 | Cash | 250 | 00 | √ | | | | 1020 | 91 |
| 349 | 2/12 | Rent | 440 | 00 | | | | | 580 | 91 |

CHECKING ACCOUNT Monthly Statement — Account Number: 924-297-8

| Date | Transaction | Amount | Balance |
|---|---|---|---|
| 1/10 | OPENING BALANCE | | 412.64 |
| 1/18 | CHECK | 34.75 | 377.89 |
| 1/23 | CHECK | 8.98 | 368.91 |
| 1/31 | DEPOSIT | 947.00 | 1315.91 |
| 2/1 | INTEREST | 4.52 | 1320.43 |
| 2/10 | CHECK | 250.00 | 1070.43 |
| 2/10 | CLOSING BALANCE | | 1070.43 |

**Example 2**

Balance the bank statement shown above.

**Solution**

| | |
|---|---|
| Current checkbook balance: | 580.91 |
| Checks: 347 | 45.00 |
| 349 | 440.00 |
| Interest: | + 4.52 |
| | 1070.43 |
| Service charge: | − 0.00 |
| | 1070.43 |
| Deposit: | − 0.00 |
| Checkbook balance: | 1070.43 |

Closing bank balance from bank statement: $1070.43

Checkbook balance: $1070.43

The bank statement and checkbook balance.

RECORD ALL CHARGES OR CREDITS THAT AFFECT YOUR ACCOUNT

| NUMBER | DATE | DESCRIPTION OF TRANSACTION | PAYMENT/DEBIT (-) | | √ T | FEE (IF ANY) (-) | DEPOSIT/CREDIT (+) | | BALANCE | |
|---|---|---|---|---|---|---|---|---|---|---|
| | | | | | | | | | $ 603 | 17 |
| | 2/15 | *Deposit* | $ | | | $ | $ 523 | 84 | 1127 | 01 |
| 234 | 2/20 | *Mortgage* | 473 | 21 | √ | | | | 653 | 80 |
| 235 | 2/27 | *Cash* | 200 | 00 | √ | | | | 453 | 80 |
| | 3/1 | *Deposit* | | | | | 523 | 84 | 977 | 64 |
| 236 | 3/12 | *Insurance* | 275 | 50 | √ | | | | 702 | 14 |
| 237 | 3/12 | *Telephone* | 48 | 73 | | | | | 653 | 41 |

CHECKING ACCOUNT Monthly Statement — Account Number: 314-271-4

| Date | Transaction | Amount | Balance |
|---|---|---|---|
| 2/14 | OPENING BALANCE | | 603.17 |
| 2/15 | DEPOSIT | 523.84 | 1127.01 |
| 2/21 | CHECK | 473.21 | 653.80 |
| 2/28 | CHECK | 200.00 | 453.80 |
| 3/1 | INTEREST | 2.11 | 455.91 |
| 3/14 | CHECK | 275.50 | 180.41 |
| 3/14 | CLOSING BALANCE | | 180.41 |

**You Try It 2**

Balance the bank statement shown above.

**Your solution**

*Solution on p. A35*

## 6.7 Exercises

***Objective A*** *Application Problems*

1. You had a checking account balance of $342.51 before making a deposit of $143.81. What is your new checking account balance?

2. Carmen had a checking account balance of $493.26 before writing a check for $48.39. What is the current checking account balance?

3. A real estate firm had a balance of $2431.76 in its rental property checking account. What is the balance in this account after a check for $1209.29 has been written?

4. The business checking account for R and R Tires showed a balance of $1536.97. What is the balance in this account after a deposit of $439.21 has been made?

5. A nutritionist had a checking account balance of $1204.63 before writing one check for $119.27 and another check for $260.09. Find the current checkbook balance.

6. Sam had a checking account balance of $3046.93 before writing a check for $1027.33 and making a deposit of $150.00. Find the current checkbook balance.

7. The business checking account for Rachael's Dry Cleaning had a balance of $3476.85 before a deposit of $1048.53 was made. The store manager then wrote checks, one for $848.37 and another for $676.19. Find the current checkbook balance.

8. Joel had a checking account balance of $427.38 before a deposit of $127.29 was made. Joel then wrote two checks, one for $43.52 and one for $249.78. Find the current checkbook balance.

9. A carpenter had a checkbook balance of $104.96 before making a deposit of $350 and writing a check for $71.29. Is there enough money in the account to purchase a refrigerator for $375?

10. A taxi driver had a checkbook balance of $149.85 before making a deposit of $245 and writing a check for $387.68. Is there enough money in the account for the bank to pay the check?

11. A sporting goods store has the opportunity to buy downhill skis and cross-country skis at a manufacturer's close-out sale. The downhill skis will cost $3500, and the cross-country skis will cost $2050. There is currently $5625.42 in the sporting goods store's checking account. Is there enough money in the account to make both purchases by check?

12. A lathe operator's current checkbook balance is $643.42. The operator wants to purchase a utility trailer for $225 and a used piano for $450. Is there enough money in the account to make the two purchases?

**13.** Balance the checkbook.

RECORD ALL CHARGES OR CREDITS THAT AFFECT YOUR ACCOUNT

| NUMBER | DATE | DESCRIPTION OF TRANSACTION | PAYMENT/DEBIT (-) | | √ T | FEE (IF ANY) (-) | DEPOSIT/CREDIT (+) | | BALANCE | |
|---|---|---|---|---|---|---|---|---|---|---|
| | | | | | | | | | $ 466 | 79 |
| 223 | 3/2 | Groceries | $ 67 | 32 | √ | $ | $ | | 399 | 47 |
| | 3/5 | Deposit | | | | | 560 | 70 | 960 | 17 |
| 224 | 3/5 | Rent | 460 | 00 | √ | | | | 500 | 17 |
| 225 | 3/7 | Gas & Electric | 42 | 35 | √ | | | | 457 | 82 |
| 226 | 3/7 | Cash | 100 | 00 | √ | | | | 357 | 82 |
| 227 | 3/7 | Insurance | 118 | 44 | | | | | 239 | 38 |
| 228 | 3/7 | Credit Card | 119 | 32 | | | | | 120 | 06 |
| 229 | 3/12 | Dentist | 42 | 00 | √ | | | | 78 | 06 |
| 230 | 3/13 | Drug Store | 17 | 03 | √ | | | | 61 | 03 |
| | 3/19 | Deposit | | | | | 560 | 70 | 621 | 73 |
| 231 | 3/22 | Car Payment | 141 | 35 | √ | | | | 480 | 38 |
| 232 | 3/25 | Cash | 100 | 00 | √ | | | | 380 | 38 |
| 233 | 3/25 | Oil Company | 66 | 40 | | | | | 313 | 98 |
| 234 | 3/28 | Plumber | 55 | 73 | √ | | | | 258 | 25 |
| 235 | 3/29 | Department Store | 88 | 39 | | | | | 169 | 86 |
| | | | | | | | | | | |

CHECKING ACCOUNT Monthly Statement — Account Number: 122-345-1

| Date | Transaction | Amount | Balance |
|---|---|---|---|
| 3/1 | OPENING BALANCE | | 466.79 |
| 3/5 | DEPOSIT | 560.70 | 1027.49 |
| 3/7 | CHECK | 67.32 | 960.17 |
| 3/8 | CHECK | 460.00 | 500.17 |
| 3/8 | CHECK | 100.00 | 400.17 |
| 3/9 | CHECK | 42.35 | 357.82 |
| 3/12 | CHECK | 118.44 | 239.38 |
| 3/14 | CHECK | 42.00 | 197.38 |
| 3/18 | CHECK | 17.03 | 180.35 |
| 3/19 | DEPOSIT | 560.70 | 741.05 |
| 3/25 | CHECK | 141.35 | 599.70 |
| 3/27 | CHECK | 100.00 | 499.70 |
| 3/29 | CHECK | 55.73 | 443.97 |
| 3/30 | INTEREST | 13.22 | 457.19 |
| 4/1 | CLOSING BALANCE | | 457.19 |

**14.** Balance the checkbook.

RECORD ALL CHARGES OR CREDITS THAT AFFECT YOUR ACCOUNT

| NUMBER | DATE | DESCRIPTION OF TRANSACTION | PAYMENT/DEBIT (-) | | √ T | FEE (IF ANY) (-) | DEPOSIT/CREDIT (+) | | BALANCE | |
|---|---|---|---|---|---|---|---|---|---|---|
| | | | | | | | | | $ 219 | 43 |
| | 5/1 | Deposit | $ | | | $ | $ 219 | 14 | 438 | 57 |
| 515 | 5/2 | Electric Bill | 22 | 35 | | | | | 416 | 22 |
| 516 | 5/2 | Groceries | 55 | 14 | | | | | 361 | 08 |
| 517 | 5/4 | Insurance | 122 | 17 | | | | | 238 | 91 |
| 518 | 5/5 | Theatre Tickets | 24 | 50 | | | | | 214 | 41 |
| | 5/8 | Deposit | | | | | 219 | 14 | 433 | 55 |
| 519 | 5/10 | Telephone | 17 | 39 | | | | | 416 | 16 |
| 520 | 5/12 | Newspaper | 12 | 50 | | | | | 403 | 66 |
| | 5/15 | Interest | | | | | 7 | 82 | 411 | 48 |
| | 5/15 | Deposit | | | | | 219 | 14 | 630 | 62 |
| 521 | 5/20 | Hotel | 172 | 90 | | | | | 457 | 72 |
| 522 | 5/21 | Credit Card | 113 | 44 | | | | | 344 | 28 |
| 523 | 5/22 | Eye Exam | 42 | 00 | | | | | 302 | 28 |
| 524 | 5/24 | Groceries | 77 | 14 | | | | | 225 | 14 |
| 525 | 5/24 | Deposit | | | | | 219 | 14 | 444 | 28 |
| 526 | 5/25 | Oil Company | 44 | 16 | | | | | 400 | 12 |
| 527 | 5/30 | Car Payment | 88 | 62 | | | | | 311 | 50 |
| 528 | 5/30 | Doctor | 37 | 42 | | | | | 274 | 08 |

CHECKING ACCOUNT Monthly Statement — Account Number: 122-345-1

| Date | Transaction | Amount | Balance |
|---|---|---|---|
| 5/1 | OPENING BALANCE | | 219.43 |
| 5/1 | DEPOSIT | 219.14 | 438.57 |
| 5/3 | CHECK | 55.14 | 383.43 |
| 5/4 | CHECK | 22.35 | 361.08 |
| 5/6 | CHECK | 24.50 | 336.58 |
| 5/8 | CHECK | 122.17 | 214.41 |
| 5/8 | DEPOSIT | 219.14 | 433.55 |
| 5/15 | INTEREST | 7.82 | 441.37 |
| 5/15 | CHECK | 17.39 | 423.98 |
| 5/15 | DEPOSIT | 219.14 | 643.12 |
| 5/23 | CHECK | 42.00 | 601.12 |
| 5/23 | CHECK | 172.90 | 428.22 |
| 5/24 | CHECK | 77.14 | 351.08 |
| 5/24 | DEPOSIT | 219.14 | 570.22 |
| 5/30 | CHECK | 88.62 | 481.60 |
| 6/1 | CLOSING BALANCE | | 481.60 |

## *Objective B* Application Problems

**15.** Balance the checkbook.

RECORD ALL CHARGES OR CREDITS THAT AFFECT YOUR ACCOUNT

| NUMBER | DATE | DESCRIPTION OF TRANSACTION | PAYMENT/DEBIT (-) | | √ T | FEE (IF ANY) (-) | DEPOSIT/CREDIT (+) | | BALANCE $ | |
|---|---|---|---|---|---|---|---|---|---|---|
| | | | | | | | | | 1035 | 18 |
| 218 | 7/2 | Mortgage | $ 284 | 60 | | $ | $ | | 750 | 58 |
| 219 | 7/4 | Telephone | 23 | 36 | | | | | 727 | 22 |
| 220 | 7/7 | Cash | 200 | 00 | | | | | 527 | 22 |
| | 7/12 | Deposit | | | | | 792 | 60 | 1319 | 82 |
| 221 | 7/15 | Insurance | 192 | 30 | | | | | 1127 | 52 |
| 222 | 7/18 | Investment | 100 | 00 | | | | | 1027 | 52 |
| 223 | 7/20 | Credit Card | 214 | 83 | | | | | 812 | 69 |
| | 7/26 | Deposit | | | | | 792 | 60 | 1605 | 29 |
| 224 | 7/27 | Department Store | 113 | 37 | | | | | 1491 | 92 |

CHECKING ACCOUNT Monthly Statement — Account Number: 122-345-1

| Date | Transaction | Amount | Balance |
|---|---|---|---|
| 7/1 | OPENING BALANCE | | 1035.18 |
| 7/1 | INTEREST | 5.15 | 1040.33 |
| 7/4 | CHECK | 284.60 | 755.73 |
| 7/6 | CHECK | 23.36 | 732.37 |
| 7/12 | DEPOSIT | 792.60 | 1524.97 |
| 7/20 | CHECK | 192.30 | 1332.67 |
| 7/24 | CHECK | 100.00 | 1232.67 |
| 7/26 | DEPOSIT | 792.60 | 2025.27 |
| 7/28 | CHECK | 200.00 | 1825.27 |
| 7/30 | CLOSING BALANCE | | 1825.27 |

### APPLYING THE CONCEPTS

**16.** When a check is written, the amount is __________ from the balance.

**17.** When a deposit is made, the amount is __________ to the balance.

**18.** In checking the bank statement, __________ to the checkbook balance all checks that have been written but not processed.

**19.** In checking the bank balance, __________ any service charge and any deposits not yet recorded.

**20.** Define the words credit and debit as they apply to checkbooks.
[W]

# Project in Mathematics

## Credit Card Finance Charges

All credit card companies charge a *fee* (finance charge) when a credit card balance is not paid within a certain number of days (usually 25 days) of the *billing date* (the date the credit card bill is sent). There may also be an annual fee. The table below shows the charges for five banks in February 1994.

| *Bank* | *Annual Interest Rate* | *Annual Fee* |
|---|---|---|
| Arkansas Federal<br>Little Rock, Arkansas | 7.75% | \$35 |
| Federal Savings Bank<br>Rogers, Arkansas | 7.92% | \$33 |
| Central Carolina Bank<br>Columbus, Georgia | 8.50% | \$29 |
| People's Bank<br>Bridgeport, Connecticut | 11.50% | \$25 |
| USAA Federal Savings<br>San Antonio, Texas | 12.50% | \$0 |

The amount of the monthly finance charge is based on the unpaid balance. For instance, using Arkansas Federal, if your unpaid balance for one month is \$250, then the monthly finance charge is calculated using the simple interest formula.

$$\text{Interest} = \text{principal} \times \text{annual interest rate} \times \text{time}$$

$$= 250 \times 0.0775 \times \frac{1}{12} \qquad \bullet\ 1 \text{ month} = \frac{1}{12} \text{ year.}$$

$$\approx 1.61458$$

The monthly finance charge is \$1.61.

***Exercises:***

1. Suppose you have an average monthly unpaid balance of \$356.00 for 12 months. With which of the banks will your *annual* finance charges be least?
2. Suppose you have an average monthly unpaid balance of \$650.00 for 12 months. With which of the banks will your *annual* finance charges be least?
3. Using your calculator, experiment with some average monthly unpaid balances and determine what average monthly unpaid balance would result in essentially the same finance charges whether you used Arkansas Federal or USAA Federal Savings.

# Chapter Summary

## Key Words

The *unit cost* is the cost of one item.

*Percent increase* is used to show how much a quantity has increased over its original value.

*Cost* is the price that a business pays for a product.

*Selling price* is the price at which a business sells a product to a customer.

*Markup* is the difference between selling price and cost.

*Markup rate* is frequently expressed as a percent of a product's cost.

*Percent decrease* is used to show how much a quantity has decreased from its original value.

*Sale price* is the price that has been reduced from the regular price.

*Discount* is the difference between the regular price and the sale price.

*Discount rate* is frequently stated as a percent of a product's regular price.

*Interest* is the amount of money paid for the privilege of using someone else's money.

*Principal* is the amount of money originally deposited or borrowed.

The percent used to determine the amount of interest is the *interest rate*.

Interest computed on the original amount is called *simple interest*.

*Compound interest* is computed not only on the original principal but also on interest already earned.

The *mortgage* is the amount that is borrowed to buy real estate.

The loan origination fee is usually a percent of the mortgage and is expressed in *points*.

*Commissions* are usually paid to the salespersons and are calculated as a percent of total sales.

An employee who receives an *hourly wage* is paid a certain amount for each hour worked.

An employee who is paid a *salary* receives payment based on a weekly, biweekly, monthly, or annual time schedule.

## Essential Rules

**To Find Unit Cost**

To find the unit cost, divide the total cost by the number of units.

**To Find Total Cost**

To find the total cost, multiply the unit cost by the number of units.

**Basic Markup Equations:**

Selling price = cost + markup

$$S = C + M$$

Markup = markup rate × cost

$$M = r \times C$$

**Basic Discount Equations:**

Sale price = regular price − discount

$$S = R - D$$

Discount = discount rate × regular price

$$D = r \times R$$

**Annual Simple Interest Equation:**

Principal × annual interest rate × time in years = interest

$$P \times r \times t = I$$

# Chapter Review

## SECTION 6.1

1. A 20-ounce box of cereal costs $2.90. Find the unit cost.

2. Twenty-four ounces of a mouthwash cost $3.49. A 60-ounce container of the same kind of mouthwash costs $8.40. Which is the better buy?

## SECTION 6.2

3. An oil stock was bought for $\$42\frac{3}{8}$ per share; 6 months later the stock was selling for $\$55\frac{1}{4}$ per share. Find the percent increase in the price of the stock for the 6 months. Round to the nearest tenth of a percent.

4. A professional baseball player received a salary of $1 million last year. This year the player signed a contract paying $12 million over 4 years. Find the yearly percent increase in the player's salary.

5. Techno-Center uses a markup rate of 35% on all computer systems. Find the selling price of a computer system that costs the store $1540.

6. Last year an oil company had earnings of $4.12 per share. This year the earnings are $4.73 per share. What is the percent increase in earnings per share? Round to the nearest percent.

7. A sporting goods store uses a markup rate of 40%. What is the markup on a ski suit that costs the store $180?

8. A suit that regularly costs $235 is on sale for 40% off the regular price. Find the sale price.

## SECTION 6.3

9. A contractor borrowed $100,000 from a credit union for 9 months at an annual interest rate of 9%. What is the simple interest due on the loan?

10. Pros' Sporting Goods borrowed $30,000 at an annual interest rate of 8% for 6 months. Find the simple interest due on the loan.

11. A computer programmer invested $25,000 in a retirement account that pays 6% interest, compounded daily. What is the value of the investment in 10 years? Use the table on page A4. Round to the nearest cent.

12. A fast-food restaurant invested $50,000 in an account that pays 7% annual interest compounded quarterly. What is the value of the investment in 1 year? Use the table on page A4.

## SECTION 6.4

13. Paula Mason purchased a home for $125,000. The lender requires a down payment of 15%. Find the amount of the down payment.

14. A credit union requires a borrower to pay $2\frac{1}{2}$ points for a loan. Find the origination fee for a loan of $75,000.

15. The monthly mortgage payment for a condominium is $523.67. The owner must pay an annual property tax of $658.32. Find the total monthly payment for the mortgage and property tax.

16. The Sweeneys bought a home for $156,000. The family made a 10% down payment and financed the remainder with a 30-year loan with an annual interest rate of 7%. Find the monthly mortgage payment. Use the monthly payment table on page A6. Round to the nearest cent.

## SECTION 6.5

17. A plumber bought a truck for $13,500. A state license of $315 and a sales tax of 6.25% of the purchase price are required. Find the total cost of the sales tax and the license fee.

18. An account executive had car expenses of $1025.58 for insurance, $605.82 for gas, $37.92 for oil, and $188.27 for maintenance during a year in which 15,320 miles were driven. Find the cost per mile for these four items. Round to the nearest tenth of a cent.

19. Mien pays a monthly car payment of $122.78. During a month in which $25.45 is principal, how much of the payment is interest?

20. A pickup truck with a slide-in camper is purchased for $14,450. A down payment of 8% is made, and the remaining cost is financed for 4 years at an annual interest rate of 9%. Find the monthly payment. Use the monthly schedule on page A6. Round to the nearest cent.

## SECTION 6.6

21. The manager of the retail store at a ski resort receives a commission of 3% on all sales at the alpine shop. Find the total commission received during a month in which the shop had $108,000 in sales.

22. Richard Valdez receives $12.60 per hour for working 40 hours a week and time and a half for working over 40 hours. Find his total income during a week in which he worked 48 hours.

## SECTION 6.7

23. Luke had a checking account balance of $1568.45 before writing checks for $123.76, $756.45, and $88.77. He then deposited a check for $344.21. Find Luke's current checkbook balance.

24. The business checking account of a donut shop showed a balance of $9567.44 before checks of $1023.55, $345.44, and $23.67 were written and checks of $555.89 and $135.91 were deposited. Find the current checkbook balance.

# Chapter Test

1. Twenty feet of lumber cost $138.40. What is the cost per foot?

2. Find the more economical purchase: 5 pounds of tomatoes for $1.65, or 8 pounds for $2.72.

3. Red snapper costs $4.15 per pound. Find the cost of $3\frac{1}{2}$ pounds. (Round to the nearest cent.)

4. An exercise bicycle increased in price from $415 to $498. Find the percent increase in the cost of the exercise bicycle.

5. Fifteen years ago a painting was priced at $6000. Today the same painting has a value of $15,000. Find the percent increase in the price of the painting during the 15 years.

6. A department store uses a 40% markup rate. Find the selling price of a compact disc player that the store purchased for $215.

7. A bookstore bought a paperback book for $5 and used a markup rate of 25%. Find the selling price of the book.

8. The price of gold dropped from $850 per ounce to $360 per ounce. What percent decrease does this amount represent? (Round to the nearest tenth of a percent.)

9. The price of a video camera dropped from $1120 to $896. What percent decrease does this price drop represent?

10. A corner hutch with a regular price of $299 is on sale for 30% off the regular price. Find the sale price.

11. A box of stationery that regularly sells for $4.50 is on sale for $2.70. Find the discount rate.

12. A construction company borrowed $75,000 for 4 months at an annual interest rate of 8%. Find the simple interest due on the loan.

13. Jorge, who is self-employed, placed $30,000 in an account that pays 6% annual interest compounded quarterly. How much interest was earned in 10 years? Use the table on page A4.

14. A savings and loan institution is giving mortgage loans that have a loan origination fee of $2\frac{1}{2}$ points. Find the loan origination fee on a home purchased with a loan of $134,000.

**15.** A new housing development offers homes with a mortgage of $222,000 for 25 years at an annual interest rate of 8%. Find the monthly mortgage payment. Use the table on page A6.

**16.** A $17,500 minivan was purchased with an 18% down payment. Find the amount financed.

**17.** A rancher bought a pickup for $11,500 with a down payment of 15% of the cost. The balance is financed for 3 years at an annual interest rate of 7%. Find the monthly car payment. Use the table on page A6.

**18.** Shaney receives an hourly wage of $13.40 an hour as an emergency room nurse. When called in at night, she receives time and a half. How much does Shaney earn in a week when working 30 hours at normal rates and 15 hours during the night?

**19.** The business checking account for a pottery store had a balance of $7349.44 before checks for $1349.67 and $344.12 were written. The store manager then made a deposit of $956.60. Find the current checkbook balance.

**20.** Balance the checkbook shown.

RECORD ALL CHARGES OR CREDITS THAT AFFECT YOUR ACCOUNT

| NUMBER | DATE | DESCRIPTION OF TRANSACTION | PAYMENT/DEBIT (-) | √ T | FEE (IF ANY) (-) | DEPOSIT/CREDIT (+) | BALANCE $ 422.13 |
|---|---|---|---|---|---|---|---|
| | 8/1 | House Payment | $ 213.72 | | $ | $ | 208.41 |
| | 8/4 | Deposit | | | | 552.60 | 761.01 |
| | 8/5 | Plane Tickets | 162.40 | | | | 598.61 |
| | 8/6 | Groceries | 66.44 | | | | 532.17 |
| | 8/10 | Car Payment | 122.37 | | | | 409.80 |
| | 8/15 | Deposit | | | | 552.60 | 962.40 |
| | 8/16 | Credit Card | 213.45 | | | | 748.95 |
| | 8/18 | Doctor | 92.14 | | | | 656.81 |
| | 8/22 | Utilities | 72.30 | | | | 584.51 |
| | 8/28 | T. V. Repair | 78.20 | | | | 506.31 |

CHECKING ACCOUNT Monthly Statement — Account Number: 122-345-1

| Date | Transaction | Amount | Balance |
|---|---|---|---|
| 8/1 | OPENING BALANCE | | 422.13 |
| 8/3 | CHECK | 213.72 | 208.41 |
| 8/4 | DEPOSIT | 552.60 | 761.01 |
| 8/8 | CHECK | 66.44 | 694.57 |
| 8/8 | CHECK | 162.40 | 532.17 |
| 8/15 | DEPOSIT | 552.60 | 1084.77 |
| 8/23 | CHECK | 72.30 | 1012.47 |
| 8/24 | CHECK | 92.14 | 920.33 |
| 9/1 | CLOSING BALANCE | | 920.33 |

# Cumulative Review

1. Simplify $12 - (10 - 8)^2 \div 2 + 3$.

2. Add: $3\frac{1}{3} + 4\frac{1}{8} + 1\frac{1}{12}$

3. Find the difference between $12\frac{3}{16}$ and $9\frac{5}{12}$.

4. Find the product of $5\frac{5}{8}$ and $1\frac{9}{15}$.

5. Divide: $3\frac{1}{2} \div 1\frac{3}{4}$

6. Simplify $\left(\frac{3}{4}\right)^2 \div \left(\frac{3}{8} - \frac{1}{4}\right) + \frac{1}{2}$.

7. Divide: $0.059\overline{)3.0792}$. Round to the nearest tenth.

8. Convert $\frac{17}{12}$ to a decimal. Round to the nearest thousandth.

9. Write "$410 in 8 hours" as a unit rate.

10. Solve the proportion $\frac{5}{n} = \frac{16}{35}$. Round to the nearest hundredth.

11. Write $\frac{5}{8}$ as a percent.

12. Find 6.5% of 420.

13. Write 18.2% as a decimal.

14. What percent of 20 is 8.4?

15. 30 is 12% of what?

16. 65 is 42% of what? Round to the nearest hundredth.

17. A series of late summer storms produced rainfall of $3\frac{3}{4}$, $8\frac{1}{2}$, and $1\frac{2}{3}$ inches during a 3-week period. Find the total rainfall during the 3 weeks.

18. The Homer family pays $\frac{1}{5}$ of its total monthly income for taxes. The family has a total monthly income of \$2850. Find the amount of the monthly income that the Homers pay in taxes.

19. In 5 years, the cost of a scientific calculator went from \$75 to \$30. What is the ratio of the decrease in price to the original price?

20. A compact car drove 417.5 miles on 12.5 gallons of gasoline. Find the number of miles driven per gallon of gasoline.

21. A 14-pound turkey costs \$12.96. Find the unit cost. (Round to the nearest cent.)

22. Eighty shares of a stock paid a dividend of \$112. At the same rate, find the dividend on 200 shares of the stock.

23. A video camera that regularly sells for \$900 is on sale for 20% off the regular price. What is the sale price?

24. A department store bought a portable disc player for \$85 and used a markup rate of 40%. Find the selling price of the disc player.

25. Sook Kim, an elementary school teacher, received an increase in salary from \$2800 per month to \$3024 per month. Find the percent increase in her salary.

26. A contractor borrowed \$120,000 for 6 months at an annual interest rate of 10%. How much simple interest is due on the loan?

27. A red sports car was purchased for \$14,000, and a down payment of \$2000 was made. The balance is financed for 3 years at an annual interest rate of 9%. Find the monthly payment. Use the table on page A6. (Round to the nearest cent.)

28. A family had a checking account balance of \$1846.78. A check of \$568.30 was deposited into the account, and checks of \$123.98 and \$47.33 were written. Find the new checking account balance.

29. Anna Gonzalez spent \$840 on gasoline and oil, \$520 on insurance, \$185 on tires, and \$432 on repairs. Find the cost per mile to drive the car 10,000 miles during the year. (Round to the nearest cent.)

30. A house has a mortgage of \$72,000 for 20 years at an annual interest rate of 11%. Find the monthly mortgage payment. Use the table on page A6. (Round to the nearest cent.)

# Appendix

## Addition Table

| + | 0 | 1 | 2 | 3 | 4 | 5 | 6 | 7 | 8 | 9 | 10 | 11 | 12 |
|---|---|---|---|---|---|---|---|---|---|---|---|---|---|
| **0** | 0 | 1 | 2 | 3 | 4 | 5 | 6 | 7 | 8 | 9 | 10 | 11 | 12 |
| **1** | 1 | 2 | 3 | 4 | 5 | 6 | 7 | 8 | 9 | 10 | 11 | 12 | 13 |
| **2** | 2 | 3 | 4 | 5 | 6 | 7 | 8 | 9 | 10 | 11 | 12 | 13 | 14 |
| **3** | 3 | 4 | 5 | 6 | 7 | 8 | 9 | 10 | 11 | 12 | 13 | 14 | 15 |
| **4** | 4 | 5 | 6 | 7 | 8 | 9 | 10 | 11 | 12 | 13 | 14 | 15 | 16 |
| **5** | 5 | 6 | 7 | 8 | 9 | 10 | 11 | 12 | 13 | 14 | 15 | 16 | 17 |
| **6** | 6 | 7 | 8 | 9 | 10 | 11 | 12 | 13 | 14 | 15 | 16 | 17 | 18 |
| **7** | 7 | 8 | 9 | 10 | 11 | 12 | 13 | 14 | 15 | 16 | 17 | 18 | 19 |
| **8** | 8 | 9 | 10 | 11 | 12 | 13 | 14 | 15 | 16 | 17 | 18 | 19 | 20 |
| **9** | 9 | 10 | 11 | 12 | 13 | 14 | 15 | 16 | 17 | 18 | 19 | 20 | 21 |
| **10** | 10 | 11 | 12 | 13 | 14 | 15 | 16 | 17 | 18 | 19 | 20 | 21 | 22 |
| **11** | 11 | 12 | 13 | 14 | 15 | 16 | 17 | 18 | 19 | 20 | 21 | 22 | 23 |
| **12** | 12 | 13 | 14 | 15 | 16 | 17 | 18 | 19 | 20 | 21 | 22 | 23 | 24 |

## Multiplication Table

| + | 0 | 1 | 2 | 3 | 4 | 5 | 6 | 7 | 8 | 9 | 10 | 11 | 12 |
|---|---|---|---|---|---|---|---|---|---|---|---|---|---|
| **0** | 0 | 0 | 0 | 0 | 0 | 0 | 0 | 0 | 0 | 0 | 0 | 0 | 0 |
| **1** | 0 | 1 | 2 | 3 | 4 | 5 | 6 | 7 | 8 | 9 | 10 | 11 | 12 |
| **2** | 0 | 2 | 4 | 6 | 8 | 10 | 12 | 14 | 16 | 18 | 20 | 22 | 24 |
| **3** | 0 | 3 | 6 | 9 | 12 | 15 | 18 | 21 | 24 | 27 | 30 | 33 | 36 |
| **4** | 0 | 4 | 8 | 12 | 16 | 20 | 24 | 28 | 32 | 36 | 40 | 44 | 48 |
| **5** | 0 | 5 | 10 | 15 | 20 | 25 | 30 | 35 | 40 | 45 | 50 | 55 | 60 |
| **6** | 0 | 6 | 12 | 18 | 24 | 30 | 36 | 42 | 48 | 54 | 60 | 66 | 72 |
| **7** | 0 | 7 | 14 | 21 | 28 | 35 | 42 | 49 | 56 | 63 | 70 | 77 | 84 |
| **8** | 0 | 8 | 16 | 24 | 32 | 40 | 48 | 56 | 64 | 72 | 80 | 88 | 96 |
| **9** | 0 | 9 | 18 | 27 | 36 | 45 | 54 | 63 | 72 | 81 | 90 | 99 | 108 |
| **10** | 0 | 10 | 20 | 30 | 40 | 50 | 60 | 70 | 80 | 90 | 100 | 110 | 120 |
| **11** | 0 | 11 | 22 | 33 | 44 | 55 | 66 | 77 | 88 | 99 | 110 | 121 | 132 |
| **12** | 0 | 12 | 24 | 36 | 48 | 60 | 72 | 84 | 96 | 108 | 120 | 132 | 144 |

## Table of Square Roots

Decimal approximations have been rounded to the nearest thousandth.

| Number | Square Root | Number | Square Root | Number | Square Root | Number | Square Root |
|---|---|---|---|---|---|---|---|
| 1 | 1 | 51 | 7.141 | 101 | 10.050 | 151 | 12.288 |
| 2 | 1.414 | 52 | 7.211 | 102 | 10.100 | 152 | 12.329 |
| 3 | 1.732 | 53 | 7.280 | 103 | 10.149 | 153 | 12.369 |
| 4 | 2 | 54 | 7.348 | 104 | 10.198 | 154 | 12.410 |
| 5 | 2.236 | 55 | 7.416 | 105 | 10.247 | 155 | 12.450 |
| 6 | 2.449 | 56 | 7.483 | 106 | 10.296 | 156 | 12.490 |
| 7 | 2.646 | 57 | 7.550 | 107 | 10.344 | 157 | 12.530 |
| 8 | 2.828 | 58 | 7.616 | 108 | 10.392 | 158 | 12.570 |
| 9 | 3 | 59 | 7.681 | 109 | 10.440 | 159 | 12.610 |
| 10 | 3.162 | 60 | 7.746 | 110 | 10.488 | 160 | 12.649 |
| 11 | 3.317 | 61 | 7.810 | 111 | 10.536 | 161 | 12.689 |
| 12 | 3.464 | 62 | 7.874 | 112 | 10.583 | 162 | 12.728 |
| 13 | 3.606 | 63 | 7.937 | 113 | 10.630 | 163 | 12.767 |
| 14 | 3.742 | 64 | 8 | 114 | 10.677 | 164 | 12.806 |
| 15 | 3.873 | 65 | 8.062 | 115 | 10.724 | 165 | 12.845 |
| 16 | 4 | 66 | 8.124 | 116 | 10.770 | 166 | 12.884 |
| 17 | 4.123 | 67 | 8.185 | 117 | 10.817 | 167 | 12.923 |
| 18 | 4.243 | 68 | 8.246 | 118 | 10.863 | 168 | 12.961 |
| 19 | 4.359 | 69 | 8.307 | 119 | 10.909 | 169 | 13 |
| 20 | 4.472 | 70 | 8.367 | 120 | 10.954 | 170 | 13.038 |
| 21 | 4.583 | 71 | 8.426 | 121 | 11 | 171 | 13.077 |
| 22 | 4.690 | 72 | 8.485 | 122 | 11.045 | 172 | 13.115 |
| 23 | 4.796 | 73 | 8.544 | 123 | 11.091 | 173 | 13.153 |
| 24 | 4.899 | 74 | 8.602 | 124 | 11.136 | 174 | 13.191 |
| 25 | 5 | 75 | 8.660 | 125 | 11.180 | 175 | 13.229 |
| 26 | 5.099 | 76 | 8.718 | 126 | 11.225 | 176 | 13.267 |
| 27 | 5.196 | 77 | 8.775 | 127 | 11.269 | 177 | 13.304 |
| 28 | 5.292 | 78 | 8.832 | 128 | 11.314 | 178 | 13.342 |
| 29 | 5.385 | 79 | 8.888 | 129 | 11.358 | 179 | 13.379 |
| 30 | 5.477 | 80 | 8.944 | 130 | 11.402 | 180 | 13.416 |
| 31 | 5.568 | 81 | 9 | 131 | 11.446 | 181 | 13.454 |
| 32 | 5.657 | 82 | 9.055 | 132 | 11.489 | 182 | 13.491 |
| 33 | 5.745 | 83 | 9.110 | 133 | 11.533 | 183 | 13.528 |
| 34 | 5.831 | 84 | 9.165 | 134 | 11.576 | 184 | 13.565 |
| 35 | 5.916 | 85 | 9.220 | 135 | 11.619 | 185 | 13.601 |
| 36 | 6 | 86 | 9.274 | 136 | 11.662 | 186 | 13.638 |
| 37 | 6.083 | 87 | 9.327 | 137 | 11.705 | 187 | 13.675 |
| 38 | 6.164 | 88 | 9.381 | 138 | 11.747 | 188 | 13.711 |
| 39 | 6.245 | 89 | 9.434 | 139 | 11.790 | 189 | 13.748 |
| 40 | 6.325 | 90 | 9.487 | 140 | 11.832 | 190 | 13.784 |
| 41 | 6.403 | 91 | 9.539 | 141 | 11.874 | 191 | 13.820 |
| 42 | 6.481 | 92 | 9.592 | 142 | 11.916 | 192 | 13.856 |
| 43 | 6.557 | 93 | 9.644 | 143 | 11.958 | 193 | 13.892 |
| 44 | 6.633 | 94 | 9.695 | 144 | 12 | 194 | 13.928 |
| 45 | 6.708 | 95 | 9.747 | 145 | 12.042 | 195 | 13.964 |
| 46 | 6.782 | 96 | 9.798 | 146 | 12.083 | 196 | 14 |
| 47 | 6.856 | 97 | 9.849 | 147 | 12.124 | 197 | 14.036 |
| 48 | 6.928 | 98 | 9.899 | 148 | 12.166 | 198 | 14.071 |
| 49 | 7 | 99 | 9.950 | 149 | 12.207 | 199 | 14.107 |
| 50 | 7.071 | 100 | 10 | 150 | 12.247 | 200 | 14.142 |

## Compound Interest Table

**Compounded Annually**

| | 4% | 5% | 6% | 7% | 8% | 9% | 10% |
|---|---|---|---|---|---|---|---|
| **1 year** | 1.04000 | 1.05000 | 1.06000 | 1.07000 | 1.08000 | 1.09000 | 1.10000 |
| **5 years** | 1.21665 | 1.27628 | 1.33823 | 1.40255 | 1.46933 | 1.53862 | 1.61051 |
| **10 years** | 1.48024 | 1.62890 | 1.79085 | 1.96715 | 2.15893 | 2.36736 | 2.59374 |
| **15 years** | 1.80094 | 2.07893 | 2.39656 | 2.75903 | 3.17217 | 3.64248 | 4.17725 |
| **20 years** | 2.19112 | 2.65330 | 3.20714 | 3.86968 | 4.66095 | 5.60441 | 6.72750 |

**Compounded Semiannually**

| | 4% | 5% | 6% | 7% | 8% | 9% | 10% |
|---|---|---|---|---|---|---|---|
| **1 year** | 1.04040 | 1.05062 | 1.06090 | 1.07123 | 1.08160 | 1.09203 | 1.10250 |
| **5 years** | 1.21899 | 1.28008 | 1.34392 | 1.41060 | 1.48024 | 1.55297 | 1.62890 |
| **10 years** | 1.48595 | 1.63862 | 1.80611 | 1.98979 | 2.19112 | 2.41171 | 2.65330 |
| **15 years** | 1.81136 | 2.09757 | 2.42726 | 2.80679 | 3.24340 | 3.74531 | 4.32194 |
| **20 years** | 2.20804 | 2.68506 | 3.26204 | 3.95926 | 4.80102 | 5.81634 | 7.03999 |

**Compounded Quarterly**

| | 4% | 5% | 6% | 7% | 8% | 9% | 10% |
|---|---|---|---|---|---|---|---|
| **1 year** | 1.04060 | 1.05094 | 1.06136 | 1.07186 | 1.08243 | 1.09308 | 1.10381 |
| **5 years** | 1.22019 | 1.28204 | 1.34686 | 1.41478 | 1.48595 | 1.56051 | 1.63862 |
| **10 years** | 1.48886 | 1.64362 | 1.81402 | 2.00160 | 2.20804 | 2.43519 | 2.68506 |
| **15 years** | 1.81670 | 2.10718 | 2.44322 | 2.83182 | 3.28103 | 3.80013 | 4.39979 |
| **20 years** | 2.21672 | 2.70148 | 3.29066 | 4.00639 | 4.87544 | 5.93015 | 7.20957 |

**Compounded Daily**

| | 4% | 5% | 6% | 7% | 8% | 9% | 10% |
|---|---|---|---|---|---|---|---|
| **1 year** | 1.04080 | 1.05127 | 1.06183 | 1.07250 | 1.08328 | 1.09416 | 1.10516 |
| **5 years** | 1.22139 | 1.28400 | 1.34983 | 1.41902 | 1.49176 | 1.56823 | 1.64861 |
| **10 years** | 1.49179 | 1.64866 | 1.82203 | 2.01362 | 2.22535 | 2.45933 | 2.71791 |
| **15 years** | 1.82206 | 2.11689 | 2.45942 | 2.85736 | 3.31968 | 3.85678 | 4.48077 |
| **20 years** | 2.22544 | 2.71810 | 3.31979 | 4.05466 | 4.95217 | 6.04830 | 7.38703 |

To use this table:

1. Locate the section which gives the desired compounding period.
2. Locate the interest rate in the top row of that section.
3. Locate the number of years in the left-hand column of that section.
4. Locate the number where the interest-rate column and the number-of-years row meet. This is the compound interest factor.

**Example:** An investment yields an annual interest rate of 10% compounded quarterly for 5 years.
The compounding period is "compounded quarterly."
The interest rate is 10%.
The number of years is 5.
The number where the row and column meet is 1.63862. This is the compound interest factor.

## Compound Interest Table

### Compounded Annually

| | 11% | 12% | 13% | 14% | 15% | 16% | 17% |
|---|---|---|---|---|---|---|---|
| **1 year** | 1.11000 | 1.12000 | 1.13000 | 1.14000 | 1.15000 | 1.16000 | 1.17000 |
| **5 years** | 1.68506 | 1.76234 | 1.84244 | 1.92542 | 2.01136 | 2.10034 | 2.19245 |
| **10 years** | 2.83942 | 3.10585 | 3.39457 | 3.70722 | 4.04556 | 4.41144 | 4.80683 |
| **15 years** | 4.78459 | 5.47357 | 6.25427 | 7.13794 | 8.13706 | 9.26552 | 10.53872 |
| **20 years** | 8.06239 | 9.64629 | 11.52309 | 13.74349 | 16.36654 | 19.46076 | 23.10560 |

### Compounded Semiannually

| | 11% | 12% | 13% | 14% | 15% | 16% | 17% |
|---|---|---|---|---|---|---|---|
| **1 year** | 1.11303 | 1.12360 | 1.13423 | 1.14490 | 1.15563 | 1.16640 | 1.17723 |
| **5 years** | 1.70814 | 1.79085 | 1.87714 | 1.96715 | 2.06103 | 2.15893 | 2.26098 |
| **10 years** | 2.91776 | 3.20714 | 3.52365 | 3.86968 | 4.24785 | 4.66096 | 5.11205 |
| **15 years** | 4.98395 | 5.74349 | 6.61437 | 7.61226 | 8.75496 | 10.06266 | 11.55825 |
| **20 years** | 8.51331 | 10.28572 | 12.41607 | 14.97446 | 18.04424 | 21.72452 | 26.13302 |

### Compounded Quarterly

| | 11% | 12% | 13% | 14% | 15% | 16% | 17% |
|---|---|---|---|---|---|---|---|
| **1 year** | 1.11462 | 1.12551 | 1.13648 | 1.14752 | 1.15865 | 1.16986 | 1.18115 |
| **5 years** | 1.72043 | 1.80611 | 1.89584 | 1.98979 | 2.08815 | 2.19112 | 2.29891 |
| **10 years** | 2.95987 | 3.26204 | 3.59420 | 3.95926 | 4.36038 | 4.80102 | 5.28497 |
| **15 years** | 5.09225 | 5.89160 | 6.81402 | 7.87809 | 9.10513 | 10.51963 | 12.14965 |
| **20 years** | 8.76085 | 10.64089 | 12.91828 | 15.67574 | 19.01290 | 23.04980 | 27.93091 |

### Compounded Daily

| | 11% | 12% | 13% | 14% | 15% | 16% | 17% |
|---|---|---|---|---|---|---|---|
| **1 year** | 1.11626 | 1.12747 | 1.13880 | 1.15024 | 1.16180 | 1.17347 | 1.18526 |
| **5 years** | 1.73311 | 1.82194 | 1.91532 | 2.01348 | 2.11667 | 2.22515 | 2.33918 |
| **10 years** | 3.00367 | 3.31946 | 3.66845 | 4.05411 | 4.48031 | 4.95130 | 5.47178 |
| **15 years** | 5.20569 | 6.04786 | 7.02625 | 8.16288 | 9.48335 | 11.01738 | 12.79950 |
| **20 years** | 9.02203 | 11.01883 | 13.45751 | 16.43582 | 20.07316 | 24.51534 | 29.94039 |

## Monthly Payment Table

| | 4% | 5% | 6% | 7% | 8% | 9% |
|---|---|---|---|---|---|---|
| **1 year** | 0.0851499 | 0.0856075 | 0.0860664 | 0.0865267 | 0.0869884 | 0.0874515 |
| **2 years** | 0.0434249 | 0.0438714 | 0.0443206 | 0.0447726 | 0.0452273 | 0.0456847 |
| **3 years** | 0.0295240 | 0.0299709 | 0.0304219 | 0.0308771 | 0.0313364 | 0.0317997 |
| **4 years** | 0.0225791 | 0.0230293 | 0.0234850 | 0.0239462 | 0.0244129 | 0.0248850 |
| **5 years** | 0.0184165 | 0.0188712 | 0.0193328 | 0.0198012 | 0.0202764 | 0.0207584 |
| **20 years** | 0.0060598 | 0.0065996 | 0.0071643 | 0.0077530 | 0.0083644 | 0.0089973 |
| **25 years** | 0.0052784 | 0.0058459 | 0.0064430 | 0.0070678 | 0.0077182 | 0.0083920 |
| **30 years** | 0.0047742 | 0.0053682 | 0.0059955 | 0.0066530 | 0.0073376 | 0.0080462 |

| | 10% | 11% | 12% | 13% |
|---|---|---|---|---|
| **1 year** | 0.0879159 | 0.0883817 | 0.0888488 | 0.0893173 |
| **2 years** | 0.0461449 | 0.0466078 | 0.0470735 | 0.0475418 |
| **3 years** | 0.0322672 | 0.0327387 | 0.0332143 | 0.0336940 |
| **4 years** | 0.0253626 | 0.0258455 | 0.0263338 | 0.0268275 |
| **5 years** | 0.0212470 | 0.0217424 | 0.0222445 | 0.0227531 |
| **20 years** | 0.0096502 | 0.0103219 | 0.0110109 | 0.0117158 |
| **25 years** | 0.0090870 | 0.0098011 | 0.0105322 | 0.0112784 |
| **30 years** | 0.0087757 | 0.0095232 | 0.0102861 | 0.0110620 |

To use this table:
1. Locate the desired interest rate in the top row.
2. Locate the number of years in the left-hand column.
3. Locate the number where the interest-rate column and the number-of-years row meet. This is the monthly payment factor.

**Example:** A home has a 30-year mortgage at an annual interest rate of 12%.
The interest rate is 12%.
The number of years is 30.
The number where the row and column meet is 0.0102861. This is the monthly payment factor.

# SOLUTIONS to Chapter 1 "You Try It"

## SECTION 1.1 *pages 3–6*

**You Try It 1**

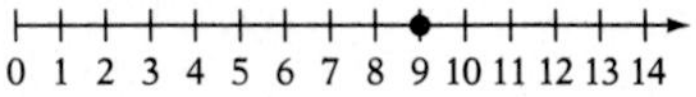


**You Try It 2** a. $45 > 29$ b. $27 > 0$

**You Try It 3** Thirty-six million four hundred sixty-two thousand seventy-five

**You Try It 4** 452,007

**You Try It 5** 60,000 + 8000 + 200 + 80 + 1

**You Try It 6** 100,000 + 9000 + 200 + 7

**You Try It 7** 370,000

**You Try It 8** 4000

## SECTION 1.2 *pages 9–12*

**You Try It 1**

$$\begin{array}{r} 347 \\ +\ 12453 \\ \hline 12{,}800 \end{array}$$

347 increased by 12,453 is 12,800.

**You Try It 2**

$$\begin{array}{r} {\scriptstyle 2}\phantom{0} \\ 95 \\ 88 \\ +\ 67 \\ \hline 250 \end{array}$$

**You Try It 3**

$$\begin{array}{r} {\scriptstyle 1\,1\ \ 2\,1}\phantom{0} \\ 392 \\ 4{,}079 \\ 89{,}035 \\ +\ \ 4{,}992 \\ \hline 98{,}498 \end{array}$$

**You Try It 4**

**Strategy** To find the total amount budgeted for the three items each month, add the three amounts ($475, $275, and $120).

**Solution**

$$\begin{array}{r} \$475 \\ 275 \\ +\ 120 \\ \hline \$870 \end{array}$$

The total amount budgeted for the three items is $870.

## SECTION 1.3 *pages 17–20*

**You Try It 1**

$$\begin{array}{r} 8925 \\ -\ 6413 \\ \hline 2512 \end{array}$$

**You Try It 2**

$$\begin{array}{r} 17{,}504 \\ -\ \ 9{,}302 \\ \hline 8{,}202 \end{array}$$

**You Try It 3**

$$\begin{array}{r} \scriptstyle 2\ 14\ 7\ 11 \\ \not3\ \not4\ \not8\ \not1 \\ -\ \ 8\ 6\ 5 \\ \hline 2\ 6\ 1\ 6 \end{array}$$

**You Try It 4**

$$\begin{array}{r} \scriptstyle 15 \\ \scriptstyle 4\ \not5\ 12 \\ 5\ 4{,}\not5\ \not6\ \not2 \\ -\ 1\ 4{,}4\ 8\ 5 \\ \hline 4\ 0{,}0\ 7\ 7 \end{array}$$

**You Try It 5**

$$\begin{array}{r} \scriptstyle 13\ 9\ 9 \\ \scriptstyle 5\ \not3\ \not{10}\ \not{10}\ 13 \\ \not6\ \not4{,}\not0\ \not0\ \not3 \\ -\ 5\ 4{,}9\ 3\ 6 \\ \hline 9\ 0\ 6\ 7 \end{array}$$

**You Try It 6**

**Strategy** To find the difference in value, subtract the value of the takeovers in 1992 ($82 billion) from the value of the takeovers in 1993 ($135 billion).

**Solution**

$$\begin{array}{r} 135 \\ -\ \ 82 \\ \hline 53 \end{array}$$

The difference in value is $53 billion.

**You Try It 7**

**Strategy** To find your take-home pay:

- Add to find the total of the deductions ($127 + $18 + $35).
- Subtract the total of the deductions from your total salary ($638).

**Solution**

$$\begin{array}{r} 127 \\ 18 \\ +\ 35 \\ \hline 180 \end{array} \text{ deductions} \qquad \begin{array}{r} 638 \\ -\ 180 \\ \hline 458 \end{array}$$

Your take-home pay is $458.

## SECTION 1.4 *pages 25–28*

**You Try It 1**

$$\begin{array}{r} \scriptstyle 35 \\ 648 \\ \times\ \ \ 7 \\ \hline 4536 \end{array}$$

**You Try It 2**

$$\begin{array}{r} 756 \\ \times\ 305 \\ \hline 3780 \\ 22680 \\ \hline 230{,}580 \end{array}$$

**You Try It 3**

**Strategy**
To find the number of cars the dealer will receive in 12 months, multiply the number of months (12) by the number of cars received each month (37).

**Solution**

$$\begin{array}{r} 37 \\ \times\ 12 \\ \hline 74 \\ 37\phantom{0} \\ \hline 444 \end{array}$$

The dealer will receive 444 cars in 12 months.

**You Try It 4**

**Strategy**
To find the total cost of the order:
- Find the cost of the sports jackets by multiplying the number of jackets (25) by the cost for each jacket ($23).
- Add the product to the cost for the suits ($4800).

**Solution**

$$\begin{array}{r} \$23 \\ \times\ 25 \\ \hline 115 \\ 46\phantom{0} \\ \hline \$575 \end{array} \quad \text{cost for jackets} \qquad \begin{array}{r} \$4800 \\ +\ \ 575 \\ \hline \$5375 \end{array}$$

The total cost of the order is $5375.

## SECTION 1.5 *pages 33–40*

**You Try It 1**

$$9\overline{)63}\ \text{quotient } 7$$

Check: $7 \times 9 = 63$

**You Try It 2**

$$\begin{array}{r} 453 \\ 9\overline{)\ 4077} \\ -36\phantom{00} \\ \hline 47\phantom{0} \\ -45\phantom{0} \\ \hline 27 \\ -27 \\ \hline 0 \end{array}$$

Check: $453 \times 9 = 4077$

**You Try It 3**

$$\begin{array}{r} 705 \\ 9\overline{)\ 6345} \\ -63\phantom{00} \\ \hline 04\phantom{0} \\ -\ 0\phantom{0} \\ \hline 45 \\ -45 \\ \hline 0 \end{array}$$

Check: $705 \times 9 = 6345$

**You Try It 4**

$$\begin{array}{r} 870\ \text{r}5 \\ 6\overline{)\ 5225} \\ -48\phantom{00} \\ \hline 42\phantom{0} \\ -42\phantom{0} \\ \hline 05 \\ -\ 0 \\ \hline 5 \end{array}$$

Check: $(870 \times 6) + 5 =$
$5220 + 5 = 5225$

**You Try It 5**

$$\begin{array}{r} 3{,}058 \text{ r}3 \\ 7\overline{)\ 21{,}409} \\ \underline{-21} \\ 0\,4 \\ \underline{-\ \ 0} \\ 40 \\ \underline{-35} \\ 59 \\ \underline{-56} \\ 3 \end{array}$$

Check: $(3058 \times 7) + 3 =$
$21{,}406 + 3 = 21{,}409$

**You Try It 6**

$$\begin{array}{r} 109 \\ 42\overline{)\ 4578} \\ \underline{-42} \\ 37 \\ \underline{-\ 0} \\ 378 \\ \underline{-378} \\ 0 \end{array}$$

Check: $109 \times 42 = 4578$

**You Try It 7**

$$\begin{array}{r} 470 \text{ r}29 \\ 39\overline{)\ 18{,}359} \\ \underline{-15\,6} \\ 2\,75 \\ \underline{-2\,73} \\ 29 \\ \underline{-\ 0} \\ 29 \end{array}$$

Check: $(470 \times 39) + 29 =$
$18{,}330 + 29 = 18{,}359$

**You Try It 8**

$$\begin{array}{r} 62 \text{ r}111 \\ 534\overline{)\ 33{,}219} \\ \underline{-32\,04} \\ 1\,179 \\ \underline{-1\,068} \\ 111 \end{array}$$

Check: $(62 \times 534) + 111 =$
$33{,}108 + 111 = 33{,}219$

**You Try It 9**

$$\begin{array}{r} 421 \text{ r}33 \\ 515\overline{)\ 216{,}848} \\ \underline{-206\,0} \\ 10\,84 \\ \underline{-10\,30} \\ 548 \\ \underline{-515} \\ 33 \end{array}$$

Check: $(421 \times 515) + 33 =$
$216{,}815 + 33 = 216{,}848$

**You Try It 10**

**Strategy** To find the number of tires to be stored on each shelf, divide the number of tires (270) by the number of shelves (15).

**Solution**

$$\begin{array}{r} 18 \\ 15\overline{)\ 270} \\ -15 \\ \hline 120 \\ -120 \\ \hline 0 \end{array}$$

Each shelf can store 18 tires.

**You Try It 11**

**Strategy** To find the number of cases produced in 8 hours:

- Find the number of cases produced in one hour by dividing the number of cans produced (12,600) by the number of cans to a case (24).
- Multiply the number of cases produced in one hour by 8.

**Solution**

$$\begin{array}{r} 525 \\ 24\overline{)\ 12{,}600} \\ -12\,0 \\ \hline 60 \\ -48 \\ \hline 120 \\ -120 \\ \hline 0 \end{array}$$

cases produced in one hour

$$\begin{array}{r} 525 \\ \times\ \ 8 \\ \hline 4200 \end{array}$$

In 8 hours, 4200 cases are produced.

## SECTION 1.6 *pages 45–46*

**You Try It 1** $2^4 \cdot 3^3$

**You Try It 2** $10^7$

**You Try It 3** $(2 \cdot 2 \cdot 2) \cdot (5 \cdot 5) = 8 \cdot 25$
$= 200$

**You Try It 4** $5 \cdot (8 - 4) \div 4 - 2$
$5 \cdot 4 \div 4 - 2$
$20 \div 4 - 2$
$5 - 2$
$3$

## SECTION 1.7 *pages 49–50*

**You Try It 1** 1, 2, 4, 5, 8, 10, 20 and 40 are factors of 40.

**You Try It 2** $44 = 2 \cdot 2 \cdot 11$

**You Try It 3** $177 = 3 \cdot 59$

# SOLUTIONS to Chapter 2 "You Try It"

## SECTION 2.1 *pages 61–62*

**You Try It 1**

| | 2 | 3 | 5 | 7 |
|---|---|---|---|---|
| 50 = | 2 | | (5 · 5) | |
| 84 = | (2 · 2) | 3 | | (7) |
| 135 = | | (3 · 3 · 3) | 5 | |

The LCM $= 2 \cdot 2 \cdot 3 \cdot 3 \cdot 3 \cdot 5 \cdot 5 \cdot 7$
$= 18{,}900$

**You Try It 2**

| | 2 | 3 | 5 |
|---|---|---|---|
| 36 = | (2 · 2) | 3 · 3 | |
| 60 = | 2 · 2 | (3) | 5 |
| 72 = | 2 · 2 · 2 | 3 · 3 | |

The GCF $= 2 \cdot 2 \cdot 3 = 12$.

**You Try It 3**

| | 2 | 3 | 5 | 11 |
|---|---|---|---|---|
| 11 = | | | | 11 |
| 24 = | 2 · 2 · 2 | 3 | | |
| 30 = | 2 | 3 | 5 | |

Since no numbers are circled, the GCF = 1.

## SECTION 2.2 *pages 65–66*

**You Try It 1** $4\frac{1}{4}$

**You Try It 2** $\frac{17}{4}$

**You Try It 3** $4\frac{2}{5}$

**You Try It 4** 4

**You Try It 5** $\frac{117}{8}$

## SECTION 2.3 *pages 69–70*

**You Try It 1** $45 \div 5 = 9$ $\quad \frac{3}{5} = \frac{3 \cdot 9}{5 \cdot 9} = \frac{27}{45}$

$\frac{27}{45}$ is equivalent to $\frac{3}{5}$.

**You Try It 2** Write 6 as $\frac{6}{1}$.

$18 \div 1 = 18 \quad 6 = \frac{6 \cdot 18}{1 \cdot 18} = \frac{108}{18}$

$\frac{108}{18}$ is equivalent to 6.

**You Try It 3** $\frac{16}{24} = \frac{\overset{1}{\cancel{2}} \cdot \overset{1}{\cancel{2}} \cdot \overset{1}{\cancel{2}} \cdot 2}{\underset{1}{\cancel{2}} \cdot \underset{1}{\cancel{2}} \cdot \underset{1}{\cancel{2}} \cdot 3} = \frac{2}{3}$

**You Try It 4** $\frac{8}{56} = \frac{\overset{1}{\cancel{2}} \cdot \overset{1}{\cancel{2}} \cdot \overset{1}{\cancel{2}}}{\underset{1}{\cancel{2}} \cdot \underset{1}{\cancel{2}} \cdot \underset{1}{\cancel{2}} \cdot 7} = \frac{1}{7}$

**You Try It 5** $\frac{15}{32} = \frac{3 \cdot 5}{2 \cdot 2 \cdot 2 \cdot 2 \cdot 2} = \frac{15}{32}$

**You Try It 6** $\frac{48}{36} = \frac{\overset{1}{\cancel{2}} \cdot \overset{1}{\cancel{2}} \cdot 2 \cdot 2 \cdot \overset{1}{\cancel{3}}}{\underset{1}{\cancel{2}} \cdot \underset{1}{\cancel{2}} \cdot \underset{1}{\cancel{3}} \cdot 3} = \frac{4}{3} = 1\frac{1}{3}$

## SECTION 2.4 *pages 73–76*

**You Try It 1**

$$\begin{array}{r} \frac{3}{8} \\ +\frac{7}{8} \\ \hline \frac{10}{8} = \frac{5}{4} = 1\frac{1}{4} \end{array}$$

**You Try It 2**

$$\begin{array}{rcl} \frac{5}{12} &=& \frac{20}{48} \\ +\frac{9}{16} &=& \frac{27}{48} \\ \hline && \frac{47}{48} \end{array}$$

**You Try It 3**

$$\begin{array}{rcl} \frac{7}{8} &=& \frac{105}{120} \\ +\frac{11}{15} &=& \frac{88}{120} \\ \hline && \frac{193}{120} = 1\frac{73}{120} \end{array}$$

**You Try It 4**

$$\begin{array}{rcl} \frac{3}{4} &=& \frac{30}{40} \\ \frac{4}{5} &=& \frac{32}{40} \\ +\frac{5}{8} &=& \frac{25}{40} \\ \hline && \frac{87}{40} = 2\frac{7}{40} \end{array}$$

**You Try It 5** $7 + \frac{6}{11} = 7\frac{6}{11}$

**You Try It 6**

$$\begin{array}{r} 29 \\ +17\frac{5}{12} \\ \hline 46\frac{5}{12} \end{array}$$

**You Try It 7**

$$\begin{array}{rcl} 7\frac{4}{5} &=& 7\frac{24}{30} \\ 6\frac{7}{10} &=& 6\frac{21}{30} \\ +13\frac{11}{15} &=& 13\frac{22}{30} \\ \hline && 26\frac{67}{30} = 28\frac{7}{30} \end{array}$$

**You Try It 8**

$$\begin{array}{rcl} 9\frac{3}{8} &=& 9\frac{45}{120} \\ 17\frac{7}{12} &=& 17\frac{70}{120} \\ +10\frac{14}{15} &=& 10\frac{112}{120} \\ \hline && 36\frac{227}{120} = 37\frac{107}{120} \end{array}$$

**You Try It 9**

**Strategy**
To find the total time spent on the activities, add the three times $\left(4\frac{1}{2}, 3\frac{3}{4}, 1\frac{1}{3}\right)$.

**Solution**

$$\begin{array}{r} 4\frac{1}{2} = 4\frac{6}{12} \\ 3\frac{3}{4} = 3\frac{9}{12} \\ +\ 1\frac{1}{3} = 1\frac{4}{12} \\ \hline 8\frac{19}{12} = 9\frac{7}{12} \end{array}$$

The total time spent on the three activities was $9\frac{7}{12}$ hours.

**You Try It 10**

**Strategy**
To find the overtime pay:

- Find the total number of overtime hours $\left(1\frac{2}{3} + 3\frac{1}{3} + 2\right)$.
- Multiply the total number of hours by the overtime hourly wage ($24).

**Solution**

$$\begin{array}{r} 1\frac{2}{3} \\ 3\frac{1}{3} \\ +\ 2 \\ \hline 6\frac{3}{3} = 7 \text{ hours} \end{array} \qquad \begin{array}{r} \$24 \\ \times\ 7 \\ \hline \$168 \end{array}$$

Jeff earned $168 in overtime pay.

## SECTION 2.5 *pages 81–84*

**You Try It 1**

$$\begin{array}{r} \frac{16}{27} \\ -\ \frac{7}{27} \\ \hline \frac{9}{27} = \frac{1}{3} \end{array}$$

**You Try It 2**

$$\begin{array}{r} \frac{13}{18} = \frac{52}{72} \\ -\ \frac{7}{24} = \frac{21}{72} \\ \hline \frac{31}{72} \end{array}$$

**You Try It 3**

$$\begin{array}{r} 17\frac{5}{9} = 17\frac{20}{36} \\ -\ 11\frac{5}{12} = 11\frac{15}{36} \\ \hline 6\frac{5}{36} \end{array}$$

**You Try It 4**

$$\begin{array}{r} 8 = 7\frac{13}{13} \\ -\ 2\frac{4}{13} = 2\frac{4}{13} \\ \hline 5\frac{9}{13} \end{array}$$

**You Try It 5**

$$\begin{array}{r} 21\frac{7}{9} = 21\frac{28}{36} = 20\frac{64}{36} \\ -\ 7\frac{11}{12} = 7\frac{33}{36} = 7\frac{33}{36} \\ \hline 13\frac{31}{36} \end{array}$$

**You Try It 6**

**Strategy**
To find the time remaining before the plane lands, subtract the number of hours already in the air $\left(2\frac{3}{4}\right)$ from the total time of the trip $\left(5\frac{1}{2}\right)$.

**Solution**

$$\begin{array}{r} 5\frac{1}{2} = 5\frac{2}{4} = 4\frac{6}{4} \\ -\ 2\frac{3}{4} = 2\frac{3}{4} = 2\frac{3}{4} \\ \hline 2\frac{3}{4} \text{ hours} \end{array}$$

The plane will land in $2\frac{3}{4}$ hours.

**You Try It 7**

**Strategy**
To find the amount of weight to be lost during the third month:
- Find the total weight loss during the first two months $\left(7\frac{1}{2} + 5\frac{3}{4}\right)$.
- Subtract the total weight loss from the goal (24 pounds).

**Solution**

$$\begin{array}{r} 7\frac{1}{2} = 7\frac{2}{4} \\ +\ 5\frac{3}{4} = 5\frac{3}{4} \\ \hline 12\frac{5}{4} = 13\frac{1}{4} \text{ pounds lost} \end{array}$$

$$\begin{array}{r} 24 = 23\frac{4}{4} \\ -\ 13\frac{1}{4} = 13\frac{1}{4} \\ \hline 10\frac{3}{4} \text{ pounds} \end{array}$$

The patient must lose $10\frac{3}{4}$ pounds to achieve the goal.

## SECTION 2.6 *pages 89–92*

**You Try It 1**

$$\frac{4}{21} \times \frac{7}{44} = \frac{4 \cdot 7}{21 \cdot 44} = \frac{\overset{1}{\cancel{2}} \cdot \overset{1}{\cancel{2}} \cdot \overset{1}{\cancel{7}}}{3 \cdot \underset{1}{\cancel{7}} \cdot \underset{1}{\cancel{2}} \cdot \underset{1}{\cancel{2}} \cdot 11} = \frac{1}{33}$$

**You Try It 2**

$$\frac{2}{21} \times \frac{10}{33} = \frac{2 \cdot 10}{21 \cdot 33} = \frac{2 \cdot 2 \cdot 5}{3 \cdot 7 \cdot 3 \cdot 11} = \frac{20}{693}$$

**You Try It 3**

$$\frac{16}{5} \times \frac{15}{24} = \frac{16 \cdot 15}{5 \cdot 24} = \frac{\overset{1}{\cancel{2}} \cdot \overset{1}{\cancel{2}} \cdot \overset{1}{\cancel{2}} \cdot 2 \cdot \overset{1}{\cancel{3}} \cdot \overset{1}{\cancel{5}}}{\underset{1}{\cancel{5}} \cdot \underset{1}{\cancel{2}} \cdot \underset{1}{\cancel{2}} \cdot \underset{1}{\cancel{2}} \cdot \underset{1}{\cancel{3}}} = 2$$

**You Try It 4**

$$5\frac{2}{5} \times \frac{5}{9} = \frac{27}{5} \times \frac{5}{9} = \frac{27 \cdot 5}{5 \cdot 9} = \frac{\overset{1}{\cancel{3}} \cdot \overset{1}{\cancel{3}} \cdot 3 \cdot \overset{1}{\cancel{5}}}{\underset{1}{\cancel{5}} \cdot \underset{1}{\cancel{3}} \cdot \underset{1}{\cancel{3}}} = 3$$

**You Try It 5**

$$3\frac{2}{5} \times 6\frac{1}{4} = \frac{17}{5} \times \frac{25}{4} = \frac{17 \cdot 25}{5 \cdot 4}$$

$$= \frac{17 \cdot \overset{1}{\cancel{5}} \cdot 5}{\underset{1}{\cancel{5}} \cdot 2 \cdot 2} = \frac{85}{4} = 21\frac{1}{4}$$

**You Try It 6**

$$3\frac{2}{7} \times 6 = \frac{23}{7} \times \frac{6}{1} = \frac{23 \cdot 6}{7 \cdot 1}$$

$$= \frac{23 \cdot 3 \cdot 2}{7 \cdot 1} = \frac{138}{7} = 19\frac{5}{7}$$

**You Try It 7**

**Strategy**
To find the value of the house today, multiply the old value of the house ($30,000) by $3\frac{1}{2}$.

**Solution**

$$\frac{30{,}000}{1} \times \frac{7}{2} = \frac{30{,}000 \cdot 7}{1 \cdot 2}$$

$$= 105{,}000$$

The value of the house today is $105,000.

**You Try It 8**

**Strategy**
To find the cost of the air compressor:

- Multiply to find the value of the drying chamber $\left(\frac{4}{5} \times \$60{,}000\right)$.
- Subtract the value of the drying chamber from the total value of the two items ($60,000).

**Solution**

$$\frac{4}{5} \times \frac{\$60{,}000}{1} = \frac{\$240{,}000}{5}$$

$$= \$48{,}000$$

value of the drying chamber

$$\begin{array}{r} \$60{,}000 \\ -\ 48{,}000 \\ \hline \$12{,}000 \end{array}$$

The cost of the air compressor was $12,000.

## SECTION 2.7 *pages 97–100*

**You Try It 1**

$$\frac{3}{7} \div \frac{2}{3} = \frac{3}{7} \times \frac{3}{2} = \frac{3 \cdot 3}{7 \cdot 2} = \frac{9}{14}$$

**You Try It 2**

$$\frac{3}{4} \div \frac{9}{10} = \frac{3}{4} \times \frac{10}{9}$$

$$= \frac{3 \cdot 10}{4 \cdot 9} = \frac{\overset{1}{\cancel{3}} \cdot \overset{1}{\cancel{2}} \cdot 5}{\underset{1}{\cancel{2}} \cdot 2 \cdot \underset{1}{\cancel{3}} \cdot 3} = \frac{5}{6}$$

**You Try It 3**

$$\frac{5}{7} \div 6 = \frac{5}{7} \div \frac{6}{1}$$

$$= \frac{5}{7} \times \frac{1}{6} = \frac{5}{7 \cdot 2 \cdot 3} = \frac{5}{42}$$

**You Try It 4**

$$12\frac{3}{5} \div 7 = \frac{63}{5} \div \frac{7}{1} = \frac{63}{5} \times \frac{1}{7}$$

$$= \frac{63 \cdot 1}{5 \cdot 7} = \frac{3 \cdot 3 \cdot \overset{1}{\cancel{7}}}{5 \cdot \underset{1}{\cancel{7}}} = \frac{9}{5} = 1\frac{4}{5}$$

**You Try It 5**

$$3\frac{2}{3} \div 2\frac{2}{5} = \frac{11}{3} \div \frac{12}{5}$$
$$= \frac{11}{3} \times \frac{5}{12} = \frac{11 \cdot 5}{3 \cdot 12}$$
$$= \frac{11 \cdot 5}{3 \cdot 2 \cdot 2 \cdot 3} = \frac{55}{36} = 1\frac{19}{36}$$

**You Try It 6**

$$2\frac{5}{6} \div 8\frac{1}{2} = \frac{17}{6} \div \frac{17}{2}$$
$$= \frac{17}{6} \times \frac{2}{17} = \frac{17 \cdot 2}{6 \cdot 17}$$
$$= \frac{\overset{1}{\cancel{17}} \cdot \overset{1}{\cancel{2}}}{\underset{1}{\cancel{2}} \cdot 3 \cdot \underset{1}{\cancel{17}}} = \frac{1}{3}$$

**You Try It 7**

$$6\frac{2}{5} \div 4 = \frac{32}{5} \div \frac{4}{1}$$
$$= \frac{32}{5} \times \frac{1}{4} = \frac{32 \cdot 1}{5 \cdot 4}$$
$$= \frac{2 \cdot 2 \cdot 2 \cdot \overset{1}{\cancel{2}} \cdot \overset{1}{\cancel{2}}}{5 \cdot \underset{1}{\cancel{2}} \cdot \underset{1}{\cancel{2}}} = \frac{8}{5} = 1\frac{3}{5}$$

**You Try It 8**

**Strategy**

To find the price of one ounce of gold, divide the total price of the coin ($195) by the number of ounces $\left(\frac{1}{2}\right)$.

**Solution**

$$195 \div \frac{1}{2} = \frac{195}{1} \div \frac{1}{2}$$
$$= \frac{195}{1} \times \frac{2}{1} = \frac{195 \cdot 2}{1 \cdot 1} = 390$$

The price of one ounce of gold is $390.

**You Try It 9**

**Strategy**

To find the length of the remaining piece:

- Divide the total length of the board (16 feet) by the length of each shelf $\left(3\frac{1}{3}\text{ feet}\right)$.
- Multiply the fractional part of the result by the length of one shelf to determine the length of the remaining piece.

**Solution**

$$16 \div 3\frac{1}{3} = 16 \div \frac{10}{3}$$
$$= \frac{16}{1} \times \frac{3}{10} = \frac{16 \cdot 3}{1 \cdot 10}$$
$$= \frac{\cancel{2} \cdot 2 \cdot 2 \cdot 2 \cdot 3}{\cancel{2} \cdot 5} = \frac{24}{5} = 4\frac{4}{5}$$

$$\frac{4}{5} \times \frac{10}{3} = \frac{40}{15} = 2\frac{2}{3}$$

The length of the piece remaining is $2\frac{2}{3}$ feet.

## SECTION 2.8 *pages 105–106*

**You Try It 1** $\frac{9}{14} > \frac{13}{21}$

**You Try It 2**

$$\left(\frac{7}{11}\right)^2 \cdot \left(\frac{2}{7}\right) = \left(\frac{7}{11} \cdot \frac{7}{11}\right) \cdot \left(\frac{2}{7}\right)$$

$$= \frac{\cancel{7} \cdot 7 \cdot 2}{11 \cdot 11 \cdot \cancel{7}} = \frac{14}{121}$$

**You Try It 3**

$$\left(\frac{1}{13}\right)^2 \cdot \left(\frac{1}{4} + \frac{1}{6}\right) \div \frac{5}{13}$$

$$\left(\frac{1}{13}\right)^2 \cdot \left(\frac{5}{12}\right) \div \frac{5}{13}$$

$$\left(\frac{1}{169}\right) \cdot \left(\frac{5}{12}\right) \div \frac{5}{13}$$

$$\left(\frac{1 \cdot 5}{13 \cdot 13 \cdot 12}\right) \div \frac{5}{13}$$

$$\frac{1 \cdot \overset{1}{\cancel{5}} \cdot \overset{1}{\cancel{13}}}{\underset{1}{\cancel{13}} \cdot 13 \cdot 12 \cdot \underset{1}{\cancel{5}}}$$

$$\frac{1}{156}$$

# SOLUTIONS to Chapter 3 "You Try It"

## SECTION 3.1 *pages 119–120*

**You Try It 1** Two hundred nine and five thousand eight hundred thirty-eight hundred-thousandths

**You Try It 2** 42,000.000207

**You Try It 3** 4.35

**You Try It 4** 3.29053

## SECTION 3.2 *pages 123–124*

**You Try It 1**

$$\begin{array}{r} {\scriptstyle 1\,2} \phantom{.62054} \\ 4.62\phantom{054} \\ 27.9\phantom{2054} \\ +\ 0.62054 \\ \hline 33.14054 \end{array}$$

**You Try It 2**

$$\begin{array}{r} {\scriptstyle 1} \phantom{.05} \\ 6.05\phantom{4} \\ 12.\phantom{374} \\ +\ 0.374 \\ \hline 18.424 \end{array}$$

**You Try It 3**

**Strategy**
To find the total income, add the four commissions (\$485.60, \$599.46, \$326.75, and \$725.42) to the salary (\$425.00).

**Solution**

$$\begin{array}{r} \$485.60 \\ 599.46 \\ 326.75 \\ 725.42 \\ +\ \ 425.00 \\ \hline \$2562.23 \end{array}$$

Anita's total income was \$2562.23.

## SECTION 3.3 *pages 127–128*

**You Try It 1**

$$\begin{array}{r} \scriptstyle 11\ 9 \\ \scriptstyle 6\ \not{1}\ \not{10}\ 13 \\ \not{7}\ \not{2}.\not{0}\ \not{3}\ 9 \\ -\quad 8.4\ 7 \\ \hline 6\ 3.5\ 6\ 9 \end{array} \qquad \text{Check: } \begin{array}{r} \scriptstyle 11\ 1 \\ 8.47 \\ +\ 63.569 \\ \hline 72.039 \end{array}$$

**You Try It 2**

$$\begin{array}{r} \scriptstyle 14\ 9 \\ \scriptstyle 2\ \not{4}\ \not{10}\ 10 \\ \not{3}\ \not{5}.\not{0}\ \not{0} \\ -\quad 9.6\ 7 \\ \hline 2\ 5.3\ 3 \end{array} \qquad \text{Check: } \begin{array}{r} \scriptstyle 11\ 1 \\ 9.67 \\ +\ 25.33 \\ \hline 35.00 \end{array}$$

**You Try It 3**

$$\begin{array}{r} \scriptstyle 16\ 9\ 9 \\ \scriptstyle 2\ 6\ \not{10}\ \not{10}\ 10 \\ \not{3}.\not{7}\ \not{0}\ \not{0}\ \not{0} \\ -\ 1.9\ 7\ 1\ 5 \\ \hline 1.7\ 2\ 8\ 5 \end{array} \qquad \text{Check: } \begin{array}{r} \scriptstyle 1\ 111 \\ 1.9715 \\ +\ 1.7285 \\ \hline 3.7000 \end{array}$$

**You Try It 4**

**Strategy**
To find the amount of change, subtract the amount paid (\$3.85) from \$5.00.

**Solution**

$$\begin{array}{r} \$5.00 \\ -\ 3.85 \\ \hline \$1.15 \end{array}$$

Your change was \$1.15.

**You Try It 5**

**Strategy**
To find the new balance:

- Add to find the total of the three checks (\$1025.60 + \$79.85 + \$162.47).
- Subtract the total from the previous balance (\$2472.69).

**Solution**

$$\begin{array}{r} \$1025.60 \\ 79.85 \\ +\ \ 162.47 \\ \hline \$1267.92 \end{array} \qquad \begin{array}{r} \$2472.69 \\ -\ 1267.92 \\ \hline \$1204.77 \end{array}$$

The new balance is \$1204.77.

## SECTION 3.4 *pages 131–134*

**You Try It 1**

$$\begin{array}{r} 870 \\ \times\ 4.6 \\ \hline 5220 \\ 3480\phantom{0} \\ \hline 4002.0 \end{array}$$

**You Try It 2**

$$\begin{array}{r} 0.000086 \\ \times\ 0.057 \\ \hline 602 \\ 430\phantom{0} \\ \hline 0.000004902 \end{array}$$

**You Try It 3**

$$\begin{array}{r} 4.68 \\ \times\ 6.03 \\ \hline 1404 \\ 28\,080 \\ \hline 28.2204 \end{array}$$

**You Try It 4** $6.9 \times 1000 = 6900$

**You Try It 5** $4.0273 \times 10^2 = 402.73$

**You Try It 6**

**Strategy**
To find the cost of water:

- Find the number of gallons of water used by multiplying the number of gallons used per day (5000) by the number of days (62).
- Then multiply the cost per 1000 gallons (1.39) by the number of gallons used.
- Add the cost of the water to the meter fee ($133.70).

**Solution**
Number of gallons $= 5000(62) = 310{,}000$

Cost of water $= \frac{310{,}000}{1000} \times 1.39 = 430.90$

Total cost $= 133.70 + 430.90 = 564.60$

The total cost is $564.60.

**You Try It 7**

**Strategy**
To find the cost of running the freezer for 210 hours, multiply the hourly cost (\$.035) by the number of hours (210).

**Solution**

$$\begin{array}{r} \$.035 \\ \times\ \ 210 \\ \hline \$7.35 \end{array}$$

The cost of running the freezer for 210 hours is \$7.35.

**You Try It 8**

**Strategy**
To find the total cost of the stereo:

- Multiply the monthly payment (\$37.18) by the number of months (18).
- Add the total to the down payment (\$175.00).

**Solution**

$$\begin{array}{r} \$37.18 \\ \times\ \ \ \ 18 \\ \hline \$669.24 \end{array} \qquad \begin{array}{r} \$699.24 \\ +\ 175.00 \\ \hline \$844.24 \end{array}$$

The total cost of the stereo is \$844.24.

## SECTION 3.5 *pages 141–144*

**You Try It 1**

$$\begin{array}{r} 2.7 \\ 0.052.\overline{)0.140.4} \\ -104 \\ \hline 36\ 4 \\ -36\ 4 \\ \hline 0 \end{array}$$

**You Try It 2**

$$\begin{array}{r} 0.4873 \approx 0.487 \\ 76\overline{)37.0420} \\ -30\ 4 \\ \hline 6\ 64 \\ -6\ 08 \\ \hline 562 \\ -532 \\ \hline 300 \\ -228 \\ \hline \end{array}$$

**You Try It 3**

$$\begin{array}{r} 72.73 \approx 72.7 \\ 5.09.\overline{)370.20.00} \\ -356\ 3 \\ \hline 13\ 90 \\ 10\ 18 \\ \hline 3\ 720 \\ -3\ 563 \\ \hline 1570 \\ -1527 \\ \hline \end{array}$$

**You Try It 4** $309.21 \div 10{,}000 = 0.030921$

**You Try It 5** $42.93 \div 10^4 = 0.004293$

**You Try It 6**

**Strategy**
To find the amount he paid in gasoline taxes.

- Find the total number of gallons of gas he used by dividing his total miles (13,400) by the number of miles traveled per gallon of gasoline (26).
- Multiply the total tax ($.4013) by the total number of gallons of gasoline used.

**Solution**
$13{,}400 \div 26 \approx 515.385$

$0.4013 \times 515.385 \approx 206.842$

Jefferson paid approximately $207 for gasoline taxes.

**You Try It 7**

**Strategy**
To find the average number of people watching TV:

- Add the number of people watching each day of the week.
- Divide the total number of people watching by 7.

**Solution**
$91.9 + 89.8 + 90.6 + 93.9 + 78.0 + 77.1 + 87.7 = 609$

$\frac{609}{7} = 87$

An average of 87 million people watch television per day.

## SECTION 3.6 *pages 149–150*

**You Try It 1** $16\overline{)9.00}$, quotient $0.56 \approx 0.6$

**You Try It 2** $4\frac{1}{6} = \frac{25}{6}$ $6\overline{)25.00}$, quotient $4.166 \approx 4.17$

**You Try It 3** $0.56 = \frac{56}{100} = \frac{14}{25}$

$5.35 = 5\frac{35}{100} = 5\frac{7}{20}$

**You Try It 4** $0.12\frac{7}{8} = \frac{12\frac{7}{8}}{100} = 12\frac{7}{8} \div 100$

$= \frac{103}{8} \times \frac{1}{100} = \frac{103}{800}$

**You Try It 5** $0.63 > \frac{5}{8}$

# SOLUTIONS to Chapter 4 "You Try It"

## SECTION 4.1 *pages 163–164*

**You Try It 1**

$$\frac{20 \text{ pounds}}{24 \text{ pounds}} = \frac{20}{24} = \frac{5}{6}$$

20 pounds:24 pounds = 20:24 = 5:6

20 pounds to 24 pounds = 20 to 24
= 5 to 6

**You Try It 2**

$$\frac{64 \text{ miles}}{8 \text{ miles}} = \frac{64}{8} = \frac{8}{1}$$

64 miles:8 miles = 64:8 = 8:1

64 miles to 8 miles = 64 to 8 = 8 to 1

**You Try It 3**

**Strategy**
To find the ratio, write the ratio of board feet of cedar (12,000) to board feet of ash (18,000) in simplest form.

**Solution**

$$\frac{12{,}000}{18{,}000} = \frac{2}{3}$$

The ratio is $\frac{2}{3}$.

**You Try It 4**

**Strategy**
To find the ratio, write the ratio of the amount spent on radio advertising ($15,000) to the amount spent on radio and television advertising ($15,000 + $20,000).

**Solution**

$$\frac{\$15{,}000}{\$15{,}000 + \$20{,}000} = \frac{\$15{,}000}{\$35{,}000} = \frac{3}{7}$$

The ratio is $\frac{3}{7}$.

## SECTION 4.2 *pages 167–168*

**You Try It 1** $\frac{15 \text{ pounds}}{12 \text{ trees}} = \frac{5 \text{ pounds}}{4 \text{ trees}}$

**You Try It 2** $\frac{260 \text{ miles}}{8 \text{ hours}}$

$$8\overline{)260.0} = 32.5$$

32.5 miles/hour

**You Try It 3**

**Strategy**
To find Erik's profit per ounce:

- Find the total profit by subtracting the cost ($1625) from the selling price ($1720).
- Find the profit per ounce by dividing the total profit by the number of ounces (5).

**Solution**

$$\begin{array}{r} 1720 \\ -\ 1625 \\ \hline 95 \end{array} \text{ total profit} \qquad 5\overline{)95} = 19$$

The profit per ounce is $19.

## SECTION 4.3 *pages 171–174*

**You Try It 1** $\frac{6}{10} \times \frac{9}{15}$ → $10 \times 9 = 90$; → $6 \times 15 = 90$

The proportion is true.

**You Try It 2** $\frac{32}{6} \times \frac{90}{8}$ → $6 \times 90 = 540$; → $32 \times 8 = 256$

The proportion is not true.

**You Try It 3**
$n \times 7 = 14 \times 3$
$n \times 7 = 42$
$n = 42 \div 7$
$n = 6$

Check: $\frac{6}{14} \times \frac{3}{7}$ → $14 \times 3 = 42$; → $6 \times 7 = 42$

**You Try It 4**
$5 \times 20 = 8 \times n$
$100 = 8 \times n$
$100 \div 8 = n$
$12.5 = n$

**You Try It 5**
$15 \times n = 20 \times 12$
$15 \times n = 240$
$n = 240 \div 15$
$n = 16$

Check: $\frac{15}{20} \times \frac{12}{16}$ → $20 \times 12 = 240$; → $15 \times 16 = 240$

**You Try It 6**
$12 \times 4 = 7 \times n$
$48 = 7 \times n$
$48 \div 7 = n$
$6.86 \approx n$

**You Try It 7**
$n \times 1 = 12 \times 4$
$n \times 1 = 48$
$n = 48 \div 1$
$n = 48$

Check: $\frac{48}{12} \times \frac{4}{1}$ → $12 \times 4 = 48$; → $48 \times 1 = 48$

**You Try It 8**
$3 \times n = 12 \times 8$
$3 \times n = 96$
$n = 96 \div 3$
$n = 32$

Check: $\frac{3}{8} \times \frac{12}{32}$ → $8 \times 12 = 96$; → $3 \times 32 = 96$

**You Try It 9**

**Strategy**
To find the number of jars that can be packed in 15 boxes, write and solve a proportion using $n$ to represent the number of jars.

**Solution**

$$\frac{24 \text{ jars}}{6 \text{ boxes}} = \frac{n \text{ jars}}{15 \text{ boxes}}$$

$$24 \times 15 = 6 \times n$$
$$360 = 6 \times n$$
$$360 \div 6 = n$$
$$60 = n$$

60 jars can be packed in 15 boxes.

**You Try It 10**

**Strategy**
To find the number of tablespoons of fertilizer needed, write and solve a proportion using $n$ to represent the number of tablespoons of fertilizer.

**Solution**

$$\frac{3 \text{ tablespoons}}{4 \text{ gallons}} = \frac{n \text{ tablespoons}}{10 \text{ gallons}}$$

$$3 \times 10 = 4 \times n$$
$$30 = 4 \times n$$
$$30 \div 4 = n$$
$$7\frac{1}{2} = n$$

For 10 gallons of water, $7\frac{1}{2}$ tablesppons of fertilizer are required.

# SOLUTIONS to Chapter 5 "You Try It"

**SECTION 5.1** *pages 189–190*

**You Try It 1** $125\% = 125 \times \frac{1}{100} = \frac{125}{100} = 1\frac{1}{4}$

$125\% = 125 \times 0.01 = 1.25$

**You Try It 2** $33\frac{1}{3}\% = 33\frac{1}{3} \times \frac{1}{100}$

$= \frac{100}{3} \times \frac{1}{100}$

$= \frac{100}{300} = \frac{1}{3}$

**You Try It 3** $0.25\% = 0.25 \times 0.01 = 0.0025$

**You Try It 4** $0.048 = 0.048 \times 100\% = 4.8\%$

**You Try It 5** $3.67 = 3.67 \times 100\% = 367\%$

**You Try It 6** $0.62\frac{1}{2} = 0.62\frac{1}{2} \times 100\%$

$= 62\frac{1}{2}\%$

**You Try It 7** $\frac{5}{6} = \frac{5}{6} \times 100\% = \frac{500\%}{6} = 83\frac{1}{3}\%$

**You Try It 8** $1\frac{4}{9} = \frac{13}{9} = \frac{13}{9} \times 100\%$

$= \frac{1300\%}{9} \approx 144.4\%$

## SECTION 5.2 *pages 193–194*

**You Try It 1** $n = 0.063 \times 150$

$n = 9.45$

**You Try It 2** $n = \frac{1}{6} \times 66$

$n = 11$

**You Try It 3**

**Strategy**

To find the new hourly wage:

- Find the amount of the raise. Write and solve a basic percent equation using $n$ to represent the amount of the raise (amount). The percent is 8%. The base is $13.50.
- Add the amount of the raise to the old wage.

**Solution**

$8\% \times \$13.50 = n$

$0.08 \times \$13.50 = n$

$\$1.08 = n$

$$\begin{array}{r} \$13.50 \\ +\ \ 1.08 \\ \hline \$14.58 \end{array}$$

The new hourly wage is $14.58.

## SECTION 5.3 *pages 197–198*

**You Try It 1**

$$\begin{aligned} n \times 32 &= 16 \\ n &= 16 \div 32 \\ n &= 0.50 \\ n &= 50\% \end{aligned}$$

**You Try It 2**

$$\begin{aligned} n \times 15 &= 48 \\ n &= 48 \div 15 \\ n &= 3.20 \\ n &= 320\% \end{aligned}$$

**You Try It 3**

$$\begin{aligned} 30 &= n \times 45 \\ 30 \div 45 &= n \\ 0.66\frac{2}{3} &= n \\ 66\frac{2}{3}\% &= n \end{aligned}$$

**You Try It 4**

**Strategy**
To find what percent of the income the income tax is, write and solve a basic percent equation using $n$ to represent the percent. The base is \$20,000 and the amount is \$3000.

**Solution**

$$n \times \$20{,}000 = \$3000$$
$$n = \$3000 \div \$20{,}000$$
$$n = 0.15 = 15\%$$

The income tax is 15% of the income.

**You Try It 5**

**Strategy**
To find the percent that did not favor Minna Oliva:

- Subtract to find the number of people that did not favor her (1000 − 667).
- Write and solve a basic percent equation using $n$ to represent the percent. The base is 1000 and the amount is the number of people who did not favor her.

**Solution**

$1000 - 667 = 333$ number who did not favor the candidate

$$n \times 1000 = 333$$
$$n = 333 \div 1000$$
$$n = 0.333 = 33.3\%$$

33.3% of the people did not favor Minna Oliva.

## SECTION 5.4 *pages 201–202*

**You Try It 1** $0.86 \times n = 215$
$$n = 215 \div 0.86$$
$$n = 250$$

**You Try It 2** $0.025 \times n = 15$
$$n = 15 \div 0.025$$
$$n = 600$$

**You Try It 3** $\frac{1}{6} \times n = 5$
$$n = 5 \div \frac{1}{6}$$
$$n = 30$$

**You Try It 4**

**Strategy**
To find the original value of the car, write and solve a basic percent equation using $n$ to represent the original value (base). The percent is 51%. The amount is \$3876.

**Solution**

$$\begin{aligned} 51\% \times n &= \$3876 \\ 0.51 \times n &= \$3876 \\ n &= \$3876 \div 0.51 \\ n &= \$7600 \end{aligned}$$

The original value of the car was \$7600.

**You Try It 5**

**Strategy**
To find the difference between the original price and the selling price:

- Find the original price. Write and solve a basic percent equation using $n$ to represent the original price (base). The percent is 80%. The amount is \$44.80.
- Subtract the sale price (\$44.80) from the original price.

**Solution**

$$\begin{aligned} 80\% \times n &= \$44.80 \\ 0.80 \times n &= \$44.80 \\ n &= \$44.80 \div 0.80 \\ n &= \$56.00 \text{ original price} \end{aligned}$$

$\$56.00 - \$44.80 = \$11.20$

The difference between the original price and the sale price is \$11.20.

## SECTION 5.5 *pages 205–206*

**You Try It 1**

$$\begin{aligned} \frac{26}{100} &= \frac{22}{n} \\ 26 \times n &= 100 \times 22 \\ 26 \times n &= 2200 \\ n &= 2200 \div 26 \\ n &\approx 84.62 \end{aligned}$$

**You Try It 2**

$$\begin{aligned} \frac{16}{100} &= \frac{n}{132} \\ 16 \times 132 &= 100 \times n \\ 2112 &= 100 \times n \\ 2112 \div 100 &= n \\ 21.12 &= n \end{aligned}$$

**You Try It 3**

**Strategy**
To find the number of days it snowed, write and solve a proportion using $n$ to represent the number of days (amount). The percent is 64%. The base is 150.

**Solution**

$$\frac{64}{100} = \frac{n}{150}$$
$$64 \times 150 = 100 \times n$$
$$9600 = 100 \times n$$
$$9600 \div 100 = n$$
$$96 = n$$

It snowed 96 days.

**You Try It 4**

**Strategy**
To find the percent of pens which were not defective:

- Subtract to find the number of pens which were not defective (200 − 5).
- Write and solve a proportion using $n$ to represent the percent of pens which were not defective. The base is 200 and the amount is the number of pens not defective.

**Solution**
$200 - 5 = 195$ number of pens not defective

$$\frac{n}{100} = \frac{195}{200}$$
$$200 \times n = 195 \times 100$$
$$200 \times n = 19500$$
$$n = 19500 \div 200$$
$$n = 97.5\%$$

97.5% of the pens were not defective.

# SOLUTIONS to Chapter 6 "You Try It"

## SECTION 6.1 *pages 219–220*

**You Try It 1**

a. $6.25 \div 5 = 1.25$
Unit cost: $1.25 per quart

b. $85 \div 4 = 21.25$
Unit cost: 21.3¢ per ear

**You Try It 2**

**Strategy**
To find the more economical purchase, compare unit costs.

**Solution**
$2.52 \div 6 = 0.42$
$1.66 \div 4 = 0.415$
$\$.415 < \$.42$

The more economical purchase is 4 cans for $1.66.

**You Try It 3**

**Strategy**
To find the total cost, multiply the unit cost ($4.96) by the number of units (7).

**Solution**
$4.96 \times 7 = 34.72$

The total cost is $34.72.

## SECTION 6.2 *pages 223–228*

**You Try It 1**

**Strategy**
To find the percent increase:
- Find the amount of increase.
- Solve the basic percent equation for *percent*.

**Solution**

$$\begin{array}{r} 2250 \\ -\ 2000 \\ \hline 250 \end{array} \qquad \begin{aligned} n \times 2000 &= 250 \\ n &= 250 \div 2000 \\ n &= 0.125 = 12.5\% \end{aligned}$$

The percent increase was 12.5%.

**You Try It 2**

**Strategy**
To find the new hourly wage:
- Solve the basic percent equation for *amount*.
- Add the amount of increase to the original wage.

**Solution**
$0.14 \times 6.50 = n$
$0.91 = n$
$6.50 + 0.91 = 7.41$

The new hourly wage is $7.41.

**You Try It 3**

**Strategy**
To find the markup, solve the basic percent equation for *amount*.

**Solution**
$0.20 \times 8 = n$
$1.60 = n$

The markup is $1.60.

**You Try It 4**

**Strategy**
To find the selling price:
- Find the markup by solving the basic percent equation for *amount*.
- Add the markup to the cost.

**Solution**
$0.55 \times 72 = n$
$39.60 = n$
$72 + 39.60 = 111.60$

The selling price is $111.60.

**You Try It 5**

**Strategy**
To find the percent decrease:
- Find the amount of decrease.
- Solve the basic percent equation for *percent*.

**Solution**
$150 - 120 = 30$
$n \times 150 = 30$
$n = 30 \div 150$
$n = 0.2 = 20\%$

The percent decrease is 20%.

**You Try It 6**

**Strategy**
To find the visibility:
- Find the amount of decrease by solving the basic percent equation for *amount*.
- Subtract the amount of decrease from the original visibility.

**Solution**
$0.40 \times 5 = n$
$2 = n$
$5 - 2 = 3$

The visibility was 3 miles.

**You Try It 7**

**Strategy**
To find the discount rate, solve the basic percent equation for *percent*.

**Solution**
$n \times 125 = 25$
$n = 0.20 = 20\%$

The discount rate is 20%.

**You Try It 8**

**Strategy**
To find the sale price:

- Find the discount by solving the basic percent equation for *amount*.
- Subtract to find the sale price.

**Solution**
$0.20 \times 10.25 = n$
$2.05 = n$
$10.25 - 2.05 = 8.20$

The sale price is \$8.20.

## SECTION 6.3 *pages 233–234*

**You Try It 1**

**Strategy**
To find the simple interest, multiply the principal by the annual interest rate by the time (in years).

**Solution**
$15{,}000 \times 0.08 \times 1.5 = 1800$

The interest due is \$1800.

**You Try It 2**

**Strategy**
To find the interest due, multiply the principal by the monthly interest rate by the time (in months).

**Solution**
$400 \times 0.012 \times 2 = 9.60$

The interest charge is \$9.60.

**You Try It 3**

**Strategy**
To find the interest earned:

- Find the new principal by multiplying the original principal by the factor found in the compound interest table.
- Subtract the original principal from the new principal.

**Solution**
$1000 \times 3.29066 = 3290.66$

The new principal is \$3290.66.

$3290.66 - 1000 = 2290.66$

The interest earned is \$2290.66.

## SECTION 6.4 *pages 237–240*

**You Try It 1**

**Strategy**
To find the mortgage:

- Find the down payment by solving the basic percent equation for *amount*.
- Subtract the down payment from the purchase price.

**Solution**
$0.25 \times 216{,}000 = n$
$54{,}000 = n$

The down payment is \$54,000.

$216{,}000 - 54{,}000 = 162{,}000$

The mortgage is \$162,000.

**You Try It 2**

**Strategy**
To find the loan origination fee, solve the basic percent equation for *amount*.

**Solution**
$0.045 \times 80{,}000 = n$
$3600 = n$

The loan origination fee was \$3600.

**You Try It 3**

**Strategy**
To find the monthly mortgage payment:

- Subtract the down payment from the purchase price to find the mortgage.
- Multiply the mortgage by the factor found in the monthly payment table.

**Solution**
$75{,}000 - 15{,}000 = 60{,}000$

The mortgage is \$60,000.

$60{,}000 \times 0.0089973 = 539.838$

The monthly mortgage payment is \$539.84.

**You Try It 4**

**Strategy**
To find the interest:

- Multipy the mortgage by the factor found in the monthly payment table to find the monthly mortgage payment.
- Subtract the principal from the monthly mortgage payment.

**Solution**
$125{,}000 \times 0.0083920 = 1049$

The monthly mortgage payment is \$1049.

$1049 - 492.65 = 556.35$

The interest on the mortgage is \$556.35.

**You Try It 5**

**Strategy**
To find the monthly payment:

- Divide the annual property tax by 12 to find the monthly property tax.
- Add the monthly property tax to the monthly mortgage payment.

**Solution**
$744 \div 12 = 62$

The monthly property tax is \$62.

$415.20 + 62 = 477.20$

The total monthly payment is \$477.20.

## SECTION 6.5 *pages 243–244*

### You Try It 1

**Strategy**
To find the amount financed:

- Find the down payment by solving the basic percent equation for *amount*.
- Subtract the down payment from the purchase price.

**Solution**
$0.20 \times 9200 = n$
$1840 = n$

The down payment is \$1840.

$9200 - 1840 = 7360$

The amount financed is \$7360.

### You Try It 2

**Strategy**
To find the license fee, solve the basic percent equation for *amount*.

**Solution**
$0.015 \times 7350 = n$
$110.25 = n$

The license fee is \$110.25.

### You Try It 3

**Strategy**
To find the cost of operating the car, multiply the cost per mile by the number of miles driven.

**Solution**
$23,000 \times 0.22 = 5060$

The cost of operating the car is \$5060.

### You Try It 4

**Strategy**
To find the cost per mile for the car insurance, divide the cost for insurance by the number of miles driven.

**Solution**
$360 \div 15,000 = 0.024$

The cost per mile for insurance is \$0.024.

### You Try It 5

**Strategy**
To find the monthly payment:

- Subtract the down payment from the purchase price to find the amount financed.
- Multiply the amount financed by the factor found in the monthly payment table.

**Solution**
$15,900 - 3975 = 11,925$

The amount financed is \$11,925.

$11,925 \times 0.0244129 = 291.123$

The monthly payment is \$291.12.

## SECTION 6.6 *pages 247–248*

**You Try It 1**

**Strategy**
To find the worker's earnings:
- Find the worker's overtime wage by multiplying the hourly wage by 2.
- Multiply the number of hours worked by the overtime hours.

**Solution**
$8.50 \times 2 = 17$

The hourly wage for overtime is $17.

$17 \times 8 = 136$

The construction worker earns $136.

**You Try It 2**

**Strategy**
To find the salary per month, divide the annual salary by the number of months in a year (12).

**Solution**
$28{,}224 \div 12 = 2352$

The contractor's monthly salary is $2352.

**You Try It 3**

**Strategy**
To find the total earnings:
- Find the commission by multiplying the commission rate by the sales over $50,000.
- Add the commission to the annual salary.

**Solution**
$175{,}000 - 50{,}000 = 125{,}000$

Sales over $50,000 totaled $125,000.

$125{,}000 \times 0.095 = 11{,}875$

Earnings from commissions totaled $11,875.

$12{,}000 + 11{,}875 = 23{,}875$

The insurance agent earned $23,875.

## SECTION 6.7 *pages 251–256*

**You Try It 1**

**Strategy**
To find the current balance:

- Subtract the amount of the check from the old balance.
- Add the amount of each deposit.

**Solution**

| | |
|---|---|
| 302.46 | |
| − 20.59 | check |
| 281.87 | |
| 176.86 | first deposit |
| + 94.73 | second deposit |
| 553.46 | |

The current checking account balance is $553.46.

**You Try It 2**

| | |
|---|---|
| Current checkbook balance: | 653.41 |
| Check: 237 | + 48.73 |
| | 702.14 |
| Interest: | + 2.11 |
| | 704.25 |
| Deposit: | − 523.84 |
| | 180.41 |

Closing bank balance from bank statement: $180.41.

Checkbook balance: $180.41.

The bank statement and checkbook balance.

# ANSWERS to Chapter 1 Odd-Numbered Exercises

## SECTION 1.1 *pages 7–8*

**1.** 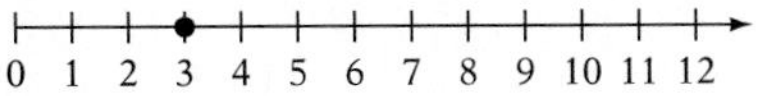

**3.** 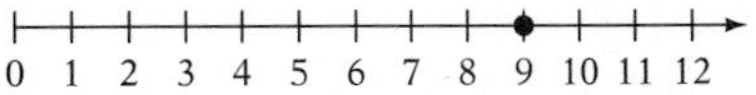


**5.** 37 < 49 **7.** 101 > 87 **9.** 245 > 158 **11.** 0 < 45 **13.** 815 < 928
**15.** Three thousand seven hundred ninety **17.** Fifty-eight thousand four hundred seventy-three
**19.** Four hundred ninety-eight thousand five hundred twelve
**21.** Six million eight hundred forty-two thousand seven hundred fifteen **23.** 357 **25.** 63,780
**27.** 7,024,709 **29.** 6000 + 200 + 90 + 5 **31.** 400,000 + 50,000 + 3000 + 900 + 20 + 1
**33.** 300,000 + 1000 + 800 + 9 **35.** 3,000,000 + 600 + 40 + 2 **37.** 850 **39.** 4000
**41.** 53,000 **43.** 250,000 **45.** False. 8270 rounded to the nearest hundred is 8300.
**47.** Yes. IX = 9 XI = 11

## SECTION 1.2 *pages 13–16*

**1.** 28 **3.** 125 **5.** 102 **7.** 154 **9.** 1489 **11.** 828 **13.** 1584 **15.** 1219
**17.** 102,317 **19.** 79,326 **21.** 1804 **23.** 1579 **25.** 19,740 **27.** 7420
**29.** 120,570 **31.** 207,453 **33.** 24,218 **35.** 11,974 **37.** 9323 **39.** 77,139
**41.** 14,383 **43.** 9473 **45.** 33,247 **47.** 5058 **49.** 1992 **51.** 68,263
**53.** Est.: 17,700 Cal.: 17,754 **55.** Est.: 2900 Cal.: 2872 **57.** Est.: 101,000 Cal.: 101,712 **59.** Est.: 158,000 Cal.: 158,763 **61.** Est.: 260,000 Cal.: 261,595
**63.** Est.: 940,000 Cal.: 946,718 **65.** Est.: 33,000,000 Cal.: 32,691,621 **67.** Est.: 34,000,000 Cal.: 34,420,922
**69.** The total attendance was 5792 people. **71.** The total number of yards gained by passing was 307. **73.** The total commission Ken received was $6973. **75a.** The total paid attendance for the fifth and sixth games was 104,031 people. **75b.** The total paid attendance for the entire series was 311,460 people. **77a.** The total amount deposited was $1664. **77b.** The new checking account balance is $3799. **79.** 90,900 **81.** No. 0 + 2 = 2 **83.** Answers will vary **85.** The total average amount for all Americans is $4527. **87.** The total average amount for Americans ages 16 to 34 is $7838.

## SECTION 1.3 *pages 21–24*

**1.** 4 **3.** 4 **5.** 10 **7.** 7 **9.** 11 **11.** 9 **13.** 22 **15.** 60 **17.** 66 **19.** 33
**21.** 501 **23.** 962 **25.** 5002 **27.** 1513 **29.** 9 **31.** 7 **33.** 31 **35.** 47
**37.** 925 **39.** 4561 **41.** 3205 **43.** 1222 **45.** 3021 **47.** 3022 **49.** 3040
**51.** 212 **53.** 60,245 **55.** 65 **57.** 17 **59.** 8 **61.** 37 **63.** 353 **65.** 57
**67.** 160 **69.** 337 **71.** 1423 **73.** 754 **75.** 2179 **77.** 5726 **79.** 3347 **81.** 6489
**83.** 889 **85.** 71,129 **87.** 698 **89.** 29,405 **91.** 49,624 **93.** 628 **95.** 6532
**97.** 4286 **99.** 4042 **101.** 5209 **103.** 10,378 **105.** 1024 **107.** 556 **109.** 126,504
**111.** 38,208 **113.** 17,438 **115.** Est.: 60,000 Cal.: 60,427 **117.** Est.: 20,000 Cal.: 21,613 **119.** Est.: 360,000 Cal.: 358,979
**121.** There is $165 left in your checking account. **123.** The amount that remains to be paid is $899. **125.** The length of the trip was 3162 miles. **127.** Florida has lost 9,286,713 acres of wetlands over the last 200 years. **129.** It took 91 months longer to withdraw principal and interest in 1992. **131. a.** True **b.** False **c.** False **133.** Answers will vary **135.** The difference in cost between the two cars is $210. **137.** The SAT scores increased the most between 1992 and 1993. The amount of the increase was 4 points.

## SECTION 1.4 *pages 29–32*

**1.** 12 **3.** 35 **5.** 25 **7.** 0 **9.** 72 **11.** 198 **13.** 335 **15.** 2492 **17.** 5463
**19.** 4200 **21.** 6327 **23.** 1896 **25.** 5056 **27.** 3450 **29.** 2674 **31.** 2096
**33.** 5046 **35.** 18,040 **37.** 21,980 **39.** 18,450 **41.** 140 **43.** 47,736
**45.** 22,456 **47.** 18,630 **49.** 336 **51.** 910 **53.** 1794 **55.** 1541 **57.** 63,063

**59.** 33,520 **61.** 380,834 **63.** 541,164 **65.** 400,995 **67.** 105,315 **69.** 428,770 **71.** 260,000 **73.** 344,463 **75.** 41,808 **77.** 189,500 **79.** 401,880 **81.** 1,052,763 **83.** 4,198,388 **85.** 259,335 **87.** 1292 **89.** 151,152 **91.** 701,634,445 **93.** Est.: 540,000 Cal.: 550,935 **95.** Est.: 1,200,000 Cal.: 1,138,134 **97.** Est.: 600,000 Cal.: 605,022 **99.** Est.: 3,000,000 Cal.: 3,174,425 **101.** Est.: 18,000,000 Cal.: 19,212,228 **103.** Est.: 72,000,000 Cal.: 73,427,310 **105.** The car could travel 516 miles on 12 gallons of gas. **107.** In 40 hours, the machine can fill 168,000 bottles of Coke. **109.** There are 307,200 pixels on the screen. **111.** The total cost of the car is $11,280. **113.** The lighting designer can save $84. **115.** Carlos paid the 4 plumbers $1380. **117.** The total cost of the 4 components is $383. **119.** 12 accidental deaths occurred each hour. 288 accidental deaths occurred each day. 105,120 accidental deaths occurred each year. **121.** Answers will vary

## SECTION 1.5 *pages 41–44*

**1.** 2 **3.** 6 **5.** 5 **7.** 16 **9.** 80 **11.** 210 **13.** 44 **15.** 206 **17.** 530 **19.** 902 **21.** 21,560 **23.** 3580 **25.** 482 **27.** 1075 **29.** 2 r1 **31.** 5 r2 **33.** 13 r1 **35.** 10 r3 **37.** 90 r2 **39.** 230 r1 **41.** 204 r3 **43.** 1347 r3 **45.** 778 r2 **47.** 391 r4 **49.** 1160 r4 **51.** 708 r2 **53.** 3825 **55.** 9044 r2 **57.** 388 r3 **59.** 11,430 **61.** 510 **63.** 3 r15 **65.** 2 r3 **67.** 21 r36 **69.** 34 r2 **71.** 8 r8 **73.** 4 r49 **75.** 200 r25 **77.** 203 r2 **79.** 35 r47 **81.** 271 **83.** 4484 r6 **85.** 608 **87.** 15 r7 **89.** 1 r563 **91.** 50 r92 **93.** 40 r7 **95.** 258 r14 **97.** 517 r70 **99.** 500 **101.** Est.: 5000 Cal.: 5129 **103.** Est.: 20,000 Cal.: 21,968 **105.** Est.: 22,500 Cal.: 24,596 **107.** Est.: 3000 Cal.: 2836 **109.** Est.: 3000 Cal.: 3024 **111.** Est.: 30,000 Cal.: 32,036 **113.** Each organization received $137,000. **115.** The monthly contribution was $298. **117.** The annual contribution in 1993 was 79 times larger. **119.** Each disk can store 368,640 bytes of information. **121.** The amount of each payment is $119. **123.** Oprah Winfrey's average hourly wage is $49,000 per hour. **125.** Steven Spielberg's annual income is 2 times larger than Madonna's. **127.** The smallest three-digit palindromic number is 212. **129.** The largest possible number divisible by 4 is 91,180. **131.** The answer to a division problem can be checked by adding the remainder to the product of the quotient and divisor.

## SECTION 1.6 *pages 47–48*

**1.** $2^3$ **3.** $6^3 \cdot 7^4$ **5.** $6^3 \cdot 11^5$ **7.** $3 \cdot 10^4$ **9.** $2^2 \cdot 3^3 \cdot 5^4$ **11.** $2 \cdot 3^2 \cdot 7^2 \cdot 11$ **13.** $2^2 \cdot 7^4 \cdot 11^4$ **15.** 8 **17.** 400 **19.** 900 **21.** 972 **23.** 120 **25.** 360 **27.** 0 **29.** 90,000 **31.** 540 **33.** 4050 **35.** 11,025 **37.** 25,920 **39.** 4,320,000 **41.** 92,160 **43.** 5 **45.** 10 **47.** 47 **49.** 8 **51.** 5 **53.** 8 **55.** 6 **57.** 53 **59.** 44 **61.** 19 **63.** 37 **65.** 168 **67.** 27 **69.** 14 **71.** 10 **73.** 9 **75.** 1024 **77.** Yes. Answers will vary

## SECTION 1.7 *pages 51–52*

**1.** 1, 2, 4 **3.** 1, 2, 5, 10 **5.** 1, 5 **7.** 1, 2, 3, 4, 6, 12 **9.** 1, 2, 4, 8 **11.** 1, 13 **13.** 1, 2, 3, 6, 9, 18 **15.** 1, 5, 25 **17.** 1, 2, 3, 4, 6, 9, 12, 18, 36 **19.** 1, 2, 4, 7, 14, 28 **21.** 1, 29 **23.** 1, 2, 11, 22 **25.** 1, 2, 4, 11, 22, 44 **27.** 1, 7, 49 **29.** 1, 37 **31.** 1, 3, 19, 57 **33.** 1, 2, 3, 4, 6, 8, 12, 16, 24, 48 **35.** 1, 3, 29, 87 **37.** 1, 2, 23, 46 **39.** 1, 2, 5, 10, 25, 50 **41.** 1, 2, 3, 6, 11, 22, 33, 66 **43.** 1, 2, 4, 5, 8, 10, 16, 20, 40, 80 **45.** 1, 2, 3, 4, 6, 7, 12, 14, 21, 28, 42, 84 **47.** 1, 5, 17, 85 **49.** 1, 101 **51.** $2 \cdot 3$ **53.** Prime **55.** $2 \cdot 2 \cdot 2 \cdot 2$ **57.** $2 \cdot 2 \cdot 3$ **59.** $3 \cdot 3$ **61.** $2 \cdot 2 \cdot 3 \cdot 3$ **63.** Prime **65.** $2 \cdot 5 \cdot 5$ **67.** $5 \cdot 13$ **69.** $2 \cdot 2 \cdot 2 \cdot 2 \cdot 5$ **71.** $2 \cdot 3 \cdot 3$ **73.** $2 \cdot 2 \cdot 7$ **75.** Prime **77.** $2 \cdot 3 \cdot 7$ **79.** $3 \cdot 3 \cdot 3 \cdot 3$ **81.** $2 \cdot 11$ **83.** Prime **85.** $2 \cdot 5 \cdot 7$ **87.** $2 \cdot 43$ **89.** $5 \cdot 19$ **91.** Prime **93.** $5 \cdot 11$ **95.** $2 \cdot 61$ **97.** $2 \cdot 59$ **99.** $2 \cdot 3 \cdot 5 \cdot 5$ **101.** $2 \cdot 2 \cdot 2 \cdot 3 \cdot 5$ **103.** $2 \cdot 2 \cdot 2 \cdot 2 \cdot 2 \cdot 5$ **105.** $7 \cdot 7 \cdot 7$ **107.** $2 \cdot 2 \cdot 2 \cdot 2 \cdot 5 \cdot 5$ **109.** $3 \cdot 3 \cdot 5 \cdot 5$ **111.** 3 and 5, 5 and 7, 11 and 13. Other answers are possible. **113.** Answers will vary

### CHAPTER REVIEW *pages 55–56*

**1.** 101 > 87 **2.** Two hundred seventy-six thousand fifty-seven **3.** 2,011,044 **4.** 10,000 + 300 + 20 + 7 **5.** 1081 **6.** 12,493 **7.** The total income from commissions was $2567. **8.** The total amount was $301. The new checking account balance is $817. **9.** 1749 **10.** 5409 **11.** A U.S. company spends $1985 per person more than an Italian company. **12.** A U.S. company spends $563 more for business travel and entertainment. **13.** No **14.** 22,761 **15.** 619,833 **16.** The total of the car payments is $1476. **17.** Vincent's total pay for the last week was $384. **18.** 2135 **19.** 1306 r59 **20.** The car drove 27 miles per gallon of gasoline. **21.** The monthly car payment is $155. **22.** $2^4 \cdot 5^3$ **23.** $5^2 \cdot 7^5$ **24.** 600 **25.** 2 **26.** 17 **27.** 8 **28.** 1, 2, 3, 6, 9, 18 **29.** 1, 2, 3, 5, 6, 10, 15, 30 **30.** $2 \cdot 3 \cdot 7$ **31.** $2 \cdot 2 \cdot 2 \cdot 3 \cdot 3$

### CHAPTER TEST *pages 57–58*

**1.** 21 > 19 (1.1A) **2.** Two hundred seven thousand sixty-eight (1.1B) **3.** 1,204,006 (1.1B) **4.** 900,000 + 6000 + 300 + 70 + 8 (1.1C) **5.** 75,000 (1.1D) **6.** 96,798 (1.2A) **7.** 135,915 (1.2A) **8a.** They drove 855 miles. (1.2C) **8b.** The odometer reading was 48,481 miles. (1.2B) **9.** 9333 (1.3A) **10.** 19,922 (1.3B) **11.** The difference is $25. (1.3C) **12.** The difference is $91. (1.3C) **13.** 726,104 (1.4A) **14.** 6,854,144 (1.4B) **15.** The investor will receive $2844 over 12 months. (1.4C) **16.** 703 (1.5A) **17.** 8710 r2 (1.5B) **18.** 1,121 r27 (1.5C) **19.** The farmer needed 3000 boxes. (1.5D) **20.** $3^3 \cdot 7^2$ (1.6A) **21.** 432 (1.6A) **22.** 9 (1.6B) **23.** 7 (1.6B) **24.** 1, 2, 4, 5, 10, 20 (1.7A) **25.** $2 \cdot 2 \cdot 3 \cdot 7$ (1.7B)

# ANSWERS to Chapter 2 Odd-Numbered Exercises

### SECTION 2.1 *pages 63–64*

**1.** 40 **3.** 24 **5.** 30 **7.** 12 **9.** 24 **11.** 60 **13.** 56 **15.** 9 **17.** 32 **19.** 36 **21.** 660 **23.** 9384 **25.** 24 **27.** 30 **29.** 24 **31.** 576 **33.** 1680 **35.** 1 **37.** 3 **39.** 5 **41.** 25 **43.** 1 **45.** 4 **47.** 4 **49.** 2 **51.** 48 **53.** 2 **55.** 1 **57.** 7 **59.** 4 **61.** 2 **63.** 1 **65.** 14 **67.** 4 **69.** 1 **71.** 12 **73.** In 12 days, Joe and his friend will have another day off together. **75.** The GCF of 3 and 5 is 1, of 7 and 11 is 1, of 29 and 43 is 1. The GCF of 3 prime numbers is 1. **77.** 4

### SECTION 2.2 *pages 67–68*

**1.** $\frac{3}{4}$ **3.** $\frac{7}{8}$ **5.** $1\frac{1}{2}$ **7.** $2\frac{5}{8}$ **9.** $3\frac{3}{5}$ **11.** $\frac{5}{4}$ **13.** $\frac{8}{3}$ **15.** $\frac{28}{8}$

**17.**  **19.** 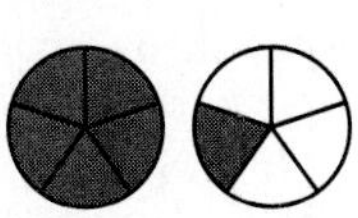 **21.** $5\frac{1}{3}$ **23.** 2 **25.** $3\frac{1}{4}$ **27.** $14\frac{1}{2}$ **29.** 17

**31.** $1\frac{7}{9}$ **33.** $1\frac{4}{5}$ **35.** 23 **37.** $1\frac{15}{16}$ **39.** $6\frac{1}{3}$ **41.** 5 **43.** 1 **45.** $\frac{14}{3}$ **47.** $\frac{26}{3}$ **49.** $\frac{59}{8}$ **51.** $\frac{25}{4}$ **53.** $\frac{121}{8}$ **55.** $\frac{41}{12}$ **57.** $\frac{34}{9}$ **59.** $\frac{38}{3}$ **61.** $\frac{38}{7}$ **63.** $\frac{63}{5}$ **65.** $\frac{41}{9}$ **67.** $\frac{117}{14}$ **69.** Answers will vary

### SECTION 2.3 *pages 71–72*

**1.** $\frac{5}{10}$ **3.** $\frac{9}{48}$ **5.** $\frac{12}{32}$ **7.** $\frac{9}{51}$ **9.** $\frac{12}{16}$ **11.** $\frac{27}{9}$ **13.** $\frac{20}{60}$ **15.** $\frac{44}{60}$ **17.** $\frac{12}{18}$ **19.** $\frac{35}{49}$ **21.** $\frac{10}{18}$ **23.** $\frac{21}{3}$ **25.** $\frac{35}{45}$ **27.** $\frac{60}{64}$ **29.** $\frac{21}{98}$ **31.** $\frac{30}{48}$ **33.** $\frac{15}{42}$ **35.** $\frac{102}{144}$ **37.** $\frac{153}{408}$

**39.** $\frac{340}{800}$ **41.** $\frac{1}{3}$ **43.** $\frac{1}{2}$ **45.** $\frac{1}{6}$ **47.** $\frac{2}{3}$ **49.** $1\frac{1}{2}$ **51.** 0 **53.** $\frac{9}{22}$ **55.** 3 **57.** $\frac{4}{21}$ **59.** $\frac{5}{11}$ **61.** $\frac{2}{9}$ **63.** $\frac{3}{4}$ **65.** $1\frac{1}{3}$ **67.** $\frac{3}{5}$ **69.** $\frac{1}{11}$ **71.** 4 **73.** $\frac{1}{3}$ **75.** $\frac{3}{5}$ **77.** $2\frac{1}{4}$ **79.** $\frac{1}{5}$ **81.**

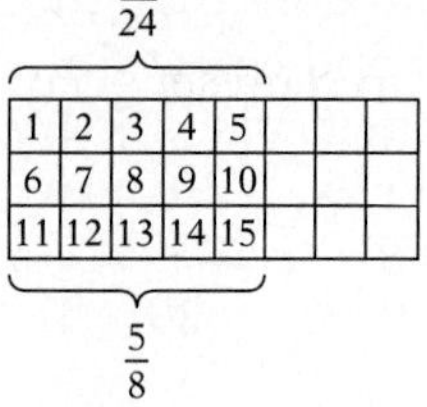


**83.** Answers will vary

## SECTION 2.4 *pages 77–80*

**1.** $\frac{3}{7}$ **3.** 1 **5.** $1\frac{4}{11}$ **7.** $3\frac{2}{5}$ **9.** $2\frac{4}{5}$ **11.** $2\frac{1}{4}$ **13.** $1\frac{3}{8}$ **15.** $1\frac{7}{15}$ **17.** $\frac{15}{16}$ **19.** $1\frac{4}{11}$ **21.** 1 **23.** $1\frac{7}{8}$ **25.** $1\frac{1}{6}$ **27.** $\frac{13}{14}$ **29.** $1\frac{1}{45}$ **31.** $\frac{17}{18}$ **33.** $\frac{35}{48}$ **35.** $\frac{23}{60}$ **37.** $\frac{56}{57}$ **39.** $\frac{13}{21}$ **41.** $2\frac{1}{12}$ **43.** $\frac{277}{315}$ **45.** $2\frac{2}{15}$ **47.** $2\frac{17}{120}$ **49.** $1\frac{89}{150}$ **51.** $1\frac{31}{72}$ **53.** $1\frac{5}{36}$ **55.** $1\frac{5}{36}$ **57.** $1\frac{29}{40}$ **59.** $1\frac{1}{42}$ **61.** $5\frac{7}{10}$ **63.** $5\frac{11}{16}$ **65.** $8\frac{23}{60}$ **67.** $18\frac{8}{9}$ **69.** $11\frac{1}{36}$ **71.** $24\frac{11}{12}$ **73.** $12\frac{17}{22}$ **75.** $15\frac{29}{160}$ **77.** $20\frac{37}{48}$ **79.** $44\frac{17}{84}$ **81.** $11\frac{7}{18}$ **83.** $8\frac{1}{12}$ **85.** $12\frac{71}{105}$ **87.** $15\frac{1}{4}$ **89.** $12\frac{7}{24}$ **91.** $29\frac{227}{240}$ **93.** $12\frac{17}{48}$ **95.** $9\frac{5}{24}$ **97.** $14\frac{1}{18}$ **99.** $12\frac{3}{4}$ **101.** The veneer is $1\frac{5}{16}$ inches thick. **103.** The length of the shaft is $1\frac{5}{16}$ inches. **105.** The price of the stock at the end of the month is $\$187\frac{1}{8}$. **107a.** Fred worked 12 hours of overtime. **107b.** Fred received $264 in overtime pay. **109.** The price of the stock at the end of the three months was $\$30\frac{5}{8}$. **111.** Answers will vary **113.** No. These activities account for only 22 hours.

## SECTION 2.5 *pages 85–88*

**1.** $\frac{2}{17}$ **3.** $\frac{1}{3}$ **5.** $\frac{1}{10}$ **7.** $\frac{7}{11}$ **9.** $\frac{1}{4}$ **11.** $\frac{2}{7}$ **13.** $\frac{4}{7}$ **15.** $\frac{1}{4}$ **17.** $\frac{9}{23}$ **19.** $\frac{1}{4}$ **21.** $\frac{1}{2}$ **23.** $\frac{19}{56}$ **25.** $\frac{1}{2}$ **27.** $\frac{4}{45}$ **29.** $\frac{11}{18}$ **31.** $\frac{5}{48}$ **33.** $\frac{37}{51}$ **35.** $\frac{17}{70}$ **37.** $\frac{49}{120}$ **39.** $\frac{8}{45}$ **41.** $\frac{11}{21}$ **43.** $\frac{23}{60}$ **45.** $\frac{1}{18}$ **47.** $5\frac{1}{5}$ **49.** $10\frac{9}{17}$ **51.** $4\frac{7}{8}$ **53.** $\frac{16}{21}$ **55.** $5\frac{1}{2}$ **57.** $5\frac{4}{7}$ **59.** $7\frac{5}{24}$ **61.** $48\frac{31}{70}$ **63.** $85\frac{2}{9}$ **65.** $9\frac{5}{13}$ **67.** $15\frac{11}{20}$ **69.** $4\frac{23}{24}$ **71.** $4\frac{37}{45}$ **73.** $1\frac{19}{24}$ **75.** $9\frac{1}{2}$ inches **77.** Ritter jumped $13\frac{7}{8}$ inches further than Coachman. Ritter jumped $10\frac{3}{4}$ inches further than McDaniel. **79a.** It is $7\frac{17}{24}$ miles to the second checkpoint. **79b.** It is $4\frac{7}{24}$ miles from the second checkpoint to the finish line. **81.** The wrestler must lose $3\frac{1}{4}$ pounds the third week. **83.** $\$27\frac{5}{8}$ **85.** $\$46\frac{1}{2}$ **87.** $6\frac{1}{8}$ **89.** $\frac{11}{15}$

## SECTION 2.6 *pages 93–96*

**1.** $\frac{7}{12}$ **3.** $\frac{7}{48}$ **5.** $\frac{5}{12}$ **7.** $\frac{1}{3}$ **9.** $\frac{11}{20}$ **11.** $\frac{1}{7}$ **13.** $\frac{1}{8}$ **15.** $\frac{5}{12}$ **17.** $\frac{9}{50}$ **19.** $\frac{5}{32}$ **21.** 6 **23.** $\frac{2}{3}$ **25.** $\frac{15}{44}$ **27.** $\frac{2}{45}$ **29.** $\frac{5}{14}$ **31.** 10 **33.** $\frac{1}{15}$ **35.** $\frac{10}{51}$ **37.** $\frac{45}{328}$ **39.** $\frac{10}{27}$ **41.** 4 **43.** $\frac{100}{357}$ **45.** $\frac{27}{100}$ **47.** $1\frac{1}{3}$ **49.** $2\frac{1}{2}$ **51.** $\frac{9}{34}$ **53.** 10 **55.** $16\frac{2}{3}$ **57.** 1 **59.** $\frac{1}{2}$ **61.** $17\frac{1}{2}$ **63.** 30 **65.** 42 **67.** $12\frac{2}{3}$ **69.** $1\frac{4}{5}$ **71.** $1\frac{2}{3}$ **73.** $12\frac{2}{9}$ **75.** $1\frac{2}{3}$ **77.** 0 **79.** $\frac{63}{184}$ **81.** $27\frac{2}{3}$ **83.** $17\frac{85}{128}$ **85.** $\frac{2}{5}$ **87.** $8\frac{1}{16}$ **89.** 18 **91.** 8 **93.** $15\frac{3}{4}$ **95.** 40 **97.** $8\frac{1}{28}$ **99.** Maria can walk $1\frac{1}{6}$ miles. **101.** $1\frac{11}{12}$ feet of the board is cut off. **103.** 535,000 gallons of propellant is used before burnout. **105a.** Ann read 288 pages. **105b.** 144 pages remain to be read. **107.** $267,000 of the monthly income remained. **109.** The weight is $54\frac{19}{36}$ pounds. **111.** $\frac{1}{2}$ **113.** Yes. **115.** $\frac{4}{9}$ **117.** $\frac{3}{16}$ **119.**

| | | |
|---|---|---|
| $\frac{2}{3}$ | $\frac{3}{4}$ | $\frac{5}{9}$ |
| $\frac{10}{9}$ | $\frac{1}{6}$ | $\frac{3}{2}$ |
| $\frac{9}{4}$ | $\frac{5}{18}$ | $\frac{4}{9}$ |

## SECTION 2.7 *pages 101–104*

**1.** $\frac{5}{6}$ **3.** $1\frac{5}{88}$ **5.** 0 **7.** $1\frac{1}{3}$ **9.** $8\frac{4}{5}$ **11.** $2\frac{73}{256}$ **13.** $1\frac{1}{9}$ **15.** $2\frac{1}{13}$ **17.** $\frac{2}{3}$ **19.** $\frac{9}{10}$ **21.** 2 **23.** 2 **25.** $\frac{1}{6}$ **27.** 6 **29.** $\frac{1}{15}$ **31.** 2 **33.** $2\frac{1}{2}$ **35.** 3 **37.** 6 **39.** $\frac{1}{2}$ **41.** $1\frac{1}{6}$ **43.** $3\frac{1}{3}$ **45.** $\frac{1}{30}$ **47.** $1\frac{4}{5}$ **49.** $8\frac{8}{9}$ **51.** $\frac{1}{6}$ **53.** $1\frac{1}{11}$ **55.** 3 **57.** $\frac{1}{5}$ **59.** 8 **61.** $\frac{7}{26}$ **63.** $\frac{9}{44}$ **65.** 120 **67.** $\frac{11}{40}$ **69.** $\frac{2}{35}$ **71.** $4\frac{4}{5}$ **73.** 24 **75.** $\frac{13}{32}$ **77.** $10\frac{2}{3}$ **79.** $6\frac{9}{10}$ **81.** $9\frac{7}{13}$ **83.** Undefined **85.** $3\frac{1}{25}$ **87.** $1\frac{139}{394}$ **89.** $8\frac{2}{7}$ **91.** $1\frac{1}{2}$ **93.** 6 **95.** $1\frac{10}{11}$ **97.** $1\frac{3}{5}$ **99.** $\frac{9}{34}$ **101.** The box contains 16 servings. **103.** The cost is $24,000 per acre. **105.** The capacity of the theater is 420. **107a.** The total weight of the fat and bone is $1\frac{5}{12}$ pounds. **107b.** There will be 28 servings. **109.** There will be 1 foot of the board remaining. **111.** There are 740 miles between the two cities. **113.** **a.** $\frac{2}{3}$ **b.** $\frac{21}{8}$ **115.** The quotient is greater. **117.** Yes. Answers will vary

## SECTION 2.8 *pages 107–108*

**1.** $\frac{11}{40} < \frac{19}{40}$ **3.** $\frac{2}{3} < \frac{5}{7}$ **5.** $\frac{5}{8} > \frac{7}{12}$ **7.** $\frac{7}{9} < \frac{11}{12}$ **9.** $\frac{13}{14} > \frac{19}{21}$ **11.** $\frac{7}{24} < \frac{11}{30}$ **13.** $\frac{9}{64}$ **15.** $\frac{8}{729}$ **17.** $\frac{2}{9}$ **19.** $\frac{3}{125}$ **21.** $\frac{8}{245}$ **23.** $\frac{1}{121}$ **25.** $\frac{16}{1225}$ **27.** $\frac{4}{49}$ **29.** $\frac{9}{125}$ **31.** $\frac{27}{88}$ **33.** $\frac{4}{121}$ **35.** $\frac{1}{30}$ **37.** $1\frac{1}{2}$ **39.** $\frac{20}{21}$ **41.** $\frac{12}{125}$ **43.** $\frac{11}{32}$ **45.** $\frac{17}{24}$ **47.** $\frac{14}{15}$ **49.** $2\frac{7}{10}$ **51.** $\frac{21}{44}$ **53.** $\frac{25}{39}$ **55.** Between $\frac{2}{3}$ and $\frac{3}{4}$.

## CHAPTER REVIEW *pages 111–112*

**1.** 36 **2.** 54 **3.** 4 **4.** 5 **5.** $\frac{13}{4}$ **6.** $1\frac{7}{8}$ **7.** $3\frac{2}{5}$ **8.** $\frac{19}{7}$ **9.** $\frac{24}{36}$ **10.** $\frac{32}{44}$ **11.** $\frac{2}{3}$ **12.** $\frac{4}{11}$ **13.** $1\frac{1}{8}$ **14.** $1\frac{13}{18}$ **15.** $5\frac{7}{8}$ **16.** $18\frac{13}{54}$ **17.** The total rainfall was $21\frac{7}{24}$ inches. **18.** $\frac{1}{3}$ **19.** $\frac{25}{48}$ **20.** $14\frac{19}{42}$ **21.** $10\frac{1}{8}$ **22.** The second checkpoint is $4\frac{3}{4}$ miles from the finish line. **23.** $\frac{1}{15}$ **24.** $\frac{1}{8}$ **25.** $9\frac{1}{24}$ **26.** $16\frac{1}{2}$ **27.** The car can travel 243 miles. **28.** 2 **29.** $\frac{3}{4}$ **30.** 2 **31.** $3\frac{1}{3}$ **32.** Each acre cost \$36,000. **33.** $\frac{11}{18} < \frac{17}{24}$ **34.** $\frac{5}{16}$ **35.** $\frac{5}{36}$ **36.** $\frac{1}{15}$

## CHAPTER TEST *pages 113–114*

**1.** 120 (2.1A) **2.** 8 (2.1B) **3.** $\frac{11}{4}$ (2.2A) **4.** $3\frac{3}{5}$ (2.2B) **5.** $\frac{49}{5}$ (2.2B) **6.** $\frac{5}{8} = \frac{45}{72}$ (2.3A) **7.** $\frac{5}{8}$ (2.3B) **8.** $1\frac{11}{12}$ (2.4A) **9.** $1\frac{61}{90}$ (2.4B) **10.** $22\frac{4}{15}$ (2.4C) **11.** The total rainfall was $21\frac{11}{24}$ inches. (2.4D) **12.** $\frac{1}{4}$ (2.5A) **13.** $\frac{7}{48}$ (2.5B) **14.** $13\frac{81}{88}$ (2.5C) **15.** The value of one share is $\$27\frac{7}{8}$. (2.5D) **16.** $\frac{4}{9}$ (2.6A) **17.** 8 (2.6B) **18.** The electrician's total earnings are \$420. (2.6C) **19.** $1\frac{3}{7}$ (2.7A) **20.** $2\frac{2}{19}$ (2.7B) **21.** There were 11 lots available for sale. (2.7C) **22.** $\frac{3}{8} < \frac{5}{12}$ (2.8A) **23.** $\frac{1}{6}$ (2.8B) **24.** $\frac{7}{24}$ (2.8C) **25.** $\frac{5}{6}$ (2.8C)

## CUMULATIVE REVIEW *pages 115–116*

**1.** 290,000 (1.1D) **2.** 291,278 (1.3B) **3.** 73,154 (1.4B) **4.** 540 R12 (1.5C) **5.** 1 (1.6B) **6.** $2 \cdot 2 \cdot 11$ (1.7B) **7.** 210 (2.1A) **8.** 20 (2.1B) **9.** $\frac{23}{3}$ (2.2B) **10.** $6\frac{1}{4}$ (2.2B) **11.** $\frac{15}{48}$ (2.3A) **12.** $\frac{2}{5}$ (2.3B) **13.** $1\frac{7}{48}$ (2.4B) **14.** $14\frac{11}{48}$ (2.4C) **15.** $\frac{13}{24}$ (2.5B) **16.** $1\frac{7}{9}$ (2.5C) **17.** $\frac{7}{20}$ (2.6A) **18.** $7\frac{1}{2}$ (2.6B) **19.** $1\frac{1}{20}$ (2.7A) **20.** $2\frac{5}{8}$ (2.7B) **21.** $\frac{1}{9}$ (2.8B) **22.** $5\frac{5}{24}$ (2.8C) **23.** There is \$862 in the checking account at the end of the week. (1.3C) **24.** The total income is \$705. (1.4C) **25.** The total weight is $12\frac{1}{24}$ pounds. (2.4D) **26.** The length of the remaining piece is $4\frac{17}{24}$ feet. (2.5D) **27.** The car travels 225 miles. (2.6C) **28.** 25 parcels can be sold from the remaining land. (2.7C)

# ANSWERS to Chapter 3 Odd-Numbered Exercises

## SECTION 3.1 *pages 121–122*

**1.** Twenty-seven hundredths **3.** One and five thousandths
**5.** Thirty-six and four tenths **7.** Thirty-five hundred-thousandths
**9.** Ten and seven thousandths **11.** Fifty-two and ninety-five hundred-thousandths
**13.** Two hundred ninety-three ten-thousandths
**15.** Six and three hundred twenty-four thousandths

**17.** Two hundred seventy-six and three thousand two hundred ninety-seven ten-thousandths
**19.** Two hundred sixteen and seven hundred twenty-nine ten-thousandths
**21.** Four thousand six hundred twenty-five and three hundred seventy-nine ten thousandths
**23.** One and one hundred-thousandths **25.** 0.762 **27.** 0.000062 **29.** 8.0304
**31.** 304.07 **33.** 362.048 **35.** 3048.2002 **37.** 7.4 **39.** 23.0 **41.** 22.68
**43.** 480.33 **45.** 1.039 **47.** 1946.375 **49.** 0.0299 **51.** 2 **53.** 0.010290
**55.** Hundredths of a second. Answers will vary
**57.** **a.** Last two zeros **b.** First zero **c.** All zeros needed **d.** First and last zero
**59.** 10.15–10.24

## SECTION 3.2 *pages 125–126*

**1.** 150.1065 **3.** 95.8446 **5.** 69.644 **7.** 92.883 **9.** 113.205 **11.** 0.69 **13.** 16.305
**15.** 110.7666 **17.** 104.4959 **19.** Est.: 234 Cal.: 234.192 **21.** Est.: 782 Cal.: 781.943
**23.** The length of the shaft is 5.65 inches. **25** The total amount of gas used is 47.8 gallons. **27.** The amount in the checking account is $3664.20. **29.** The total number of television viewers is 366.2 million. **31.** No, the customer does not have enough money. **33.** No, the rope cannot be wrapped around the box.

## SECTION 3.3 *pages 129–130*

**1.** 5.627 **3.** 113.6427 **5.** 6.7098 **7.** 215.697 **9.** 53.8776 **11.** 72.7091 **13.** 4.685
**15.** 10.0365 **17.** 0.7727 **19.** 3.273 **21.** 791.247 **23.** 547.951 **25.** 403.8557
**27.** 22.479 **29.** Est.: 600 Cal.: 590.25 **31.** Est.: 30 Cal.: 35.194 **33.** Est.: 3 Cal.: 2.74506 **35.** Est.: 7 Cal.: 7.14925
**37.** The amount of sales was $469.61. **39.** The missing dimension is 2.59 feet. **41.** The price of gasoline was $1.03 per gallon. **43.** The wage earner would need an additional $20,085.58. **45.** 0.05, 0.005, 0.0005 **47.** The difference in average speeds was 41.983 mph.

## SECTION 3.4 *pages 135–140*

**1.** 0.36 **3.** 0.30 **5.** 0.25 **7.** 0.45 **9.** 6.93 **11.** 1.84 **13.** 4.32 **15.** 0.74
**17.** 39.5 **19.** 2.72 **21.** 0.603 **23.** 0.096 **25.** 13.50 **27.** 79.80 **29.** 4.316
**31.** 1.794 **33.** 0.06 **35.** 0.072 **37.** 0.1323 **39.** 0.03568 **41.** 0.0784 **43.** 0.076
**45.** 34.48 **47.** 580.5 **49.** 20.148 **51.** 0.04255 **53.** 0.17686 **55.** 0.19803
**57.** 14.8657 **59.** 0.0006608 **61.** 53.9961 **63.** 0.536335 **65.** 0.429 **67.** 2.116
**69.** 0.476 **71.** 1.022 **73.** 37.96 **75.** 2.318 **77.** 3.2 **79.** 6.5 **81.** 6285.6
**83.** 3200 **85.** 35,700 **87.** 6.3 **89.** 3.9 **91.** 49,000 **93.** 6.7 **95.** 0.012075
**97.** 0.117796 **99.** 0.31004 **101.** 0.082845 **103.** 5.175
**105.** Est.: 90 Cal.: 91.2 **107.** Est.: 0.8 Cal.: 1.0472 **109.** Est.: 4.5 Cal.: 3.897 **111.** Est.: 12 Cal.: 11.2406
**113.** Est.: 0.32 Cal.: 0.371096 **115.** Est.: 30 Cal.: 31.8528 **117.** Est.: 2000 Cal.: 1941.0695 **119.** Est.: 0.00005 Cal.: 0.0000435
**121.** The cost to rent the car is $20.52. **123.** The amount received from recycling is $14.06. **125.** The broker's fee is $173.25. **127a.** The amount of the payments is $4590. **127b.** The total cost of the car is $6590. **129.** The total cost to rent the car is $64.20. **131.** The value of the shares is $2047.39. **133.** Grade 1 of steel costs $21.12, Grade 2 costs $29.84, and Grade 3 costs $201.66. **135a.** The cost will be $52.90. **135b.** The cost will be $79.60. **135c.** The cost will be $61.45. **137.** The total cost of the parts is $260.60. **139.** The total cost for the parts and labor is $1290.48. **141.** Answers will vary

## SECTION 3.5 *pages 145–148*

**1.** 0.82 **3.** 4.8 **5.** 89 **7.** 60 **9.** 84.3 **11.** 32.3 **13.** 5.06 **15.** 1.3
**17.** 0.11 **19.** 3.8 **21.** 6.3 **23.** 0.6 **25.** 2.5 **27.** 1.1 **29.** 130.6 **31.** 0.81
**33.** 42.40 **35.** 40.70 **37.** 0.46 **39.** 0.019 **41.** 0.087 **43.** 0.360 **45.** 0.103
**47.** 0.009 **49.** 1 **51.** 3 **53.** 1 **55.** 57 **57.** 0.407 **59.** 4.267 **61.** 0.01037

**63.** 0.008295 **65.** 0.82537 **67.** 0.032 **69.** 0.23627 **71.** 0.000053 **73.** 0.0018932 **75.** 18.42 **77.** 16.07 **79.** 0.0135 **81.** 0.023678 **83.** 0.112 **85.** Est.: 10 Cal.: 11.1632 **87.** Est.: 1000 Cal.: 884.0909 **89.** Est.: 1.5 Cal.: 1.8269 **91.** Est.: 50 Cal.: 58.8095 **93.** Est.: 100 Cal.: 72.3053 **95.** Est.: 0.0025 Cal.: 0.0023 **97.** Ramon averaged 6.23 yards per carry. **99.** Each car insurance payment was \$236.90. **101.** Three shelves can be cut from the board. **103.** The amount of the dividend is \$1.72. **105a.** The amount to be paid in monthly payments is \$842.58. **105b.** The monthly payment is \$46.81. **107.** The possible answers are 1¢, 3¢, 5¢, 9¢, 15¢, or 45¢. **109.** Answers will vary **111.** The average yogurt consumption is 5.16 pounds per person. **113.** × **115.** × **117.** ÷ **119.** 2.53

## SECTION 3.6 *pages 151–152*

**1.** 0.625 **3.** 0.667 **5.** 0.167 **7.** 0.417 **9.** 1.750 **11.** 1.500 **13.** 4.000 **15.** 0.003 **17.** 7.080 **19.** 37.500 **21.** 0.375 **23.** 0.208 **25.** 3.333 **27.** 5.444 **29.** 0.313 **31.** $\frac{4}{5}$ **33.** $\frac{8}{25}$ **35.** $\frac{1}{8}$ **37.** $1\frac{1}{4}$ **39.** $16\frac{9}{10}$ **41.** $8\frac{2}{5}$ **43.** $8\frac{437}{1000}$ **45.** $2\frac{1}{4}$ **47.** $\frac{23}{150}$ **49.** $\frac{703}{800}$ **51.** $1\frac{17}{25}$ **53.** $\frac{9}{200}$ **55.** $16\frac{18}{25}$ **57.** $\frac{33}{100}$ **59.** $\frac{1}{3}$ **61.** $0.15 < 0.5$ **63.** $6.65 > 6.56$ **65.** $2.504 > 2.054$ **67.** $\frac{3}{8} > 0.365$ **69.** $\frac{2}{3} > 0.65$ **71.** $\frac{5}{9} > 0.55$ **73.** $0.62 > \frac{7}{15}$ **75.** $0.161 > \frac{1}{7}$ **77.** $0.86 > 0.855$ **79.** $1.005 > 0.5$ **81.** **a.** False **b.** False **c.** True **83.** No. 0.0402 rounded to hundredths is 0.04, to thousandths is 0.040. **85.** Answers will vary

## CHAPTER REVIEW *pages 155–156*

**1.** Twenty-two and ninety-two ten-thousandths **2.** Three hundred forty-two and thirty-seven hundredths **3.** 34.025 **4.** 3.06753 **5.** 7.94 **6.** 0.05678 **7.** 36.714 **8.** 833.958 **9.** The checking account balance is \$478.02. **10.** 22.8635 **11.** 4.8785 **12.** The new checking account balance is \$661.51. **13.** 8.932 **14.** 25.7446 **15.** The amount of income tax paid was \$5600. **16.** 54.5 **17.** 6.594 **18.** The amount of each monthly payment is \$123.45. **19.** 0.778 **20.** 2.33 **21.** $\frac{3}{8}$ **22.** $\frac{2}{3}$ **23.** $0.055 < 0.1$ **24.** $\frac{5}{8} > 0.62$

## CHAPTER TEST *pages 157–158*

**1.** Forty-five and three hundred two ten-thousandths (3.1A) **2.** 209.07086 (3.1A) **3.** 7.095 (3.1B) **4.** 0.0740 (3.1B) **5.** 255.957 (3.2A) **6.** 458.581 (3.2A) **7.** The total income was \$1543.57. (3.2B) **8.** 4.087 (3.3A) **9.** 27.76626 (3.3A) **10.** 1.37 inch (3.3B) **11.** 0.00548 (3.4A) **12.** Los Angeles had the highest rate for a gallon of gasoline. (3.2B) **13.** Boston had the lowest tax for a gallon of gasoline. (3.2B) **14.** The cost of the long distance call is \$4.63. (3.4B) **15.** 232 (3.5A) **16.** 1.538 (3.5A) **17.** The amount of each monthly payment is \$142.85. (3.5B) **18.** 0.692 (3.6A) **19.** $\frac{33}{40}$ (3.6B) **20.** $0.66 < 0.666$ (3.6C)

## CUMULATIVE REVIEW *pages 159–160*

**1.** 235 r17 (1.5C) **2.** 128 (1.6A) **3.** 3 (1.6B) **4.** 72 (2.1A) **5.** $4\frac{2}{5}$ (2.2B) **6.** $\frac{37}{8}$ (2.2B) **7.** $\frac{25}{60}$ (2.3A) **8.** $1\frac{17}{48}$ (2.4B) **9.** $8\frac{35}{36}$ (2.4C) **10.** $5\frac{23}{36}$ (2.5C) **11.** $\frac{1}{12}$ (2.6A) **12.** $9\frac{1}{8}$ (2.6B) **13.** $1\frac{2}{9}$ (2.7A) **14.** $\frac{19}{20}$ (2.7B) **15.** $\frac{3}{16}$ (2.8B) **16.** $2\frac{5}{18}$ (2.8C) **17.** Sixty-five and three hundred nine ten-thousandths (3.1A)

**18.** 504.6991 (3.2A) **19.** 21.0764 (3.3A) **20.** 55.26066 (3.4A) **21.** 2.154 (3.5A) **22.** 0.733 (3.6A) **23.** $\frac{1}{6}$ (3.6B) **24.** $\frac{8}{9} < 0.98$ (3.6C) **25.** There were 234 passengers on the continuing flight. (1.3C) **26.** The value of each share is $\$32\frac{7}{8}$. (2.5D) **27.** The checking account balance was $617.38. (3.3B) **28.** The resulting thickness of the bushing was 1.395 inches. (3.3B) **29.** The amount of income tax paid last year was $6008.80. (3.4B) **30.** The monthly payment is $23.87. (3.5B)

# ANSWERS to Chapter 4 Odd-Numbered Exercises

## SECTION 4.1 *pages 165–166*

**1.** $\frac{1}{5}$ 1:5 1 TO 5 **3.** $\frac{2}{1}$ 2:1 2 TO 1 **5.** $\frac{3}{8}$ 3:8 3 TO 8 **7.** $\frac{37}{24}$ 37:24 37 TO 24 **9.** $\frac{1}{1}$ 1:1 1 TO 1 **11.** $\frac{7}{10}$ 7:10 7 TO 10 **13.** $\frac{1}{2}$ 1:2 1 TO 2 **15.** $\frac{2}{1}$ 2:1 2 TO 1 **17.** $\frac{3}{4}$ 3:4 3 TO 4 **19.** $\frac{5}{3}$ 5:3 5 TO 3 **21.** $\frac{2}{3}$ 2:3 2 TO 3 **23.** $\frac{2}{1}$ 2:1 2 TO 1 **25.** The ratio of housing costs to total income is $\frac{1}{3}$. **27.** The ratio of utilities cost to food cost is $\frac{3}{8}$. **29.** The ratio of College Senior football to High School Senior football players is $\frac{1}{30}$. **31.** The ratio of turns in the primary coil to the turns in the secondary coil is $\frac{1}{12}$. **33a.** The amount of increase is $20,000. **33b.** The ratio of the increase to the original value is $\frac{2}{9}$. **35.** The ratio of the increase in price to the original price is $\frac{5}{16}$. **37.** No. $\frac{265}{778}$ is greater than $\frac{1}{3}$. **39.** No. Answers will vary

## SECTION 4.2 *pages 169–170*

**1.** $\frac{3\text{ pounds}}{4\text{ people}}$ **3.** $\frac{\$20}{3\text{ boards}}$ **5.** $\frac{20\text{ miles}}{1\text{ gallon}}$ **7.** $\frac{5\text{ children}}{2\text{ families}}$ **9.** $\frac{8\text{ gallons}}{1\text{ hour}}$ **11.** 2.5 feet/second **13.** $325/week **15.** 110 trees/acre **17.** $4.71/hour **19.** 52.4 miles/hour **21.** 28 miles/gallon **23.** $1.65/pound **25.** The student drives 28.4 miles per gallon of gas. **27.** The rocket uses 213,600 gallons of fuel per minute. **29.** The dividend is $1.60 per share. **31a.** There were 4878 compact discs meeting company standards. **31b.** The cost was $5.44 per disc. **33.** The profit was $0.60 per box of strawberries. **35.** The cost was $8.33 per thousand viewers. **37.** 60 minutes had the lowest cost per thousand viewers.

## SECTION 4.3 *pages 175–178*

**1.** True **3.** Not true **5.** Not true **7.** True **9.** True **11.** Not true **13.** True **15.** True **17.** True **19.** Not true **21.** True **23.** Not true **25.** $n = 3$ **27.** $n = 6$ **29.** $n = 20$ **31.** $n = 2$ **33.** $n = 8$ **35.** $n = 4$ **37.** $n = 4.38$ **39.** $n = 9.6$ **41.** $n = 10.67$ **43.** $n = 6.67$ **45.** $n = 26.25$ **47.** $n = 96$ **49.** $n = 9.78$ **51.** $n = 3.43$ **53.** $n = 1.34$ **55.** $n = 50.4$ **57.** There are 50 calories in the serving of cereal. **59.** There were 50 pounds of fertilizer used. **61.** The property tax is $2500. **63.** It would take 2496 bricks. **65.** The distance between the two cities is 16 miles. **67.** 1.25 ounces of medication are required. **69.** There would be 160,000 people voting in the election. **71.** There would be 960 braking defects. **73.** You own 400 shares of stock. **75.** Students used the computer 300 hours. **77.** The national debt would have been $6.76 trillion in 1992. The difference is $2.76 trillion. **79.** The bowling ball would weigh $2\frac{2}{3}$ pounds on the moon. **81.** Answers will vary

## CHAPTER REVIEW *pages 181–182*

**1.** $\frac{2}{7}$, 2:7, 2 TO 7 **2.** $\frac{2}{5}$, 2:5, 2 TO 5 **3.** $\frac{1}{1}$, 1:1, 1 TO 1 **4.** $\frac{2}{5}$, 2:5, 2 TO 5 **5.** The ratio of the high temperature to the low temperature is 2:1. **6.** The ratio of increase to original value is $\frac{1}{2}$. **7.** The ratio of decrease in price to original price is $\frac{2}{5}$. **8.** The ratio of TV advertising to newspaper advertising is $\frac{5}{2}$. **9.** $\frac{\text{100 miles}}{\text{3 hours}}$ **10.** $\frac{\$15}{\text{4 hours}}$ **11.** 62.5 miles/hour **12.** $7.50/hour **13.** 27.2 miles/gallon **14.** $1.75/pound **15.** The cost is $44.75 per share. **16.** Mahesh drove an average of 56.8 miles/hour. **17.** The cost is $.68/pound. **18.** The cost per radio is $37.50. **19.** True **20.** True **21.** True **22.** Not true **23.** $n = 36$ **24.** $n = 68$ **25.** $n \approx 65.45$ **26.** $n \approx 19.44$ **27.** There were 22.5 pounds of fertilizer used. **28.** The cost of the insurance is $193.50. **29.** The property tax is $2400. **30.** It would take 1344 blocks to build the wall.

## CHAPTER TEST *pages 183–184*

**1.** $\frac{3}{2}$, 3:2, 3 TO 2 (4.1A) **2.** $\frac{3}{5}$, 3:5, 3 TO 5 (4.1A) **3.** $\frac{1}{3}$, 1:3, 1 TO 3 (4.1A) **4.** $\frac{1}{6}$, 1:6, 1 TO 6 (4.1A) **5.** The ratio of the city temperature to the desert temperature is $\frac{43}{56}$. (4.1B) **6.** The ratio of the cost of radio advertising to the total cost of advertising is $\frac{8}{13}$. (4.1B) **7.** $\frac{\text{9 supports}}{\text{4 feet}}$ (4.2A) **8.** $\frac{\$27}{\text{4 boards}}$ (4.2A) **9.** $1836.40/month (4.2B) **10.** 30.5 miles/gallon (4.2B) **11.** The cost of the lumber is $1.73/foot. (4.2C) **12.** The plane's speed is 538 miles/hour. (4.2C) **13.** True (4.3A) **14.** Not true (4.3A) **15.** $n = 40.5$ (4.3B) **16.** $n = 144$ (4.3B) **17.** The student's body contains 132 pounds of water. (4.3C) **18.** The person requires 0.875 ounce of medication. (4.3C) **19.** The property tax is $2800. (4.3C) **20.** The dividend would be $625. (4.3C)

## CUMULATIVE REVIEW *pages 185–186*

**1.** 9158 (1.3B) **2.** $2^4 \cdot 3^3$ (1.6A) **3.** 3 (1.6B) **4.** $2 \cdot 2 \cdot 2 \cdot 2 \cdot 2 \cdot 5$ (1.7B) **5.** 36 (2.1A) **6.** 14 (2.1B) **7.** $\frac{5}{8}$ (2.3B) **8.** $8\frac{3}{10}$ (2.4C) **9.** $5\frac{11}{18}$ (2.5C) **10.** $2\frac{5}{6}$ (2.6B) **11.** $4\frac{2}{3}$ (2.7B) **12.** $\frac{23}{30}$ (2.8C) **13.** Four and seven hundred nine ten-thousandths (3.1A) **14.** 2.10 (3.1B) **15.** 1.990 (3.5A) **16.** $\frac{1}{15}$ (3.6B) **17.** $\frac{1}{8}$ (4.1A) **18.** $\frac{29¢}{\text{2 bars}}$ (4.2A) **19.** 33.4 miles/gallon (4.2B) **20.** $n = 4.25$ (4.3B) **21.** 57.2 miles/hour (4.2C) **22.** $n = 36$ (4.3B) **23.** The new balance is $744. (1.3C) **24.** The monthly payment is $370. (1.5D) **25.** There are 105 pages remaining to be read. (2.6C) **26.** The cost for each acre is $36,000. (2.7C) **27.** The amount of change is $19.62. (3.3B) **28.** The player's batting average is 0.271. (3.5B) **29.** There will be 25 inches eroded. (4.3C) **30.** The person needs 1.6 ounces of medication. (4.3C)

# ANSWERS to Chapter 5 Odd-Numbered Exercises

## SECTION 5.1 *pages 191–192*

**1.** $\frac{1}{4}$, 0.25 **3.** $1\frac{3}{10}$, 1.30 **5.** 1, 1.00 **7.** $\frac{73}{100}$, 0.73 **9.** $3\frac{83}{100}$, 3.83 **11.** $\frac{7}{10}$, 0.70 **13.** $\frac{22}{25}$, 0.88 **15.** $\frac{8}{25}$, 0.32 **17.** $\frac{2}{3}$ **19.** $\frac{5}{6}$ **21.** $\frac{1}{9}$ **23.** $\frac{5}{11}$ **25.** $\frac{3}{70}$ **27.** $\frac{1}{15}$

**29.** 0.065 **31.** 0.0055 **33.** 0.0825 **35.** 0.0675 **37.** 0.0045 **39.** 0.804 **41.** 16% **43.** 5% **45.** 1% **47.** 70% **49.** 124% **51.** 0.4% **53.** 0.6% **55.** 310.6% **57.** 54% **59.** 33.3% **61.** 62.5% **63.** 16.7% **65.** 17.5% **67.** 177.8% **69.** 30% **71.** $23\frac{1}{3}\%$ **73.** $237\frac{1}{2}\%$ **75.** $216\frac{2}{3}\%$

**77.** **a.** False; 4(200%) = 8 **b.** False; $\frac{4}{200\%} = 2$ **c.** True **d.** False; 125% = 1.25

**79.** Answers will vary **81.** $\frac{1}{2}$ **83.** No; 0.495 **85.** The fraction is less than $\frac{1}{100}$.

## SECTION 5.2 *pages 195–196*

**1.** 8 **3.** 10.8 **5.** 0.075 **7.** 80 **9.** 51.895 **11.** 7.5 **13.** 13 **15.** 3.75 **17.** 20 **19.** 210 **21.** 5% of 95 **23.** 82% of 16 **25.** 84% of 32 **27.** 25,805.0324 **29.** The amount deducted for income tax is $403.20. **31.** The profit is $4000. **33.** The rebate is $1680. **35a.** The sales tax is $570. **35b.** The total cost including sales tax is $10,070. **37.** The total number of employees needed is 671. **39.** There were 23 permanent injuries obtained playing offense. **41.** The difference between the number of viewers is 752,000. **43.** The increase in price is $204.26.

## SECTION 5.3 *pages 199–200*

**1.** 32% **3.** $16\frac{2}{3}\%$ **5.** 200% **7.** 37.5% **9.** 18% **11.** 0.25% **13.** 20% **15.** 400% **17.** 2.5% **19.** 37.5% **21.** 0.25% **23.** 2.4% **25.** 9.6% **27.** The company spent 5% of its budget for advertising. **29.** The dividend is 3.2%. **31.** 62.5% of the families liked the TV show. **33a.** The fat is 20% of the roast. **33b.** The side of beef would weigh 110 pounds. **35.** 98.5% of the concrete slabs met safety requirements. **37.** Answers will vary **39.** The government spent 20.9% of its income on public debt. **41.** Home heating consisted of 56.7% of the utility bill.

## SECTION 5.4 *pages 203–204*

**1.** 75 **3.** 50 **5.** 100 **7.** 85 **9.** 1200 **11.** 19.2 **13.** 7.5 **15.** 32 **17.** 200 **19.** 80 **21.** 9 **23.** 504 **25.** 108 **27.** 7122.1548 **29.** The estimated life of the brakes is 50,000 miles. **31.** The city's population was 56,000 people. **33.** The price at the competitor's store is $46.50. **35a.** The manufacturer tested 3000 computer boards. **35b.** There were 2976 computer boards that were not defective. **37.** 6.6 grams **39.** No.

## SECTION 5.5 *pages 207–208*

**1.** 65 **3.** 25% **5.** 75 **7.** 12.5% **9.** 400 **11.** 19.5 **13.** 14.8% **15.** 62.62 **17.** 5 **19.** 45 **21.** 15 **23.** $24,500 **25.** $71.25 **27.** Enrico typed 98% of the words correctly. **29.** The winner's share increased by 5.6%. **31.** The harvest decreased by 19.1%. **33.** The current yield is 7.6%. **35.** The projected increase is 13.7%. **37.** The potency of the solution is reduced by 87.5%.

## CHAPTER REVIEW *pages 211–212*

**1.** $\frac{3}{25}$ **2.** $\frac{1}{6}$ **3.** 0.42 **4.** 0.076 **5.** 38% **6.** 150% **7.** $n = 60$ **8.** $n = 5.4$ **9.** $n = 19.36$ **10.** $n = 77.5$ **11.** The company spent $4500 for TV advertising. **12.** The total cost is $1041.25. **13.** $n = 150\%$. **14.** $n = 20\%$ **15.** $n = 7.3\%$ **16.** $n = 613.3\%$ **17.** The dividend is 4.6% of the stock price. **18.** The percent increase in value was 64%. **19.** $n = 75$ **20.** $n = 10.9$ **21.** $n = 157.5$ **22.** $n = 504$ **23.** The city's population was 70,000 people. **24.** The previous year's batting average was 0.252. **25.** $n = 198.4$ **26.** $n = 160\%$ **27.** The computer cost $3000. **28.** Trent answered 85% of the questions correctly.

## CHAPTER TEST *pages 213–214*

**1.** 0.973 (5.1A) **2.** $\frac{1}{6}$ (5.1A) **3.** 30% (5.1B) **4.** 163% (5.1B) **5.** 150% (5.1B) **6.** $66\frac{2}{3}\%$ (5.1B) **7.** $n = 50.05$ (5.2A) **8.** $n = 61.36$ (5.2A) **9.** 76% of 13 (5.2A) **10.** 212% of 12 (5.2A) **11.** The company spends $4500 for advertising. (5.2B) **12.** There were 1170 pounds of vegetables unspoiled. (5.2B) **13.** One serving provides 14.7% of the daily recommended amount of potassium. (5.3B) **14.** One serving provides 9.1% of the daily recommended number of calories. (5.3B) **15.** The store's temporary employees were 16% of the total number of permanent employees. (5.3B) **16.** Conchita answered 91.3% of the questions correctly. (5.3B) **17.** $n = 80$ (5.4A) **18.** $n = 28.3$ (5.4A) **19.** There were 32,000 transistors tested. (5.4B) **20.** The increase is 60% of the original price. (5.4B) **21.** $n = 143.0$ (5.5A) **22.** 1000% (5.5A) **23.** The increase in the hourly wage is $1.02. (5.5B) **24.** The population now is 220% of the population 10 years ago. (5.5B) **25.** The value of the car is $6500. (5.5B)

## CUMULATIVE REVIEW *pages 215–216*

**1.** 4 (1.6B) **2.** 240 (2.1A) **3.** $10\frac{11}{24}$ (2.4C) **4.** $12\frac{41}{48}$ (2.5C) **5.** $12\frac{4}{7}$ (2.6B) **6.** $\frac{7}{24}$ (2.7B) **7.** $\frac{1}{3}$ (2.8B) **8.** $\frac{13}{36}$ (2.8C) **9.** 3.08 (3.1B) **10.** 1.1196 (3.3A) **11.** 34.2813 (3.5A) **12.** 3.625 (3.6A) **13.** $1\frac{3}{4}$ (3.6B) **14.** $\frac{3}{8} < 0.87$ (3.6C) **15.** $n = 53.3$ (4.3B) **16.** $9.60/hour (4.2B) **17.** $\frac{11}{60}$ (5.1A) **18.** $83\frac{1}{3}\%$ (5.1B) **19.** 19.56 (5.2A) **20.** $133\frac{1}{3}\%$ (5.3A) **21.** 9.92 (5.4A) **22.** 342.9% (5.5A) **23.** Sergio's take-home pay is $592. (2.6C) **24.** The monthly payment is $92.25. (3.5B) **25.** There were 420 gallons of gasoline used during the month. (3.5B) **26.** The real estate tax is $3000. (4.3C) **27.** The sales tax is 6% of the purchase price. (5.3B) **28.** 45% of the people surveyed did not favor the candidate. (5.3B) **29.** The 1990 value is 250% of the 1985 value. (5.5B) **30.** 18% of the children tested had levels of lead that exceeded federal standards. (5.5B)

# ANSWERS to Chapter 6 Odd-Numbered Exercises

## SECTION 6.1 *pages 221–222*

**1.** The unit cost is $0.056 per ounce. **3.** The unit cost is $0.149 per ounce. **5.** The unit cost is $0.033 per screw. **7.** The unit cost is $0.195 per foot. **9.** The unit cost is $0.037 per ounce. **11.** The unit cost is $2.995 per clamp. **13.** The more economical purchase is 12 ounces for $1.79. **15.** The more economical purchase is 12 ounces for $2.64. **17.** The more economical purchase is 46 ounces for $0.69. **19.** The more economical purchase is 2 quarts for $4.79. **21.** The more economical purchase is 200 tablets for $9.98. **23.** The more economical purchase is 16 ounces for $1.32. **25.** The total cost is $8.37. **27.** The total cost is $2.94. **29.** The total cost is $3.51. **31.** The total cost is $0.40. **33.** The total cost is $2.01. **35.** Answers will vary

## SECTION 6.2 *pages 229–232*

**1.** The percent increase is 8%. **3.** The percent increase is $33\frac{1}{3}\%$. **5a.** The amount of increase is $2925. **5b.** The new salary is $35,425. **7.** The dividend per share of stock is $1.60. **9.** The home is 240 square feet larger. **11.** The markup is $71.25. **13.** The markup is $12.60. **15.** The markup rate is 30%. **17a.** The markup is $77.76. **17b.** The selling price is $239.76. **19a.** The markup is $21.70. **19b.** The selling price is $83.70. **21.** The selling price is $227.20. **23.** The percent decrease is 40%. **25.** The percent decrease is 40%.

**27.** The car loses $3360 in value. **29a.** The amount of the decrease is 12 cameras. **29b.** The percent decrease is 60%. **31a.** The amount of decrease was $15.20. **31b.** The average monthly bill is $60.80. **33.** The percent decrease is 20%. **35.** The discount rate is $33\frac{1}{3}\%$. **37.** The discount is $67.50. **39.** The discount rate is 25%. **41a.** The discount is $0.17 per pound. **41b.** The sale price is $0.68 per pound. **43a.** The discount is $4. **43b.** The discount rate is 25%. **45.** The percent increase was 21.1%. **47.** WstwOn **49a.** The total cost is $172. **49b.** The total selling price is $258. **49c.** The florist expects to sell 186 roses. **49d.** The selling price is $1.39 per rose. **51.** Answers will vary

## SECTION 6.3 *pages 235–236*

**1.** The simple interest due is $6750. **3.** The simple interest due is $16. **5a.** The interest on the loan is $72,000. **5b.** The monthly payment is $6187.50. **7a.** The simple interest is $1080. **7b.** The monthly payment is $545. **9.** The monthly payment is $1332.50. **11.** The value will be $1414.78. **13.** The value will be $12,380.43. **15.** The value will be $28,212. **17a.** The value of the investment will be $6040.86. **17b.** The interest earned on the investment is $3040.86. **19.** Answers will vary **21a.** The value after the deposit will be $200.50. **21b.** The value on March 1 will be $301.50.

## SECTION 6.4 *pages 241–242*

**1.** The mortgage is $82,450. **3.** The down payment is $7500. **5.** The down payment is $212,500. **7.** The loan origination fee is $3750. **9a.** The down payment is $4750. **9b.** The mortgage is $90,250. **11.** The mortgage is $86,400. **13.** The monthly mortgage payment is $1157.73. **15.** No, the couple cannot afford to buy the home. **17.** The monthly property tax is $112.35. **19a.** The monthly mortgage payment is $1678.40. **19b.** The interest is $736.68 per month. **21.** The total monthly payment (mortgage and property tax) is $982.70. **23.** The monthly mortgage payment is $1430.83. **25.** $88,917.60 of interest can be saved.

## SECTION 6.5 *pages 245–246*

**1.** Amanda does not have enough money for the down payment. **3.** The sales tax is $742.50. **5.** The license fee is $250. **7a.** The sales tax is $420. **7b.** The total cost of the sales tax and the license fee is $595. **9a.** The down payment is $550. **9b.** The amount financed is $1650. **11.** The amount financed is $28,000. **13.** The monthly truck payment is $348.39. **15.** The cost to operate the car is $5120. **17.** The cost per mile is $0.11. **19.** The interest is $74.75. **21a.** The amount financed is $76,600. **21b.** The monthly truck payment is $1590.09. **23.** The monthly car payment is $709.88. **25.** The 10% loan with the application fee has a lower loan cost.

## SECTION 6.6 *pages 249–250*

**1.** Lewis earns $220. **3.** The commission earned is $2955. **5.** The commission received was $84. **7.** The teacher received $3244 per month. **9.** The electrician's overtime hourly wage was $31.60. **11.** The commission earned is $112.50. **13.** The typist earned $393.75. **15.** The consultant's hourly wage is $85. **17a.** Mark's hourly wage is $12.75. **17b.** Mark earned $102 on Saturday. **19a.** The increase in pay is $0.85 per hour. **19b.** The nurse's hourly pay is $9.35. **21.** The commission earned is $375. **23.** The difference in monthly pay is $7408.33. **25.** The projected increase is 500,500 analysts.

## SECTION 6.7 *pages 257–260*

**1.** The new balance is $486.32. **3.** The balance in the account is $1222.47. **5.** The current balance is $825.27. **7.** The current balance is $3000.82. **9.** Yes, there is enough money to purchase a refrigerator. **11.** Yes, there is enough money to make both purchases. **13.** The bank statement and checkbook balance. **15.** The bank statement and checkbook balance. **17.** Added **19.** Subtract

## CHAPTER REVIEW *pages 263–264*

**1.** 14.5¢/ounce **2.** 60 ounces for $8.40 **3.** The percent increase was 30.4%. **4.** The yearly percent increase was 200%. **5.** The selling price is $2079. **6.** The percent increase in earnings is 15%. **7.** The markup is $72. **8.** The sale price is $141. **9.** The simple interest due is $6750. **10.** The simple interest due is $1200. **11.** The value of the investment is $45,550.75. **12.** The value of the investment will be $53,593. **13.** The down payment is $18,750. **14.** The loan origination fee is $1875. **15.** The total monthly payment is $578.53. **16.** The monthly mortgage payment is $934.08. **17.** The total cost is $1158.75. **18.** The cost for the 4 items is 12.1¢/mile. **19.** The amount of interest is $97.33. **20.** The monthly payment is $330.82. **21.** The commission received was $3240. **22.** Richard's total income was $655.20. **23.** The current balance is $943.68. **24.** The current balance is $8866.58.

## CHAPTER TEST *pages 265–266*

**1.** The cost per foot is $6.92. (6.1A) **2.** The more economical purchase is 5 pounds for $1.65. (6.1B) **3.** The cost is $14.53. (6.1C) **4.** The percent increase is 20%. (6.2A) **5.** The percent increase was 150%. (6.2A) **6.** The selling price is $301. (6.2B) **7.** The selling price is $6.25. (6.2B) **8.** The percent decrease was 57.6%. (6.2C) **9.** The percent decrease was 20%. (6.2C) **10.** The sale price is $209.30. (6.2D) **11.** The discount rate is 40%. (6.2D) **12.** The simple interest due is $2000. (6.3A) **13.** The interest earned was $24,420.60. (6.3B) **14.** The loan origination fee is $3350. (6.4A) **15.** The monthly mortgage payment is $1713.44. (6.4B) **16.** The amount financed is $14,350. (6.5A) **17.** The monthly car payment is $301.82. (6.5B) **18.** The nurse earns $703.50. (6.6A) **19.** The current balance is $6612.25. (6.7A) **20.** The bank statement and checkbook balance. (6.7B)

## CUMULATIVE REVIEW *pages 267–268*

**1.** 13 (1.6B) **2.** $8\frac{13}{24}$ (2.4C) **3.** $2\frac{37}{48}$ (2.5C) **4.** 9 (2.6B) **5.** 2 (2.7B) **6.** 5 (2.8C) **7.** 52.2 (3.5A) **8.** 1.417 (3.6A) **9.** $51.25/hour (4.2B) **10.** $n = 10.94$ (4.3B) **11.** 62.5% (5.1B) **12.** 27.3 (5.2A) **13.** 0.182 (5.1A) **14.** 42% (5.3A) **15.** 250 (5.4A) **16.** 154.76 (5.5A) **17.** The total rainfall was $13\frac{11}{12}$ inches. (2.4D) **18.** The amount the family paid in taxes was $570. (2.6C) **19.** The ratio of decrease in price to original price is $\frac{3}{5}$. (4.1B) **20.** The car drove 33.4 miles per gallon. (4.2C) **21.** The unit cost is $0.93. (4.2C) **22.** The dividend is $280. (4.3C) **23.** The sale price is $720. (5.2B) **24.** The selling price is $119. (6.2B) **25.** The percent increase is 8%. (6.2A) **26.** The simple interest due is $6000. (6.3A) **27.** The monthly payment is $381.60. (6.5B) **28.** The new balance is $2243.77. (6.7A) **29.** The cost per mile is $0.20. (6.5B) **30.** The monthly mortgage payment is $743.18. (6.4B)

# Glossary

**addend** In addition, one of the numbers added. (Sec. 1.2)

**addition** The process of finding the total of two numbers. (Sec. 1.2)

**addition property of zero** Zero added to a number does not change the number. (Sec. 1.2)

**associative property of addition** Numbers to be added can be grouped (with parentheses, for example) in any order; the sum will be the same. (Sec. 1.2)

**associative property of multiplication** Numbers to be multiplied can be grouped (with parentheses, for example) in any order; the product will be the same. (Sec. 1.4)

**balancing a checkbook** Determining if the checking account balance is accurate. (Sec. 6.7)

**basic percent equation** Percent times base equals amount. (Sec. 5.2)

**commission** That part of the pay earned by a salesperson that is calculated as a percent of the salesperson's sales. (Sec. 6.6)

**common factor** A number that is a factor of two or more numbers is a common factor of those numbers. (Sec. 2.1)

**common multiple** A number that is a multiple of two or more numbers is a common multiple of those numbers. (Sec. 2.1)

**commutative property of addition** Two numbers can be added in either order; the sum will be the same. (Sec. 1.2)

**commutative property of multiplication** Two numbers can be multiplied in either order; the product will be the same. (Sec. 1.4)

**composite number** A number that has whole-number factors besides 1 and itself. For instance, 18 is a composite number. (Sec. 1.7)

**compound interest** Interest computed not only on the original amount but also on interest already earned. (Sec. 6.3)

**cost** The price that a business pays for a product. (Sec. 6.2)

**decimal** A number written in decimal notation. (Sec. 3.1)

**decimal notation** Notation in which a number consists of a whole-number part, a decimal point, and a decimal part. (Sec. 3.1)

**decimal part** In decimal notation, that part of the number that follows the decimal point. (Sec. 3.1)

**decimal point** In decimal notation, the point that separates the whole-number part from the decimal part. (Sec. 3.1)

**denominator** The part of a fraction that appears below the fraction bar. (Sec. 2.2)

**difference** In subtraction, the result of subtracting two numbers. (Sec. 1.3)

**discount** The difference between the regular price and the sale price. (Sec. 6.2)

**discount rate** The percent of a product's regular price that is represented by the discount. (Sec. 6.2)

**dividend** In division, the number into which the divisor is divided to yield the quotient. (Sec. 1.5)

**division** The process of finding the quotient of two numbers. (Sec. 1.5)

**divisor** In division, the number that is divided into the dividend to yield the quotient. (Sec. 1.5)

**equivalent fractions** Equal fractions with different denominators. (Sec. 2.3)

**expanded form** The number 46,208 can be written in expanded form as 40,000 + 6000 + 0 + 8. (Sec. 1.1)

**exponent** In exponential notation, the raised number that indicates how many times the number to which it is attached is taken as a factor. (Sec. 1.6)

**exponential notation** The expression of a number to some power, indicated by an exponent. (Sec. 1.6)

**factors** In multiplication, the numbers that are multiplied. (Sec. 1.4)

**fraction** Notation used to represent the number of equal parts of a whole. (Sec. 2.2)

**fraction bar** Bar that separates the numerator of a fraction from the denominator. (Sec. 2.2)

**graph of a whole number** A heavy dot placed directly above that number on the number line. (Sec. 1.1)

**greater than** A number that appears to the right of a given number on the number line is greater than that given number. (Sec. 1.1)

**greatest common factor** The largest common factor of two or more numbers. (Sec. 2.1)

**hourly wage** Pay calculated on the basis of a certain amount for each hour worked. (Sec. 6.6)

**improper fraction** A fraction greater than or equal to 1. (Sec. 2.2)

**interest** The amount of money paid for the privilege of using someone else's money. (Sec. 6.3)

**interest rate** The percent used to determine the amount of interest. (Sec. 6.3)

**inverting a fraction** Interchanging the numerator and denominator. (Sec. 2.7)

**least common denominator** The least common multiple of denominators. (Sec. 2.4)

**least common multiple** The smallest common multiple of two or more numbers. (Sec. 2.1)

**less than** A number that appears to the left of a given number on the number line is less than that given number. (Sec. 1.1)

**markup** The difference between selling price and cost. (Sec. 6.2)

**markup rate** The percent of a product's cost that is represented by the markup. (Sec. 6.2)

**minuend** In subtraction, the number from which another number (the subtrahend) is subtracted. (Sec. 1.3)

**mixed number** A number greater than 1 that has a whole-number part and a fractional part. (Sec. 2.2)

**mortgage** The amount borrowed to buy real estate. (Sec. 6.4)

**multiples of a number** The products of that number and the numbers 1, 2, 3, .... (Sec. 2.1)

**multiplication** The process of finding the product of two numbers. (Sec. 1.4)

**multiplication property of one** The product of a number and one is the number. (Sec. 1.4)

**multiplication property of zero** The product of a number and zero is zero. (Sec. 1.4)

**number line** A line on which a number can be graphed. (Sec. 1.1)

**numerator** The part of a fraction that appears above the fraction bar. (Sec. 2.2)

**Order of Operations Agreement** A set of rules that tell us in what order to perform the operations that occur in a numerical expression. (Sec. 1.6)

**percent** Parts per hundred. (Sec. 5.1)

**percent decrease** A decrease of a quantity expressed as a portion of its original value. (Sec. 6.2)

**percent increase** An increase of a quantity expressed as a portion of its original value. (Sec. 6.2)

**period** In a number written in standard form, each group of digits separated by a comma. (Sec. 1.1)

**place value** The position of each digit in a number in standard form determines that digit's place value. (Sec. 1.1)

**points** A term banks use to mean percent of a mortgage; used to express the loan origination fee. (Sec. 6.4)

**prime factorization** The expression of a number as the product of its prime factors. (Sec. 1.7)

**prime number** A number whose only whole-number factors are 1 and itself. For instance, 13 is a prime number. (Sec. 1.7)

**principal** The amount of money originally deposited or borrowed. (Sec. 6.3)

**product** In multiplication, the result of multiplying two numbers. (Sec. 1.4)

**proper fraction** A fraction less than 1. (Sec. 2.2)

**proportion** An expression of the equality of two ratios or rates. (Sec. 4.3)

**quotient** In division, the result of dividing the divisor into the dividend. (Sec. 1.5)

**rate** A comparison of two quantities that have different units. (Sec. 4.2)

**rate in simplest form** A rate in which the numbers that form the rate have no common factors. (Sec. 4.2)

**ratio** A comparison of two quantities that have the same units. (Sec. 4.1)

**ratio in simplest form** A ratio in which the two numbers that form the ratio do not have a common factor. (Sec. 4.1)

**reciprocal of a fraction** The fraction with the numerator and denominator interchanged. (Sec. 2.7)

**remainder** In division, the quantity left over when it is not possible to separate objects or numbers into a whole number of even groups. (Sec. 1.5)

**rounding** Giving an approximate value of an exact number. (Sec. 1.1)

**salary** Pay based on a weekly, biweekly, monthly, or annual time schedule. (Sec. 6.6)

**sale price** The reduced price. (Sec. 6.2)

**selling price** The price for which a business sells a product to a customer. (Sec. 6.2)

**simple interest** Interest computed on the original amount. (Sec. 6.3)

**simplest form** A fraction is in simplest form when there are no common factors in the numerator and the denominator. (Sec. 2.3)

**solving a proportion** Finding a number to replace the unknown so that the proportion is true. (Sec. 4.3)

**standard form** A whole number is in standard form when it is written using the digits 0, 1, 2, ..., 9. An example is 46,208. (Sec. 1.1)

**subtraction** The process of finding the difference between two numbers. (Sec. 1.3)

**subtrahend** In subtraction, the number that is subtracted from another number (the minuend). (Sec. 1.3)

**sum** In addition, the total of the numbers added. (Sec. 1.2)

**true proportion** A proportion in which the fractions are equal when written in lowest terms. (Sec. 4.3)

**unit cost** The cost of one item. (Sec. 6.1)

**unit rate** A rate in which the number in the denominator is 1. (Sec. 4.2)

**units** In the quantity 3 feet, feet are the units in which the measurement is made. (Sec. 4.1)

**whole numbers** The whole numbers are 0, 1, 2, 3, .... (Sec. 1.1)

**whole-number part** In decimal notation, that part of the number that appears before the decimal point. (Sec. 3.1)

# Index